TRIGONOMETRY SUPPLEMENT

Montgomery College

Taken from:

Algebra and Trigonometry, Second Edition
by Robert F. Blitzer

PEARSON Custom Publishing

PEARSON Prentice Hall

Taken from:

Algebra and Trigonometry, Second Edition
by Robert F. Blitzer

Published by Prentice Hall
Upper Saddle River, New Jersey 07458

This special edition published in cooperation with Pearson Custom Publishing.

Printed in the United States of America

10 9 8 7 6 5 4 3 2 1

ISBN 0-536-86537-X

2004360482

AP

Please visit our web site at *www.pearsoncustom.com*

PEARSON Custom Publishing

PEARSON CUSTOM PUBLISHING
75 Arlington Street, Suite 300, Boston, MA 02116
A Pearson Education Company

Contents

Chapter 5

Chapter 6

Trigonometric Functions

Chapter 5

Have you had days where your physical, intellectual, and emotional potentials were all at their peak? Then there are those other days when we feel we should not even bother getting out of bed. Do our potentials run in oscillating cycles like the tides? Can they be described mathematically? In this chapter you will encounter functions that enable us to model phenomena that occur in cycles.

What a day! It started when you added two miles to your morning run. You've experienced a feeling of peak physical well-being ever since. College was wonderful: You actually enjoyed two difficult lectures and breezed through a math test that had you worried. Now you're having dinner with an old group of friends. You experience the warmth from bonds of friendship filling the room.

SECTION 5.1 *Angles and Their Measure*

Objectives

1. Recognize and use the vocabulary of angles.
2. Use degree measure.
3. Draw angles in standard position.
4. Find coterminal angles.
5. Find complements and supplements.
6. Use radian measure.
7. Convert between degrees and radians.
8. Find the length of a circular arc.
9. Use linear and angular speed to describe motion on a circular path.

The San Francisco Museum of Modern Art was constructed in 1995 to illustrate how art and architecture can enrich one another. The exterior involves geometric shapes, symmetry, and unusual facades. Although there are no windows, natural light streams in through a truncated cylindrical skylight that crowns the building. The architect worked with a scale model of the museum at the site and observed how light hit it during different times of the day. These observations were used to cut the cylindrical skylight at an angle that maximizes sunlight entering the interior.

Angles play a critical role in creating modern architecture. They are also fundamental in trigonometry. In this section, we begin our study of trigonometry by looking at angles and methods for measuring them.

1 Recognize and use the vocabulary of angles.

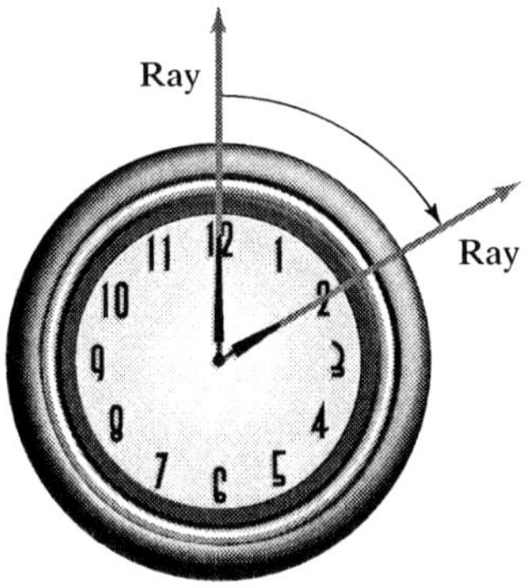

Figure 5.1 Clock with hands forming an angle

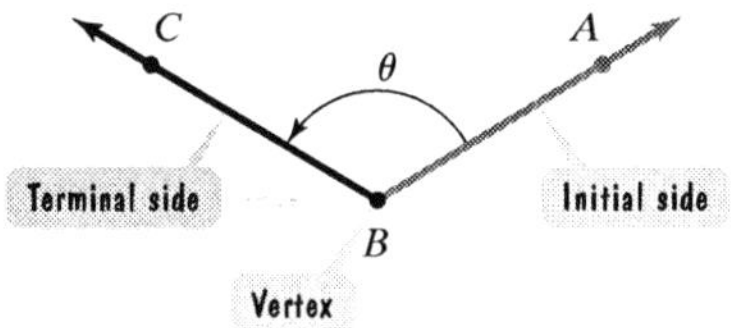

Figure 5.2 An angle; two rays with a common endpoint

Angles

The hour hand of a clock suggests a **ray,** a part of a line that has only one endpoint and extends forever in the opposite direction. An **angle** is formed by two rays that have a common endpoint. One ray is called the **initial side** and the other the **terminal side.**

A rotating ray is often a useful way to think about angles. The ray in Figure 5.1 rotates from 12 to 2. The ray pointing to 12 is the **initial side** and the ray pointing to 2 is the **terminal side.** The common endpoint of an angle's initial side and terminal side is the **vertex** of the angle.

Figure 5.2 shows an angle. The arrow near the vertex shows the direction and the amount of rotation from the initial side to the terminal side. Several methods can be used to name an angle. Lowercase Greek letters, such as α (alpha), β (beta), γ (gamma), and θ (theta), are often used.

An angle is in **standard position** if

- its vertex is at the origin of a rectangular coordinate system

and

- its initial side lies along the positive x-axis.

The angles in Figure 5.3 are both in standard position.

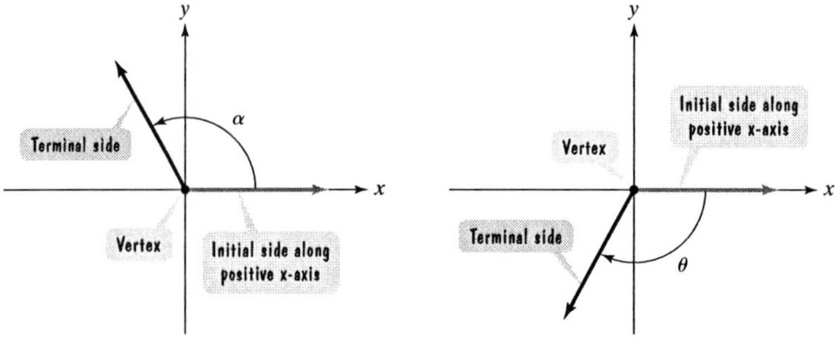

(a) α in standard position; α positive

(b) θ in standard position; θ negative

Figure 5.3 Two angles in standard position

When we see an initial side and a terminal side in place, there are two kinds of rotation that could have generated it. The arrow in Figure 5.3(a) indicates that the rotation from the initial side to the terminal side is in the counterclockwise direction. **Positive angles** are generated by counterclockwise rotation. Thus, angle α is positive. By contrast, the arrow in Figure 5.3(b) shows that the rotation from the initial side to the terminal side is in the clockwise direction. **Negative angles** are generated by clockwise rotation. Thus, angle θ is negative.

When an angle is in standard position, its terminal side can lie in a quadrant. We say that the angle **lies in that quadrant.** For example, in Figure 5.3(a), the terminal side of angle α lies in quadrant II. Thus, angle α lies in quadrant II. By contrast, in Figure 5.3(b), the terminal side of angle θ lies in quadrant III. Thus, angle θ lies in quadrant III.

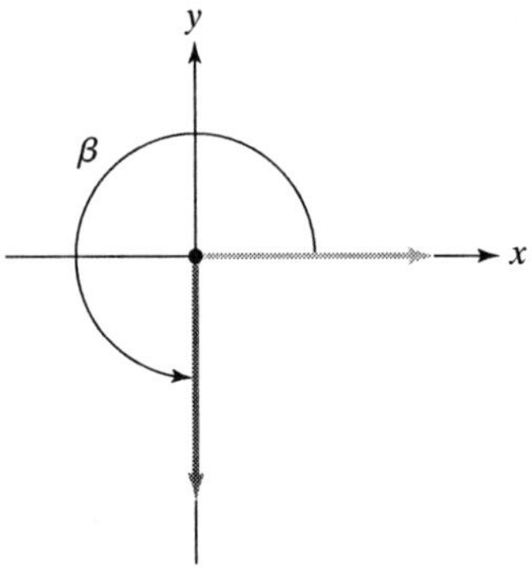

Figure 5.4 β is a quadrantal angle.

Must all angles in standard position lie in a quadrant? The answer is no. The terminal side can lie on the x-axis or the y-axis. For example, angle β in Figure 5.4 has a terminal side that lies on the negative y-axis. An angle is called a **quadrantal angle** if its terminal side lies on the x-axis or the y-axis. Angle β in Figure 5.4 is an example of a quadrantal angle.

2 Use degree measure.

Measuring Angles Using Degrees

Angles are measured by determining the amount of rotation from the initial side to the terminal side. One way to measure angles is in **degrees,** symbolized by a small, raised circle °. Think of the hour hand of a clock. From 12 noon to 12 midnight, the hour hand moves around in a complete circle. By definition, the ray has rotated through 360 degrees, or 360°. Using 360° as the amount of rotation of a ray back onto itself, a degree, 1°, is $\frac{1}{360}$ of a complete rotation.

A complete 360° rotation

Figure 5.5 shows angles classified by their degree measurement. An **acute angle** measures less than 90° [see Figure 5.5(a)]. A **right angle,** one quarter of a complete rotation, measures 90° [Figure 5.5(b)]. Examine the right angle—do you see a small square at the vertex? This symbol is used to indicate a right angle. An **obtuse angle** measures more than 90°, but less than 180° [Figure 5.5(c)]. Finally, a **straight angle,** one-half a complete rotation, measures 180° [Figure 5.5(d)].

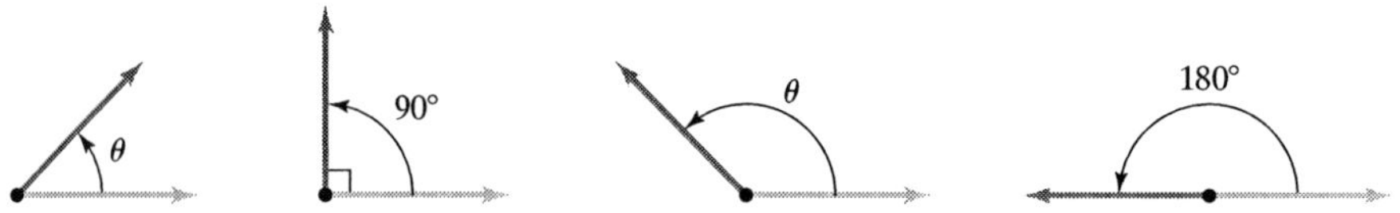

(a) **Acute angle** ($0° < \theta < 90°$)

(b) **Right angle** ($\frac{1}{4}$ rotation)

(c) **Obtuse angle** ($90° < \theta < 180°$)

(d) **Straight angle** ($\frac{1}{2}$ rotation)

Figure 5.5 Classifying angles by their degree measurement

3 Draw angles in standard position.

We will be using notation such as $\theta = 60°$ to refer to an angle θ whose measure is 60°. We also refer to *an angle of 60°* or a *60° angle*, rather than using the more precise (but cumbersome) phrase *an angle whose measure is 60°*.

Technology

Fractional parts of degrees are measured in minutes and seconds. One minute, written 1′, is $\frac{1}{60}$ degree: $1' = \frac{1}{60}°$.

One second, written 1″, is $\frac{1}{3600}$ degree: $1'' = \frac{1}{3600}°$.

For example,

$31° 47' 12''$

$= \left(31 + \frac{47}{60} + \frac{12}{3600}\right)°$

$\approx 31.787°$.

Many calculators have keys for changing an angle from degree, minute, second notation (D°M′S″) to a decimal form and vice versa.

EXAMPLE 1 Drawing Angles in Standard Position

Draw each angle in standard position:

a. a 45° angle **b.** a 225° angle **c.** a −135° angle **d.** a 405° angle.

Solution Because we are drawing angles in standard position, each vertex is at the origin and each initial side lies along the positive x-axis.

a. A 45° angle is half of a right angle. The angle lies in quadrant I and is shown in Figure 5.6(a).

b. A 225° angle is a positive angle. It has a counterclockwise rotation of 180° followed by a counterclockwise rotation of 45°. The angle lies in quadrant III and is shown in Figure 5.6(b).

c. A −135° angle is negative angle. It has a clockwise rotation of 90° followed by a clockwise rotation of 45°. The angle lies in quadrant III and is shown in Figure 5.6(c).

d. A 405° angle is a positive angle. It has a counterclockwise rotation of 360°, one complete rotation, followed by a counterclockwise rotation of 45°. The angle lies in quadrant I and is shown in Figure 5.6(d).

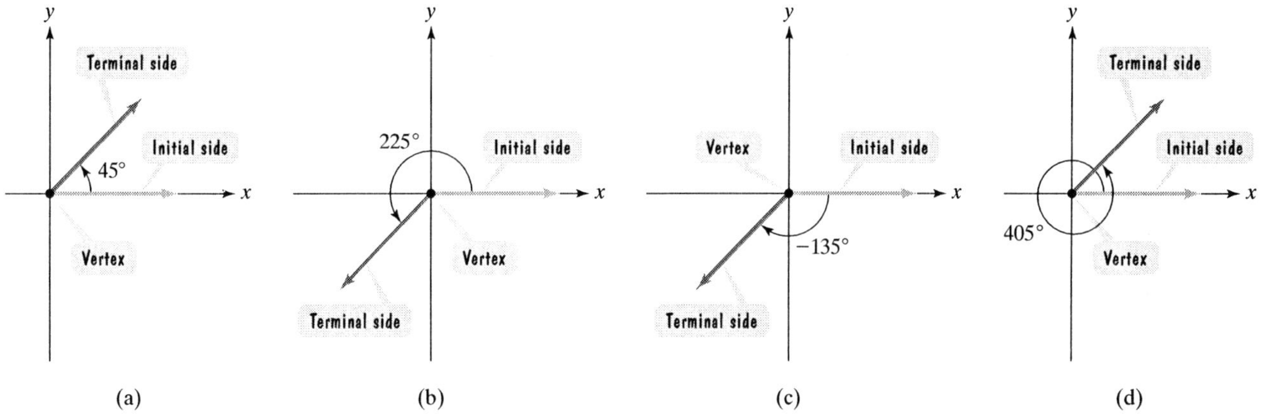

Figure 5.6 Four angles in standard position

Check Point 1 Draw each angle in standard position:

a. a 30° angle **b.** a 210° angle

c. a −120° angle **d.** a 390° angle.

4 Find coterminal angles.

Look at Figure 5.6 again. The 45° and 405° angles in parts (a) and (d) have the same initial and terminal sides. Similarly, the 225° and −135° angles in parts (b) and (c) have the same initial and terminal sides. Two angles with the same initial and terminal sides are called **coterminal angles.**

Every angle has infinitely many coterminal angles. Why? Think of an angle in standard position. One or more complete rotations of 360°, clockwise or counterclockwise, result in angles with the same initial and terminal sides as the original angle.

Coterminal Angles

An angle of $x°$ is coterminal with angles of

$$x° + k \cdot 360°$$

where k is an integer.

Two coterminal angles for an angle of $x°$ can be found by adding 360° to $x°$ and subtracting 360° from $x°$.

Counterclockwise Clocks

The counterclockwise rotation associated with positive angles was used in England to manufacture counterclockwise clocks. They ran backward but told the time perfectly correctly.

EXAMPLE 2 Finding Coterminal Angles

Assume the following angles are in standard position. Find a positive angle less than 360° that is coterminal with:

a. a 420° angle **b.** a −120° angle.

Solution We obtain the coterminal angle by adding or subtracting 360°. The requirement to obtain a positive angle less than 360° determines whether we should add or subtract.

a. For a 420° angle, subtract 360° to find a positive coterminal angle.

$$420° - 360° = 60°$$

A 60° angle is coterminal with a 420° angle. Figure 5.7(a) illustrates that these angles have the same initial and terminal sides.

b. For a −120° angle, add 360° to find a positive coterminal angle.

$$-120° + 360° = 240°$$

A 240° angle is coterminal with a −120° angle. Figure 5.7(b) illustrates that these angles have the same initial and terminal sides.

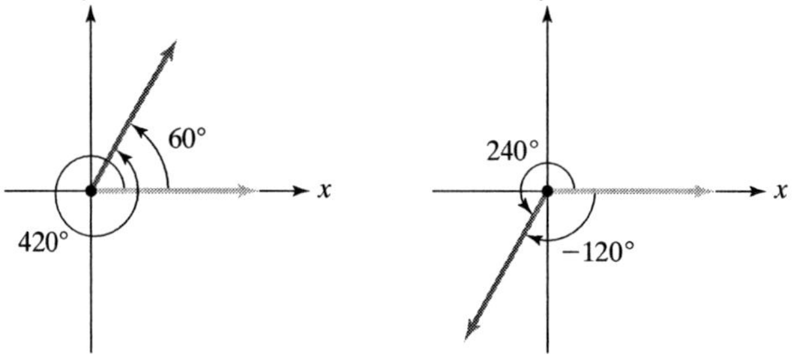

(a) Angles of 420° and 60° are coterminal.

(b) Angles of −120° and 240° are coterminal.

Figure 5.7 Pairs of coterminal angles

Check Point 2 Find a positive angle less than 360° that is coterminal with:

a. a 400° angle **b.** a −135° angle.

5 Find complements and supplements.

Two positive angles are **complements** if their sum is 90°. For example, angles of 70° and 20° are complements because $70° + 20° = 90°$.

Two positive angles are **supplements** if their sum is 180°. For example, angles of 130° and 50° are supplements because $130° + 50° = 180°$.

Finding Complements and Supplements

- For an $x°$ angle, the complement is a $90° - x°$ angle. Thus, the complement's measure is found by subtracting the angle's measure from 90°.
- For an $x°$ angle, the supplement is a $180° - x°$ angle. Thus, the supplement's measure is found by subtracting the angle's measure from 180°.

Because we use only positive angles for complements and supplements, some angles do not have complements and supplements.

EXAMPLE 3 Complements and Supplements

If possible, find the complement and the supplement of the given angle:

a. $\theta = 62°$ **b.** $\alpha = 123°$.

Solution We find the complement by subtracting the angle's measure from 90°. We find the supplement by subtracting the angle's measure from 180°.

a. We begin with $\theta = 62°$.

$$\text{complement} = 90° - 62° = 28°$$
$$\text{supplement} = 180° - 62° = 118°$$

For a 62° angle, the complement is a 28° angle and the supplement is a 118° angle.

b. Now we turn to $\alpha = 123°$. For the angle's complement, we consider subtracting 123° from 90°. The difference is negative. Because we use only positive angles for complements, a 123° angle has no complement. It does, however, have a supplement.

$$\text{supplement} = 180° - 123° = 57°$$

The supplement of a 123° angle is a 57° angle.

Check Point 3 If possible, find the complement and the supplement of the given angle:

a. $\theta = 78°$ **b.** $\alpha = 150°$.

6 Use radian measure.

Measuring Angles Using Radians

Another way to measure angles is in *radians*. Let's first define an angle measuring **1 radian.** We use a circle of radius r. In Figure 5.8, we've constructed an angle whose vertex is at the center of the circle. Such an angle is called a **central angle.** Notice that this central angle intercepts an arc along the circle measuring r units. The radius of the circle is also r units. The measure of such an angle is 1 radian.

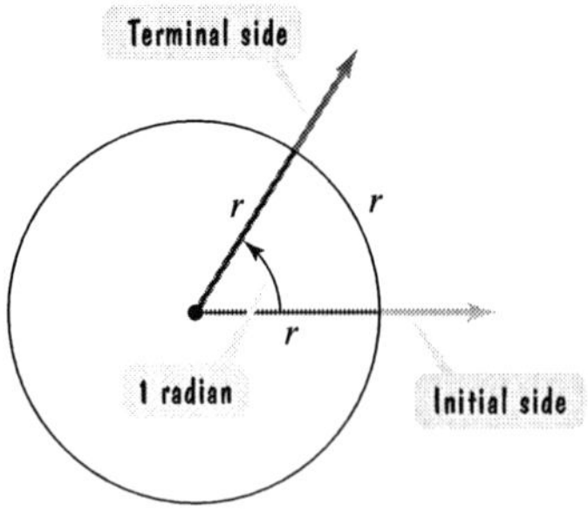

Figure 5.8 For a 1-radian angle, the intercepted arc and the radius are equal.

Definition of a Radian

One radian is the measure of the central angle of a circle that intercepts an arc equal in length to the radius of the circle.

The **radian measure** of any central angle is the length of the intercepted arc divided by the circle's radius. In Figure 5.9(a), the length of the arc intercepted by angle β is double the radius, r. We find the measure of angle β in radians by dividing the length of the intercepted arc by the radius.

$$\beta = \frac{\text{length of the intercepted arc}}{\text{radius}} = \frac{2r}{r} = 2$$

Thus, angle β measures 2 radians.

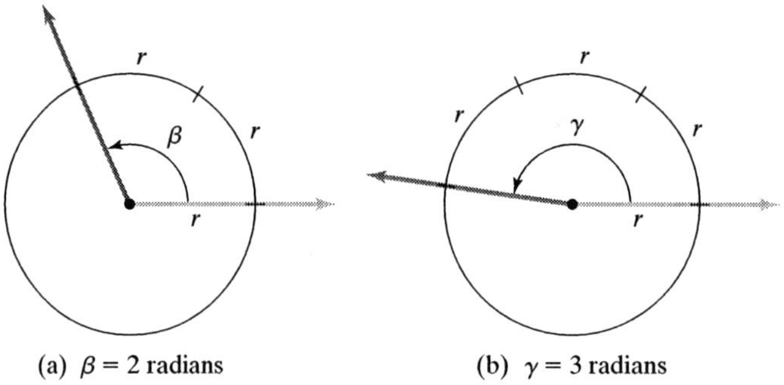

Figure 5.9 Two central angles measured in radians

In Figure 5.9(b), the length of the intercepted arc is triple the radius, r. Let us find the measure of angle γ:

$$\gamma = \frac{\text{length of the intercepted arc}}{\text{radius}} = \frac{3r}{r} = 3.$$

Thus, angle γ measures 3 radians.

Radian Measure

Consider an arc of length s on a circle of radius r. The measure of the central angle, θ, that intercepts the arc is

$$\theta = \frac{s}{r} \text{ radians.}$$

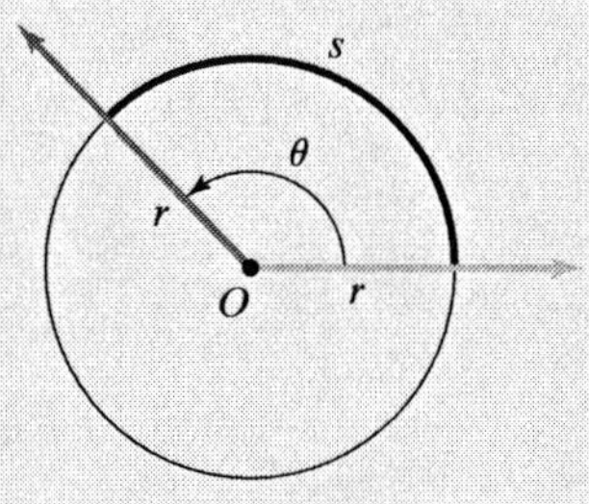

EXAMPLE 4 Computing Radian Measure

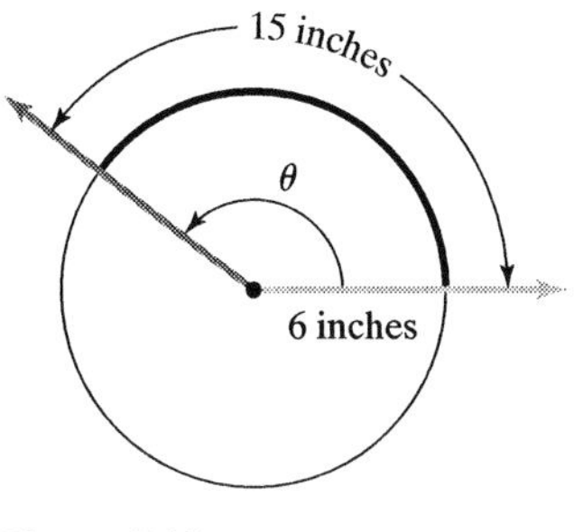

Figure 5.10

A central angle, θ, in a circle of radius 6 inches intercepts an arc of length 15 inches. What is the radian measure of θ?

Solution Angle θ is shown in Figure 5.10. The radian measure of a central angle is the length of the intercepted arc, s, divided by the circle's radius, r. The length of the intercepted arc is 15 inches: $s = 15$ inches. The circle's radius is 6 inches: $r = 6$ inches. Now we use the formula for radian measure to find the radian measure of θ.

$$\theta = \frac{s}{r} = \frac{15\ \text{inches}}{6\ \text{inches}} = 2.5$$

Thus, the radian measure of θ is 2.5.

Study Tip

Before applying the formula for radian measure, be sure that the same unit of length is used for the intercepted arc, s, and the radius, r.

In Example 4, notice that the units (inches) cancel when we use the formula for radian measure. We are left with a number with no units. Thus, if an angle θ has a measure of 2.5 radians, we can write $\theta = 2.5$ radians or $\theta = 2.5$. We will often include the word *radians* simply for emphasis. There should be no confusion as to whether radian or degree measure is being used. Why is this so? If θ has a degree measure of, say, 2.5°, we must include the degree symbol and write $\theta = 2.5°$, and *not* $\theta = 2.5$.

Check Point 4 A central angle, θ, in a circle of radius 12 feet intercepts an arc of length 42 feet. What is the radian measure of θ?

7 Convert between degrees and radians.

Relationship between Degrees and Radians

How can we obtain a relationship between degrees and radians? We compare the number of degrees and the number of radians in one complete rotation, shown in Figure 5.11. We know that 360° is the amount of rotation of a ray back onto itself. The length of the intercepted arc is equal to the circumference of the circle. Thus, the radian measure of this central angle is the circumference of the circle divided by the circle's radius, r. The circumference of a circle of radius r is $2\pi r$. We use the formula for radian measure to find the radian measure of the 360° angle.

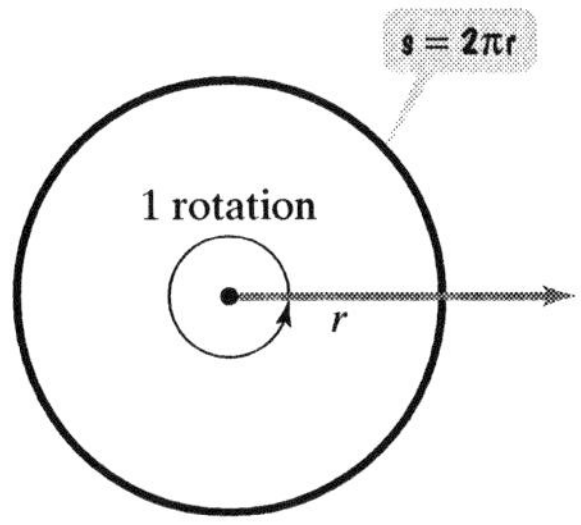

Figure 5.11 A complete rotation

$$\theta = \frac{s}{r} = \frac{\text{the circle's circumference}}{r} = \frac{2\pi r}{r} = 2\pi$$

Because one complete rotation measures 360° and 2π radians,

$$360° = 2\pi \text{ radians.}$$

Dividing both sides by 2, we have

$$180° = \pi \text{ radians.}$$

Dividing this last equation by 180° or π gives the conversion rules that appear on the next page.

Study Tip

The unit you are converting *to* appears in the *numerator* of the conversion factor.

Conversion between Degrees and Radians

Using the basic relationship π radians $= 180°$,

1. To convert degrees to radians, multiply degrees by $\frac{\pi \text{ radians}}{180°}$.
2. To convert radians to degrees, multiply radians by $\frac{180°}{\pi \text{ radians}}$.

Angles that are fractions of a complete rotation are usually expressed in radian measure as fractional multiples of π, rather than as decimal approximations. For example, we write $\theta = \frac{\pi}{2}$ rather than using the decimal approximation $\theta \approx 1.57$.

EXAMPLE 5 Converting from Degrees to Radians

Convert each angle in degrees to radians:

a. 30° **b.** 90° **c.** −135°.

Solution To convert degrees to radians, multiply by $\frac{\pi \text{ radians}}{180°}$. Observe how the degree units cancel.

a. $30° = 30° \cdot \frac{\pi \text{ radians}}{180°} = \frac{30\pi}{180} \text{ radians} = \frac{\pi}{6} \text{ radians}$

b. $90° = 90° \cdot \frac{\pi \text{ radians}}{180°} = \frac{90\pi}{180} \text{ radians} = \frac{\pi}{2} \text{ radians}$

c. $-135° = -135° \cdot \frac{\pi \text{ radians}}{180°} = -\frac{135\pi}{180} \text{ radians} = -\frac{3\pi}{4} \text{ radians}$

Divide the numerator and denominator by 45.

Check Point 5 Convert each angle in degrees to radians:

a. 60° **b.** 270° **c.** −300°.

EXAMPLE 6 Converting from Radians to Degrees

Convert each angle in radians to degrees:

a. $\frac{\pi}{3}$ radians **b.** $-\frac{5\pi}{3}$ radians **c.** 1 radian.

Solution To convert radians to degrees, multiply by $\frac{180°}{\pi \text{ radians}}$. Observe how the radian units cancel.

Study Tip

In Example 6(c), we see that 1 radian is approximately 57°. Keep in mind that a radian is much larger than a degree.

a. $\frac{\pi}{3}\text{ radians} = \frac{\pi \text{ radians}}{3} \cdot \frac{180°}{\pi \text{ radians}} = \frac{180°}{3} = 60°$

b. $-\frac{5\pi}{3}\text{ radians} = -\frac{5\pi \text{ radians}}{3} \cdot \frac{180°}{\pi \text{ radians}} = -\frac{5 \cdot 180°}{3} = -300°$

c. $1 \text{ radian} = 1 \text{ radian} \cdot \frac{180°}{\pi \text{ radians}} = \frac{180°}{\pi} \approx 57.3°$

Check Point 6 Convert each angle in radians to degrees:

a. $\frac{\pi}{4}$ radians **b.** $-\frac{4\pi}{3}$ radians **c.** 6 radians.

Figure 5.12 illustrates the degree and radian measures of angles that you will commonly see in trigonometry. Each angle is in standard position, so that the initial side lies along the positive x-axis. We will be using both degree and radian measure for these angles.

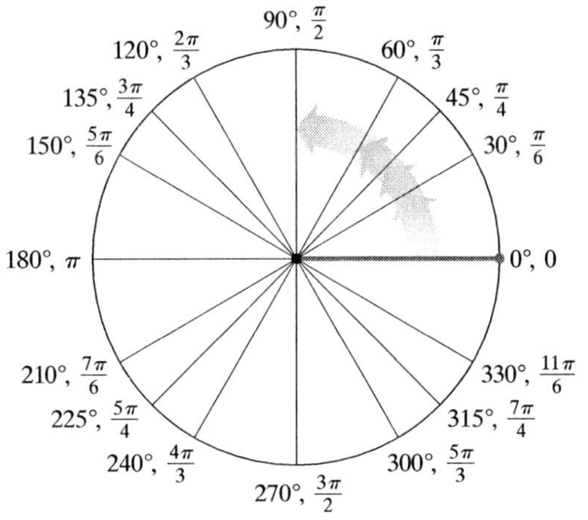

Figure 5.12 Degree and radian measures of selected angles

8 Find the length of a circular arc.

The Length of a Circular Arc

We can use the radian measure formula, $\theta = \frac{s}{r}$, to find the length of the arc of a circle. How do we do this? Remember that s represents the length of the arc intercepted by the central angle θ. Thus, by solving the formula for s, we have an equation for arc length.

The Length of a Circular Arc

Let r be the radius of a circle and θ the nonnegative radian measure of a central angle of the circle. The length of the arc intercepted by the central angle is

$$s = r\theta.$$

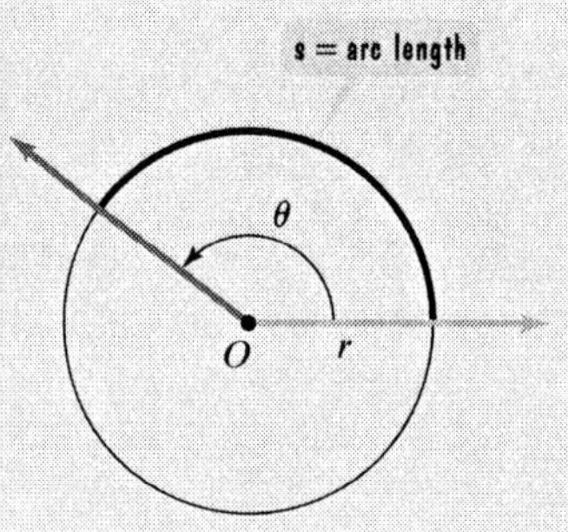

EXAMPLE 7 Finding the Length of a Circular Arc

A circle has a radius of 10 inches. Find the length of the arc intercepted by a central angle of 120°.

Study Tip

The unit used to describe the length of a circular arc is the same unit that is given in the circle's radius.

Solution The formula $s = r\theta$ can be used only when θ is expressed in radians. Thus, we begin by converting 120° to radians. Multiply by $\frac{\pi \text{ radians}}{180°}$.

$$120° = 120° \cdot \frac{\pi \text{ radians}}{180°} = \frac{120\pi}{180} \text{ radians} = \frac{2\pi}{3} \text{ radians}$$

Now we can use the formula $s = r\theta$ to find the length of the arc. The circle's radius is 10 inches: $r = 10$ inches. The measure of the central angle, in radians, is $\frac{2\pi}{3}$: $\theta = \frac{2\pi}{3}$. The length of the arc intercepted by this central angle is

$$s = r\theta = (10 \text{ inches})\left(\frac{2\pi}{3}\right) = \frac{20\pi}{3} \text{ inches} \approx 20.94 \text{ inches.}$$

Check Point 7 A circle has a radius of 6 inches. Find the length of the arc intercepted by a central angle of 45°. Express arc length in terms of π. Then round your answer to two decimal places.

9 Use linear and angular speed to describe motion on a circular path.

Linear and Angular Speed

A carousel contains four circular rows of animals. As the carousel revolves, the animals in the outer row travel a greater distance per unit of time than those in the inner rows. These animals have a greater *linear speed* than those in the inner rows. By contrast, all animals, regardless of the row, complete the same number of revolutions per unit of time. All animals in the four circular rows travel at the same *angular speed.*

Using v for linear speed and ω (omega) for angular speed, we define these two kinds of speeds along a circular path as follows:

Definitions of Linear and Angular Speed

If a point is in motion on a circle of radius r through an angle of θ radians in time t, then its **linear speed** is

$$v = \frac{s}{t}$$

where s is the arc length given by $s = r\theta$, and its **angular speed** is

$$\omega = \frac{\theta}{t}.$$

The hard drive in a computer rotates at 3600 revolutions per minute. This angular speed, expressed in revolutions per minute, can also be expressed in revolutions per second, radians per minute, and radians per second. Using 2π radians $= 1$ revolution, we express the angular speed of a hard drive in radians per minute as follows:

3600 revolutions per minute

$$= \frac{3600 \cancel{\text{revolutions}}}{1 \text{ minute}} \cdot \frac{2\pi \text{ radians}}{1 \cancel{\text{revolution}}} = \frac{7200\pi \text{ radians}}{1 \text{ minute}}$$

$$= 7200\pi \text{ radians per minute.}$$

We can establish a relationship between the two kinds of speed by dividing both sides of the arc length formula, $s = r\theta$, by t:

$$\frac{s}{t} = \frac{r\theta}{t} = r\frac{\theta}{t}.$$

This expression defines linear speed. (pointing to $\frac{s}{t}$)

This expression defines angular speed. (pointing to $\frac{\theta}{t}$)

Thus, linear speed is the product of the radius and the angular speed.

Linear Speed in Terms of Angular Speed

The linear speed, v, of a point a distance r from the center of rotation is given by

$$v = r\omega$$

where ω is the angular speed in radians per unit of time.

EXAMPLE 8 Finding Linear Speed

A wind machine used to generate electricity has blades that are 10 feet in length (see Figure 5.13). The propeller is rotating at four revolutions per second. Find the linear speed, in feet per second, of the tips of the blades.

Figure 5.13

Figure 5.13, repeated

Solution We are given ω, the angular speed.

$$\omega = 4 \text{ revolutions per second}$$

We use the formula $v = r\omega$ to find v, the linear speed. Before applying the formula, we must express ω in radians per second.

$$\omega = \frac{4 \cancel{\text{revolutions}}}{1 \text{ second}} \cdot \frac{2\pi \text{ radians}}{1 \cancel{\text{revolution}}} = \frac{8\pi \text{ radians}}{1 \text{ second}} \text{ or } \frac{8\pi}{1 \text{ second}}$$

The angular speed of the propeller is 8π radians per second. The linear speed is

$$v = r\omega = 10 \text{ feet} \cdot \frac{8\pi}{1 \text{ second}} = \frac{80\pi \text{ feet}}{\text{second}}.$$

The linear speed of the tips of the blades is 80π feet per second, which is approximately 251 feet per second.

Check Point 8 A 45-rpm record has an angular speed of 45 revolutions per minute. Find the linear speed, in inches per minute, at the point where the needle is 1.5 inches from the record's center.

EXERCISE SET 5.1

Practice Exercises

In Exercises 1–6, each angle is in standard position. Determine the quadrant in which the angle lies.

1. 145° **2.** 285°

3. −100° **4.** −110°

5. 362° **6.** 364°

In Exercises 7–10, classify the angle as acute, right, straight, or obtuse.

7.

8.

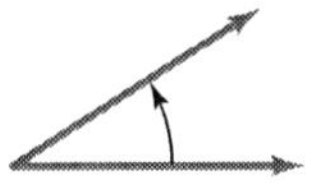

9.

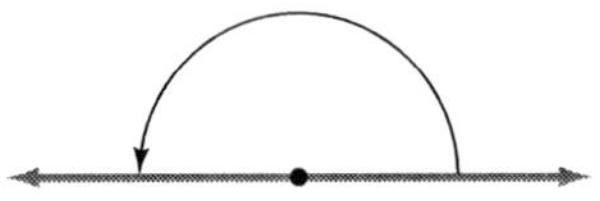

10.

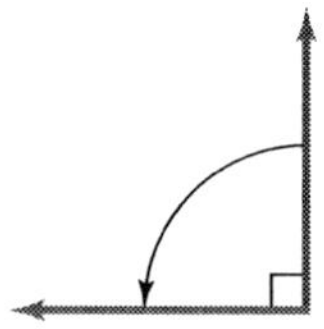

In Exercises 11–18, draw each angle in standard position.

11. 135° **12.** 120°

13. −150° **14.** −240°

15. 420° **16.** 450°

17. −90° **18.** −270°

In Exercises 19–24, find a positive angle less than 360° that is coterminal with the given angle.

19. 395° **20.** 415°

21. −150° **22.** −160°

23. −45° **24.** −40°

In Exercises 25–30, if possible, find the complement and the supplement of the given angle.

25. 52° **26.** 85°

27. 37.4° **28.** 47.6°

29. 111° **30.** 95°

In Exercises 31–36, find the radian measure of the central angle of a circle of radius r that intercepts an arc of length s.

	Radius, r	Arc length, s
31.	10 inches	40 inches
32.	5 feet	30 feet
33.	6 yards	8 yards
34.	8 yards	18 yards
35.	1 meter	400 centimeters
36.	1 meter	600 centimeters

In Exercises 37–44, convert each angle in degrees to radians. Express your answer as a multiple of π.

37. 45° **38.** 18°

39. 135° **40.** 150°

41. 300° **42.** 330°

43. −225° **44.** −270°

In Exercises 45–52, convert each angle in radians to degrees.

45. $\frac{\pi}{2}$ **46.** $\frac{\pi}{9}$

47. $\frac{2\pi}{3}$ **48.** $\frac{3\pi}{4}$

49. $\frac{7\pi}{6}$ **50.** $\frac{11\pi}{6}$

51. -3π **52.** -4π

In Exercises 53–58, convert each angle in degrees to radians. Round to two decimal places.

53. 18° **54.** 76°

55. −40° **56.** −50°

57. 200° **58.** 250°

In Exercises 59–64, convert each angle in radians to degrees. Round to two decimal places.

59. 2 radians **60.** 3 radians

61. $\frac{\pi}{13}$ radians **62.** $\frac{\pi}{17}$ radians

63. −4.8 radians **64.** −5.2 radians

In Exercises 65–68, find the length of the arc on a circle of radius r intercepted by a central angle θ. Express arc length in terms of π. Then round your answer to two decimal places.

	Radius, r	Central angle, θ
65.	12 inches	$\theta = 45°$
66.	16 inches	$\theta = 60°$
67.	8 feet	$\theta = 225°$
68.	9 yards	$\theta = 315°$

In Exercises 69–70, express each angular speed in radians per second.

69. 6 revolutions per second

70. 20 revolutions per second

Application Exercises

71. The minute hand of a clock moves from 12 to 2 o'clock, or $\frac{1}{6}$ of a complete revolution. Through how many degrees does it move? Through how many radians does it move?

72. The minute hand of a clock moves from 12 to 4 o'clock, or $\frac{1}{3}$ of a complete revolution. Through how many degrees does it move? Through how many radians does it move?

73. The minute hand of a clock is 8 inches long and moves from 12 to 2 o'clock. How far does the tip of the minute hand move? Express your answer in terms of π and then round to two decimal places.

74. The minute hand of a clock is 6 inches long and moves from 12 to 4 o'clock. How far does the tip of the minute hand move? Express your answer in terms of π and then round to two decimal places.

75. The figure shows a highway sign that warns of a railway crossing. The lines that form the cross pass through the circle's center and intersect at right angles. If the radius of the circle is 24 inches, find the length of each of the four arcs formed by the cross. Express your answer in terms of π and then round to two decimal places.

76. The radius of a wheel is 80 centimeters. If the wheel rotates through an angle of 60°, how many centimeters does it move? Express your answer in terms of π and then round to two decimal places.

How do we measure the distance between two points, A and B, on Earth? We measure along a circle with a center, C, at the center of Earth. The radius of the circle is equal to the distance from C to the surface. Use the fact that Earth is a sphere of radius equal to approximately 4000 miles to solve Exercises 77–80.

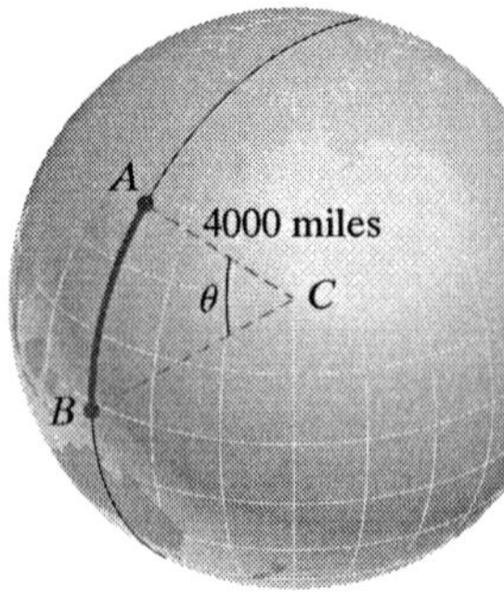

77. If two points, A and B, are 8000 miles apart, express angle θ in radians and in degrees.

78. If two points, A and B, are 10,000 miles apart, express angle θ in radians and in degrees.

79. If $\theta = 30°$, find the distance between A and B to the nearest mile.

80. If $\theta = 10°$, find the distance between A and B to the nearest mile.

81. The angular speed of a point on Earth is $\frac{\pi}{12}$ radians per hour. The Equator lies on a circle of radius approximately 4000 miles. Find the linear velocity, in miles per hour, of a point on the Equator.

82. A ferris wheel has a radius of 25 feet. The wheel is rotating at three revolutions per minute. Find the linear speed, in feet per minute, of this ferris wheel.

83. A water wheel has a radius of 12 feet. The wheel is rotating at 20 revolutions per minute. Find the linear speed, in feet per minute, of the water.

84. On a carousel, the outer row of animals is 20 feet from the center. The inner row of animals is 10 feet from the center. The carousel is rotating at 2.5 revolutions per minute. What is the difference, in feet per minute, in the linear speeds of the animals in the outer and inner rows? Round to the nearest foot per minute.

Writing in Mathematics

85. What is an angle?

86. What determines the size of an angle?

87. Describe an angle in standard position.

88. Explain the difference between positive and negative angles. What are coterminal angles?

89. Explain what is meant by one radian.

90. Explain how to find the radian measure of a central angle.

91. Describe how to convert an angle in degrees to radians.

92. Explain how to convert an angle in radians to degrees.

93. Explain how to find the length of a circular arc.

94. If a carousel is rotating at 2.5 revolutions per minute, explain how to find the linear speed of a child seated on one of the animals.

95. The angular velocity of a point on Earth is $\frac{\pi}{12}$ radians per hour. Describe what happens every 24 hours.

96. Have you ever noticed that we use the vocabulary of angles in everyday speech? Here is an example:

> My opinion about art museums took a 180° turn after visiting the San Francisco Museum of Modern Art.

Explain what this means. Then give another example of the vocabulary of angles in everyday use.

Technology Exercises

In Exercises 97–100, use the keys on your calculator or graphing utility for converting an angle in degrees, minutes, and seconds (D°M′S″) into decimal form, and vice versa.

In Exercises 97–98, convert each angle to a decimal in degrees. Round your answer to two decimal places.

97. 30°15′10″

98. 65°45′20″

In Exercises 99–100, convert each angle to D°M′S″ form. Round your answer to the nearest second.

99. 30.42°

100. 50.42°

Critical Thinking Exercises

101. If $\theta = \frac{3}{2}$, is this angle larger or smaller than a right angle?

102. A railroad curve is laid out on a circle. What radius should be used if the track is to change direction by 20° in a distance of 100 miles? Round your answer to the nearest mile.

103. Assuming Earth to be a sphere of radius 4000 miles, how many miles north of the Equator is Miami, Florida, if it is 26° north from the Equator? Round your answer to the nearest mile.

SECTION 5.2 Right Triangle Trigonometry

Objectives

1. Use right triangles to evaluate trigonometric functions.
2. Find function values for $30°\left(\frac{\pi}{6}\right)$, $45°\left(\frac{\pi}{4}\right)$, and $60°\left(\frac{\pi}{3}\right)$.
3. Recognize and use fundamental identities.
4. Use equal cofunctions of complements.
5. Evaluate trigonometric functions with a calculator.
6. Use right triangle trigonometry to solve applied problems.

In the last century, Ang Rita Sherpa climbed Mount Everest ten times, all without the use of bottled oxygen.

Mountain climbers have forever been fascinated by reaching the top of Mount Everest, sometimes with tragic results. The mountain, on Asia's Tibet-Nepal border, is Earth's highest, peaking at an incredible 29,035 feet. The heights of mountains can be found using **trigonometry.** The word *trigonometry* means *measurement of triangles*. Trigonometry is used in navigation, building, and engineering. For centuries, Muslims used trigonometry and the stars to navigate across the Arabian desert to Mecca, the birthplace of the prophet Muhammad, the founder of Islam. The ancient Greeks used trigonometry to record the locations of thousands of stars and worked out the motion of the Moon relative to Earth. Today, trigonometry is used to study the structure of DNA, the master molecule that determines how we grow from a single cell to a complex, fully developed adult.

1 Use right triangles to evaluate trigonometric functions.

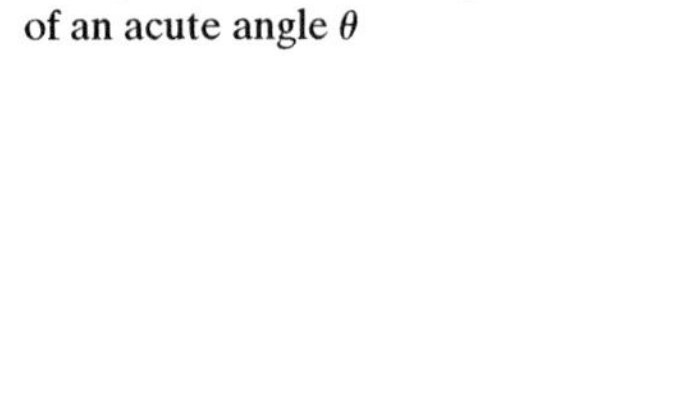

Figure 5.14 Naming a right triangle's sides from the point of view of an acute angle θ

The Six Trigonometric Functions

We begin the study of trigonometry by defining six functions, the six *trigonometric functions*. The inputs for these functions are measures of acute angles in right triangles. The outputs are the ratios of the lengths of the sides of right triangles.

Figure 5.14 shows a right triangle with one of its acute angles labeled θ. The side opposite the right angle is known as the **hypotenuse.** The other sides of the triangle are described by their position relative to the acute angle θ. One side is opposite θ and one is adjacent to θ.

The trigonometric functions have names that are words, rather than single letters such as f, g, and h. For example, the **sine of** θ is the length of the side opposite θ divided by the length of the hypotenuse:

$$\sin\theta = \frac{\text{length of side opposite } \theta}{\text{length of hypotenuse}}.$$

Input is the measure of an acute angle.

Output is the ratio of the lengths of the sides.

The ratio of lengths depends on angle θ and thus is a function of θ. The expression $\sin\theta$ really means $\sin(\theta)$, where sine is the name of the function and θ, the measure of an acute angle, is an input.

Here are the names of the six trigonometric functions, along with their abbreviations:

Name	Abbreviation	Name	Abbreviation
sine	sin	cosecant	csc
cosine	cos	secant	sec
tangent	tan	cotangent	cot

Now, let θ be an acute angle in a right triangle, shown in Figure 5.15. The length of the side opposite θ is a, the length of the side adjacent to θ is b, and the length of the hypotenuse is c.

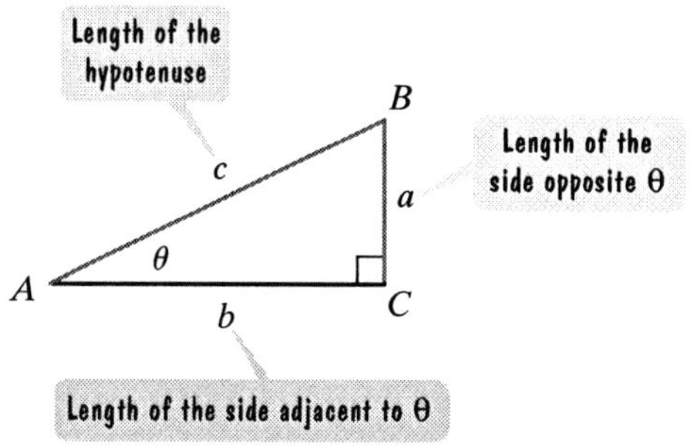

Figure 5.15

Right Triangle Definitions of Trigonometric Functions

See Figure 5.15. The six **trigonometric functions of the acute angle θ** are defined as follows:

$$\sin\theta = \frac{\text{length of side opposite angle } \theta}{\text{length of hypotenuse}} = \frac{a}{c} \qquad \csc\theta = \frac{\text{length of hypotenuse}}{\text{length of side opposite angle } \theta} = \frac{c}{a}$$

$$\cos\theta = \frac{\text{length of side adjacent to angle } \theta}{\text{length of hypotenuse}} = \frac{b}{c} \qquad \sec\theta = \frac{\text{length of hypotenuse}}{\text{length of side adjacent to angle } \theta} = \frac{c}{b}$$

$$\tan\theta = \frac{\text{length of side opposite angle } \theta}{\text{length of side adjacent to angle } \theta} = \frac{a}{b} \qquad \cot\theta = \frac{\text{length of side adjacent to angle } \theta}{\text{length of side opposite angle } \theta} = \frac{b}{a}$$

Each of the trigonometric functions of the acute angle θ is positive. Observe that the functions in the second column in the box are the reciprocals of the corresponding functions in the first column.

Figure 5.16 on the next page shows four right triangles of varying sizes. In each of the triangles, θ is the same acute angle, measuring approximately 56.3°. All four of these similar triangles have the same shape and the lengths of corresponding sides are in the same ratio. In each triangle, the tangent function has the same value: $\tan\theta = \frac{3}{2}$.

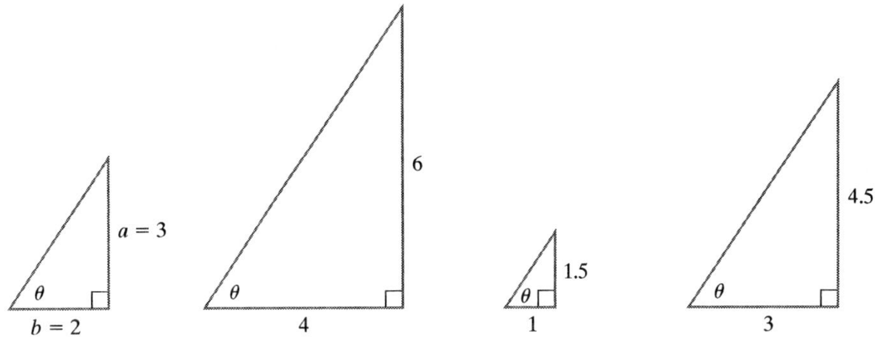

$\tan\theta = \frac{a}{b} = \frac{3}{2}$ $\quad \tan\theta = \frac{6}{4} = \frac{3}{2}$ $\quad \tan\theta = \frac{1.5}{1} = \frac{3}{2}$ $\quad \tan\theta = \frac{4.5}{3} = \frac{3}{2}$

Figure 5.16 A particular acute angle always gives the same ratio of opposite to adjacent sides.

In general, **the trigonometric function values of θ depend only on the size of angle θ, and not on the size of the triangle.**

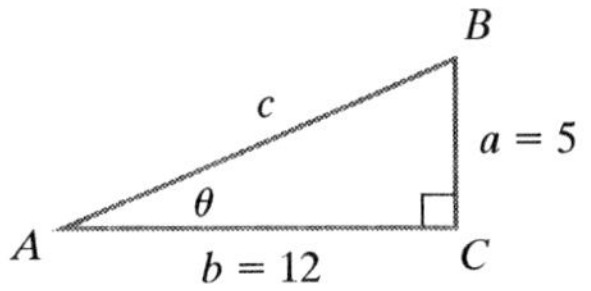

Figure 5.17

EXAMPLE 1 Evaluating Trigonometric Functions

Find the value of each of the six trigonometric functions of θ in Figure 5.17.

Solution We need to find the values of the six trigonometric functions of θ. However, we must know the lengths of all three sides of the triangle (a, b, and c) to evaluate all six functions. The values of a and b are given. We can use the Pythagorean Theorem, $c^2 = a^2 + b^2$, to find c.

$a = 5$ $\quad b = 12$

$$c^2 = a^2 + b^2 = 5^2 + 12^2 = 25 + 144 = 169$$

$$c = \sqrt{169} = 13$$

Now that we know the lengths of the three sides of the triangle, we apply the definitions of the six trigonometric functions of θ. Referring to these lengths as opposite, adjacent, and hypotenuse, we have

$$\sin\theta = \frac{\text{opposite}}{\text{hypotenuse}} = \frac{5}{13} \qquad \csc\theta = \frac{\text{hypotenuse}}{\text{opposite}} = \frac{13}{5}$$

$$\cos\theta = \frac{\text{adjacent}}{\text{hypotenuse}} = \frac{12}{13} \qquad \sec\theta = \frac{\text{hypotenuse}}{\text{adjacent}} = \frac{13}{12}$$

$$\tan\theta = \frac{\text{opposite}}{\text{adjacent}} = \frac{5}{12} \qquad \cot\theta = \frac{\text{adjacent}}{\text{opposite}} = \frac{12}{5}.$$

Study Tip

The functions in the second column are reciprocals of those in the first column. You can obtain their values by exchanging the numerator and denominator of the corresponding ratios in the first column.

Check Point 1 Find the value of each of the six trigonometric functions of θ in the figure.

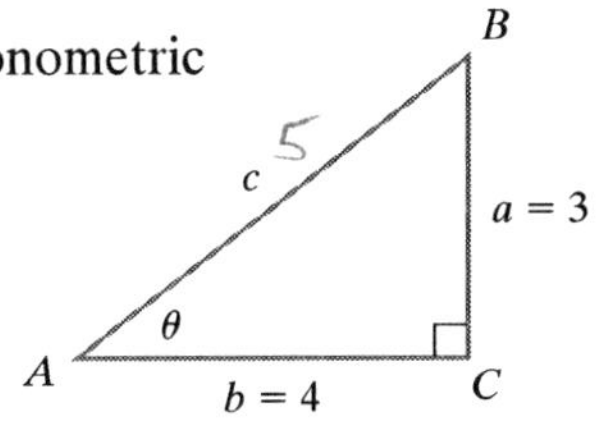

2 Find function values for $30°\left(\frac{\pi}{6}\right)$, $45°\left(\frac{\pi}{4}\right)$, and $60°\left(\frac{\pi}{3}\right)$.

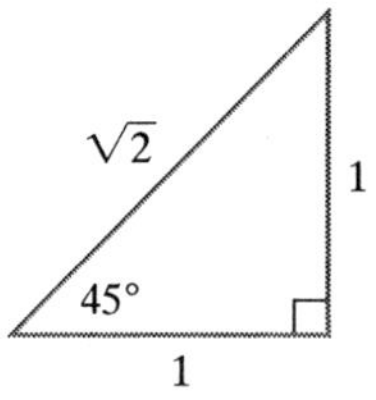

Figure 5.18 An isosceles right triangle

Function Values for Some Special Angles

A 45°, or $\frac{\pi}{4}$ radian, angle occurs frequently in trigonometry. How do we find the values of the trigonometric functions of 45°? We construct a right triangle with a 45° angle, shown in Figure 5.18. The triangle actually has two 45° angles. Thus, the triangle is isosceles—that is, it has two sides of the same length. Assume that each leg of the triangle has a length equal to 1. We can find the length of the hypotenuse using the Pythagorean Theorem.

$$(\text{length of hypotenuse})^2 = 1^2 + 1^2 = 2$$
$$\text{length of hypotenuse} = \sqrt{2}$$

With Figure 5.18, we can determine the trigonometric function values for 45°.

EXAMPLE 2 Evaluating Trigonometric Functions of 45°

Use Figure 5.18 to find sin 45°, cos 45°, and tan 45°.

Solution We apply the definitions of these three trigonometric functions.

$$\sin 45° = \frac{\text{length of side opposite } 45°}{\text{length of hypotenuse}} = \frac{1}{\sqrt{2}}$$
$$\cos 45° = \frac{\text{length of side adjacent to } 45°}{\text{length of hypotenuse}} = \frac{1}{\sqrt{2}}$$
$$\tan 45° = \frac{\text{length of side opposite } 45°}{\text{length of side adjacent to } 45°} = \frac{1}{1} = 1$$

Check Point 2 Use Figure 5.18 to find csc 45°, sec 45°, and cot 45°.

When you worked Check Point 2, did you actually use Figure 5.18 or did you use reciprocals to find the values?

$$\csc 45° = \sqrt{2} \qquad \sec 45° = \sqrt{2} \qquad \cot 45° = 1$$

Take the reciprocal of $\sin 45° = \frac{1}{\sqrt{2}}$.

Take the reciprocal of $\cos 45° = \frac{1}{\sqrt{2}}$.

Take the reciprocal of $\tan 45° = \frac{1}{1}$.

We found that $\sin 45° = \frac{1}{\sqrt{2}}$ and $\cos 45° = \frac{1}{\sqrt{2}}$. This value is often expressed by rationalizing the denominator:

$$\frac{1}{\sqrt{2}} = \frac{1}{\sqrt{2}} \cdot \frac{\sqrt{2}}{\sqrt{2}} = \frac{\sqrt{2}}{2}.$$

We are multiplying by 1 and not changing the value of $\frac{1}{\sqrt{2}}$.

Thus, $\sin 45° = \frac{\sqrt{2}}{2}$ and $\cos 45° = \frac{\sqrt{2}}{2}$.

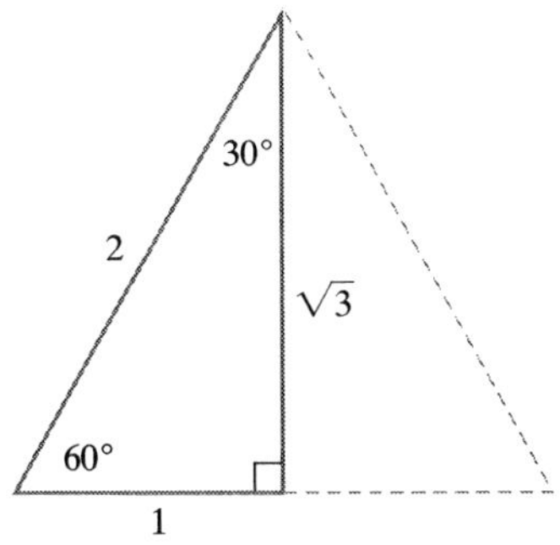

Figure 5.19 30°–60°–90° triangle

Two other angles that occur frequently in trigonometry are 30°, or $\frac{\pi}{6}$ radian, and 60°, or $\frac{\pi}{3}$ radian, angles. We can find the values of the trigonometric functions of 30° and 60° by using a right triangle. To form this right triangle, draw an equilateral triangle—that is a triangle with all sides the same length. Assume that each side has a length equal to 2. Now take half of the equilateral triangle. We obtain the right triangle in Figure 5.19. This right triangle has a hypotenuse of length 2 and a leg of length 1. The other leg has length a, which can be found using the Pythagorean Theorem.

$$a^2 + 1^2 = 2^2$$
$$a^2 + 1 = 4$$
$$a^2 = 3$$
$$a = \sqrt{3}$$

With the right triangle in Figure 5.19, we can determine the trigonometric functions for 30° and 60°.

EXAMPLE 3 Evaluating Trigonometric Functions of 30° and 60°

Use Figure 5.19 to find sin 60°, cos 60°, sin 30°, and cos 30°.

Solution We begin with 60°. Use the angle on the lower left in Figure 5.19.

$$\sin 60° = \frac{\text{length of side opposite } 60°}{\text{length of hypotenuse}} = \frac{\sqrt{3}}{2}$$

$$\cos 60° = \frac{\text{length of side adjacent to } 60°}{\text{length of hypotenuse}} = \frac{1}{2}$$

To find sin 30° and cos 30°, use the angle on the upper right in Figure 5.19.

$$\sin 30° = \frac{\text{length of side opposite } 30°}{\text{length of hypotenuse}} = \frac{1}{2}$$

$$\cos 30° = \frac{\text{length of side adjacent to } 30°}{\text{length of hypotenuse}} = \frac{\sqrt{3}}{2}$$

Check Point 3 Use Figure 5.19 to find tan 60° and tan 30°. If a radical appears in a denominator, rationalize the denominator.

Because we will often use the function values of 30°, 45°, and 60°, you should learn to construct the right triangles shown in Figures 5.18 and 5.19. With sufficient practice, you will memorize the values in the box on the next page.

Sines, Cosines, and Tangents of Special Angles

$$\sin 30° = \sin \frac{\pi}{6} = \frac{1}{2} \qquad \cos 30° = \cos \frac{\pi}{6} = \frac{\sqrt{3}}{2} \qquad \tan 30° = \tan \frac{\pi}{6} = \frac{\sqrt{3}}{3}$$

$$\sin 45° = \sin \frac{\pi}{4} = \frac{\sqrt{2}}{2} \qquad \cos 45° = \cos \frac{\pi}{4} = \frac{\sqrt{2}}{2} \qquad \tan 45° = \tan \frac{\pi}{4} = 1$$

$$\sin 60° = \sin \frac{\pi}{3} = \frac{\sqrt{3}}{2} \qquad \cos 60° = \cos \frac{\pi}{3} = \frac{1}{2} \qquad \tan 60° = \tan \frac{\pi}{3} = \sqrt{3}$$

3 Recognize and use fundamental identities.

Fundamental Identities

Many relationships exist among the six trigonometric functions. These relationships are described using **trigonometric identities.** For example, $\csc\theta$ is defined as the reciprocal of $\sin\theta$. This relationship can be expressed by the identity

$$\csc\theta = \frac{1}{\sin\theta}.$$

This identity is one of six **reciprocal identities.**

Reciprocal Identities

$$\sin\theta = \frac{1}{\csc\theta} \qquad \cos\theta = \frac{1}{\sec\theta} \qquad \tan\theta = \frac{1}{\cot\theta}$$

$$\csc\theta = \frac{1}{\sin\theta} \qquad \sec\theta = \frac{1}{\cos\theta} \qquad \cot\theta = \frac{1}{\tan\theta}$$

Two other relationships that follow from the definitions of the trigonometric functions are called the **quotient identities.**

Quotient Identities

$$\tan\theta = \frac{\sin\theta}{\cos\theta} \qquad \cot\theta = \frac{\cos\theta}{\sin\theta}$$

If $\sin\theta$ and $\cos\theta$ are known, a quotient identity and three reciprocal identities make it possible to find the value of each of the four remaining trigonometric functions.

EXAMPLE 4 Using Quotient and Reciprocal Identities

Given $\sin\theta = \frac{1}{2}$ and $\cos\theta = \frac{\sqrt{3}}{2}$, find the value of each of the four remaining trigonometric functions.

Solution We can find $\tan\theta$ by using the quotient identity that describes $\tan\theta$ as the quotient of $\sin\theta$ and $\cos\theta$.

$$\tan\theta = \frac{\sin\theta}{\cos\theta} = \frac{\frac{1}{2}}{\frac{\sqrt{3}}{2}} = \frac{1}{2}\cdot\frac{2}{\sqrt{3}} = \frac{1}{\sqrt{3}} = \frac{1}{\sqrt{3}}\cdot\frac{\sqrt{3}}{\sqrt{3}} = \frac{\sqrt{3}}{3}$$

Rationalize the denominator.

We use the reciprocal identities to find the value of each of the remaining three functions.

$$\csc\theta = \frac{1}{\sin\theta} = \frac{1}{\frac{1}{2}} = 2$$

$$\sec\theta = \frac{1}{\cos\theta} = \frac{1}{\frac{\sqrt{3}}{2}} = \frac{2}{\sqrt{3}} = \frac{2}{\sqrt{3}}\cdot\frac{\sqrt{3}}{\sqrt{3}} = \frac{2\sqrt{3}}{3}$$

Rationalize the denominator.

$$\cot\theta = \frac{1}{\tan\theta} = \frac{1}{\frac{1}{\sqrt{3}}} = \sqrt{3}$$

We found $\tan\theta = \frac{1}{\sqrt{3}}$. We could use $\tan\theta = \frac{\sqrt{3}}{3}$, but then we would have to rationalize the denominator.

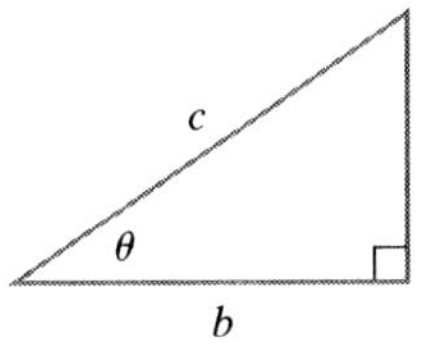

Figure 5.20

Check Point 4 Given $\sin\theta = \frac{2}{3}$ and $\cos\theta = \frac{\sqrt{5}}{3}$, find the value of each of the four remaining trigonometric functions.

Other relationships among trigonometric functions follow from the Pythagorean Theorem. Using Figure 5.20, the Pythagorean Theorem states that

$$a^2 + b^2 = c^2.$$

To obtain ratios that correspond to trigonometric functions, divide both sides of this equation by c^2.

$$\frac{a^2}{c^2} + \frac{b^2}{c^2} = 1 \quad \text{or} \quad \left(\frac{a}{c}\right)^2 + \left(\frac{b}{c}\right)^2 = 1$$

In Figure 5.20, $\sin\theta = \frac{a}{c}$, so this is $(\sin\theta)^2$.

In Figure 5.20, $\cos\theta = \frac{b}{c}$, so this is $(\cos\theta)^2$.

Based on the observations in the voice balloons, we see that

$$(\sin\theta)^2 + (\cos\theta)^2 = 1.$$

We will use the notation $\sin^2\theta$ for $(\sin\theta)^2$ and $\cos^2\theta$ for $(\cos\theta)^2$. With this notation, we can write the identity as

$$\sin^2\theta + \cos^2\theta = 1.$$

Two additional identities can be obtained from $a^2 + b^2 = c^2$ by dividing both sides by b^2 and a^2, respectively. The three identities are called the **Pythagorean identities.**

Pythagorean Identities

$$\sin^2\theta + \cos^2\theta = 1 \qquad 1 + \tan^2\theta = \sec^2\theta \qquad 1 + \cot^2\theta = \csc^2\theta$$

EXAMPLE 5 Using a Pythagorean Identity

Given that $\sin\theta = \frac{3}{5}$ and θ is an acute angle, find the value of $\cos\theta$ using a trigonometric identity.

Solution We can find the value of $\cos\theta$ by using the Pythagorean identity

$$\sin^2\theta + \cos^2\theta = 1.$$

$$\left(\frac{3}{5}\right)^2 + \cos^2\theta = 1 \qquad \text{We are given that } \sin\theta = \frac{3}{5}.$$

$$\frac{9}{25} + \cos^2\theta = 1 \qquad \text{Square } \frac{3}{5}: \left(\frac{3}{5}\right)^2 = \frac{3^2}{5^2} = \frac{9}{25}.$$

$$\cos^2\theta = 1 - \frac{9}{25} \qquad \text{Subtract } \frac{9}{25} \text{ from both sides.}$$

$$\cos^2\theta = \frac{16}{25} \qquad \text{Simplify: } 1 - \frac{9}{25} = \frac{25}{25} - \frac{9}{25} = \frac{16}{25}.$$

$$\cos\theta = \sqrt{\frac{16}{25}} = \frac{4}{5} \qquad \text{Because } \theta \text{ is an acute angle, } \cos\theta \text{ is positive.}$$

Thus, $\cos\theta = \frac{4}{5}$.

Check Point 5 Given that $\sin\theta = \frac{1}{2}$ and θ is an acute angle, find the value of $\cos\theta$ using a trigonometric identity.

4 Use equal cofunctions of complements.

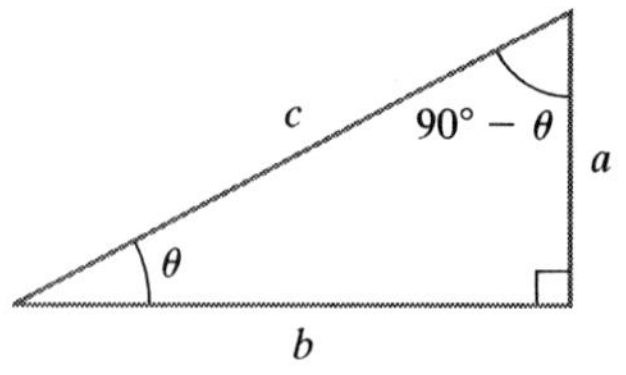

Figure 5.21

Trigonometric Functions and Complements

Another relationship among trigonometric functions is based on angles that are complements. Refer to Figure 5.21. Because the sum of the angles of any triangle is 180°, in a right triangle the sum of the acute angles is 90°. Thus, the acute angles are complements. If the degree measure of one acute angle is θ, then the degree measure of the other acute angle is $(90° - \theta)$. This angle is shown on the upper right in Figure 5.21.

Let's use Figure 5.21 to compare $\sin\theta$ and $\cos(90° - \theta)$.

$$\sin\theta = \frac{\text{length of side opposite } \theta}{\text{length of hypotenuse}} = \frac{a}{c}$$

$$\cos(90° - \theta) = \frac{\text{length of side adjacent to } (90° - \theta)}{\text{length of hypotenuse}} = \frac{a}{c}$$

Thus, $\sin\theta = \cos(90° - \theta)$. If two angles are complements, the sine of one equals the cosine of the other. Because of this relationship, the sine and cosine are called *cofunctions* of each other. The name *cosine* is a shortened form of the phrase *complement's sine*.

Any pair of trigonometric functions f and g for which

$$f(\theta) = g(90^\circ - \theta) \quad \text{and} \quad g(\theta) = f(90^\circ - \theta)$$

are called **cofunctions.** Using Figure 5.21, we can show that the tangent and cotangent are also cofunctions of each other. So are the secant and cosecant.

Cofunction Identities

The value of a trigonometric function of θ is equal to the cofunction of the complement of θ.

$$\sin\theta = \cos(90^\circ - \theta) \qquad \cos\theta = \sin(90^\circ - \theta)$$
$$\tan\theta = \cot(90^\circ - \theta) \qquad \cot\theta = \tan(90^\circ - \theta)$$
$$\sec\theta = \csc(90^\circ - \theta) \qquad \csc\theta = \sec(90^\circ - \theta)$$

If θ is in radians, replace 90° with $\frac{\pi}{2}$.

EXAMPLE 6

Find a cofunction with the same value as the given expression:

a. $\sin 72^\circ$ **b.** $\csc\frac{\pi}{3}$.

Solution Because the value of a trigonometric function of θ is equal to the cofunction of the complement of θ, we need to find the complement of each angle. We do this by subtracting the angle's measure from 90° or its radian equivalent, $\frac{\pi}{2}$.

a. $\sin 72^\circ = \cos(90^\circ - 72^\circ) = \cos 18^\circ$

We have a function and its cofunction.

b. $\csc\frac{\pi}{3} = \sec\left(\frac{\pi}{2} - \frac{\pi}{3}\right) = \sec\left(\frac{3\pi}{6} - \frac{2\pi}{6}\right) = \sec\frac{\pi}{6}$

We have a cofunction and its function.

Perform the subtraction using the least common denominator, 6.

Check Point 6 Find a cofunction with the same value as the given expression:

a. $\sin 46^\circ$ **b.** $\cot\frac{\pi}{12}$.

5 Evaluate trigonometric functions with a calculator.

Using a Calculator to Evaluate Trigonometric Functions

The values of the trigonometric functions obtained with the special triangles are exact values. For most angles other than 30°, 45°, and 60°, we approximate the value of each of the trigonometric functions using a calculator. The first step is to set the calculator to the correct *mode*, degrees or radians, depending on how the acute angle is measured.

Most calculators have keys marked [SIN], [COS], and [TAN]. For example, to find the value of sin 30°, set the calculator to the degree mode and enter 30 [SIN] on most scientific calculators and [SIN] 30 [ENTER] on most graphing calculators. Consult the manual for your calculator.

To evaluate the cosecant, secant, and cotangent functions, use the key for the respective reciprocal function, [SIN], [COS], or [TAN], and then use the reciprocal key. The reciprocal key is [$1/x$] on many scientific calculators and [x^{-1}] on many graphing calculators. For example, we can evaluate $\sec\frac{\pi}{12}$ using the following reciprocal relationship:

$$\sec\frac{\pi}{12} = \frac{1}{\cos\frac{\pi}{12}}.$$

Using the radian mode, enter one of the following keystroke sequences:

Many Scientific Calculators

[π] [÷] 12 [=] [COS] [$1/x$]

Many Graphing Calculators

[(] [COS] [(] [π] [÷] 12 [)] [)] [x^{-1}] [ENTER].

Rounding the display to four decimal places, we obtain $\sec\frac{\pi}{12} = 1.0353$.

EXAMPLE 7 Evaluating Trigonometric Functions with a Calculator

Use a calculator to find the value to four decimal places:

a. cos 48.2° **b.** cot 1.2.

Solution

Scientific Calculator Solution

Function	**Mode**	**Keystrokes**	**Display, rounded to four decimal places**
a. cos 48.2°	Degree	48.2 [COS]	0.6665
b. cot 1.2	Radian	1.2 [TAN] [$1/x$]	0.3888

Graphing Calculator Solution

Function	**Mode**	**Keystrokes**	**Display, rounded to four decimal places**
a. cos 48.2°	Degree	[COS] 48.2 [ENTER]	0.6665
b. cot 1.2	Radian	[(] [TAN] 1.2 [)] [)] [x^{-1}] [ENTER]	0.3888

Check Point 7 Use a calculator to find the value to four decimal places:

a. $\sin 72.8°$ **b.** $\csc 1.5$.

6 Use right triangle trigonometry to solve applied problems.

Applications

Many applications of right triangle trigonometry involve the angle made with an imaginary horizontal line. As shown in Figure 5.22, an angle formed by a horizontal line and the line of sight to an object that is above the horizontal line is called the **angle of elevation.** The angle formed by a horizontal line and the line of sight to an object that is below the horizontal line is called the **angle of depression.** Transits and sextants are instruments used to measure such angles.

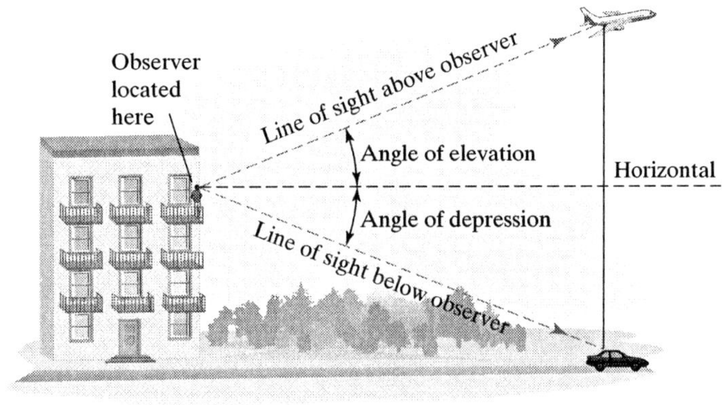

Figure 5.22

EXAMPLE 8 Problem Solving Using an Angle of Elevation

Sighting the top of a building, a surveyor measured the angle of elevation to be 22°. The transit is 5 feet above the ground and 300 feet from the building. Find the building's height.

Solution The situation is illustrated in Figure 5.23. Let a be the height of the portion of the building that lies above the transit. The height of the building is the transit's height, 5 feet, plus a. Thus, we need to identify a trigonometric function that will make it possible to find a. In terms of the 22° angle, we are looking for the side opposite the angle. The transit is 300 feet from the building, so the side adjacent to the 22° angle is 300 feet. Because we have a known angle, an unknown opposite side, and a known adjacent side, we select the tangent function.

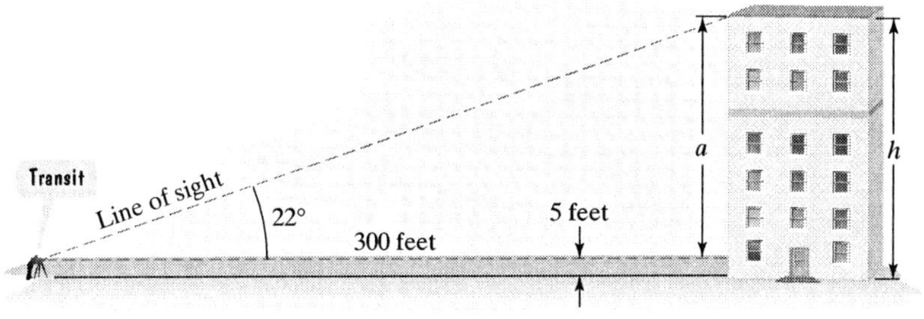

Figure 5.23

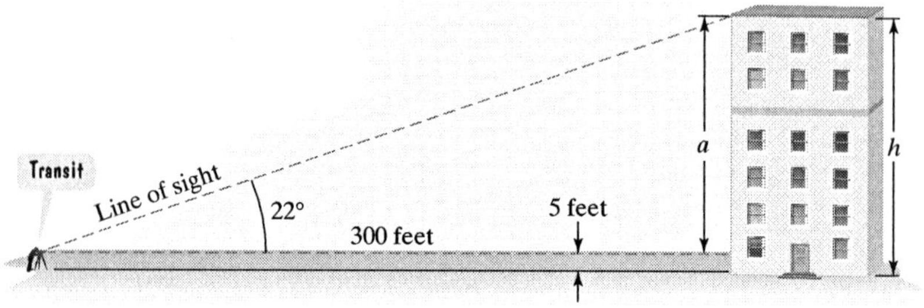

Figure 5.23, repeated

$$\tan 22° = \frac{a}{300}$$

Length of side opposite the 22° angle

Length of side adjacent to the 22° angle

$a = 300 \tan 22°$ — Multiply both sides of the equation by 300.

$a \approx 300(0.4040) \approx 121$ — Find $\tan 22°$ with a calculator in the degree mode.

The height of the part of the building above the transit is approximately 121 feet. Thus, the height of the building is determined by adding the transit's height, 5 feet, to 121 feet.

$$h \approx 5 + 121 = 126$$

The building's height is approximately 126 feet.

Check Point 8 The irregular blue shape in Figure 5.24 represents a lake. The distance across the lake, a, is unknown. To find this distance, a surveyor took the measurements shown in the figure. What is the distance across the lake?

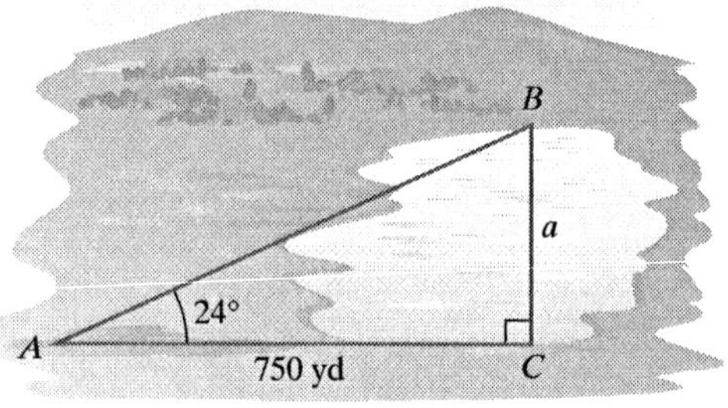

Figure 5.24

If two sides of a right triangle are known, an appropriate trigonometric function can be used to find an acute angle θ in the triangle. You will also need to use the *inverse key* on a calculator. This key uses a function value to display the acute angle θ. For example, suppose that $\sin\theta = 0.866$. We can find θ in the degree mode by using the secondary *inverse sine* key, usually labeled [SIN⁻¹].

Study Tip

[SIN^{-1}] is not a button you will actually press. It is the secondary function for the button labeled [SIN].

Many Scientific Calculators:

.866 [2nd] [SIN^{-1}]

Many Graphing Calculators:

[2nd] [SIN^{-1}] .866 [ENTER]

The display should show approximately 59.99, which can be rounded to 60. Thus, if $\sin\theta = 0.866$, then $\theta \approx 60°$.

EXAMPLE 9 Determining the Angle of Elevation

A building that is 21 meters tall casts a shadow 25 meters long. Find the angle of elevation of the sun to the nearest degree.

Solution The situation is illustrated in Figure 5.25. We are asked to find θ. We begin with the tangent function.

$$\tan\theta = \frac{\text{side opposite } \theta}{\text{side adjacent to } \theta} = \frac{21}{25}$$

We use a calculator in the degree mode to find θ.

Many Scientific Calculators:

[( ] 21 [÷] 25 [ )] [2nd] [TAN^{-1}]

Many Graphing Calculators:

[2nd] [TAN^{-1}] [( ] 21 [÷] 25 [ )] [ENTER]

The display should show approximately 40. Thus, the angle of elevation of the sun is approximately 40°.

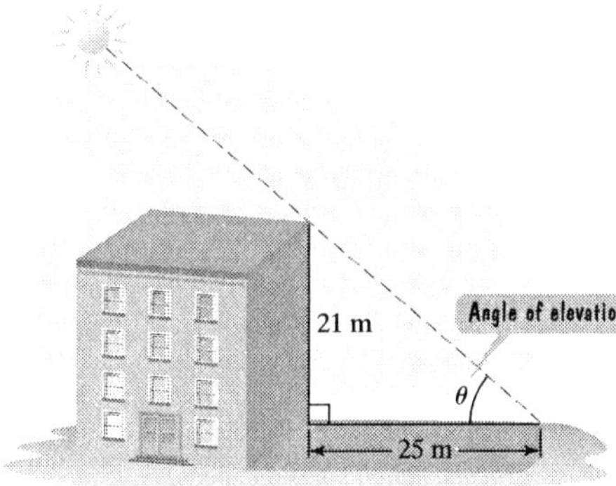

Figure 5.25

Check Point 9 A flagpole that is 14 meters tall casts a shadow 10 meters long. Find the angle of elevation of the sun to the nearest degree.

The Mountain Man

In the 1930s, a *National Geographic* team headed by Brad Washburn used trigonometry to create a map of the 5000-square-mile region of the Yukon, near the Canadian border. The team started with aerial photography. By drawing a network of angles on the photographs, the approximate locations of the major mountains and their rough heights were determined. The expedition then spent three months on foot to find the exact heights. Team members established two base points a known distance apart, one directly under the mountain's peak. By measuring the angle of elevation from one of the base points to the peak, the tangent function was used to determine the peak's height. The Yukon expedition was a major advance in the way maps are made.

EXERCISE SET 5.2

Practice Exercises

In Exercises 1–8, use the Pythagorean Theorem to find the length of the missing side of each right triangle. Then find the value of each of the six trigonometric functions of θ.

1.

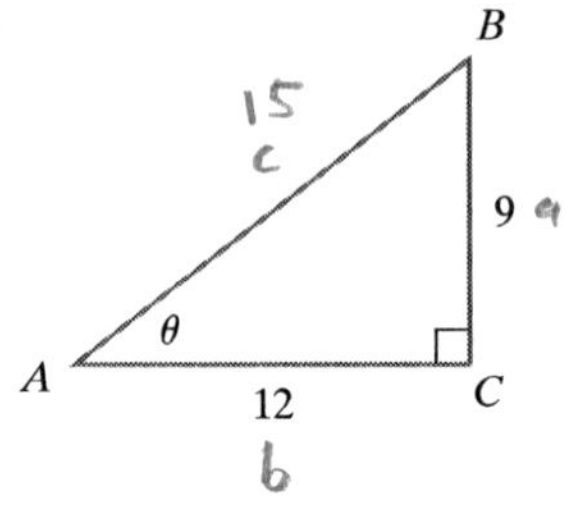

2.

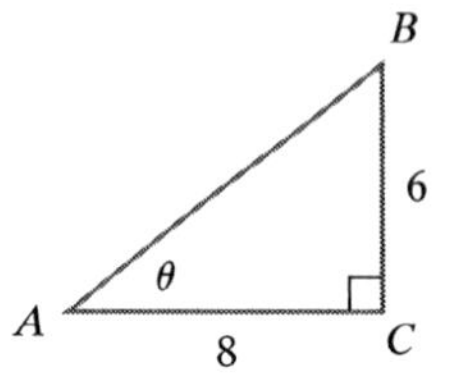

3.

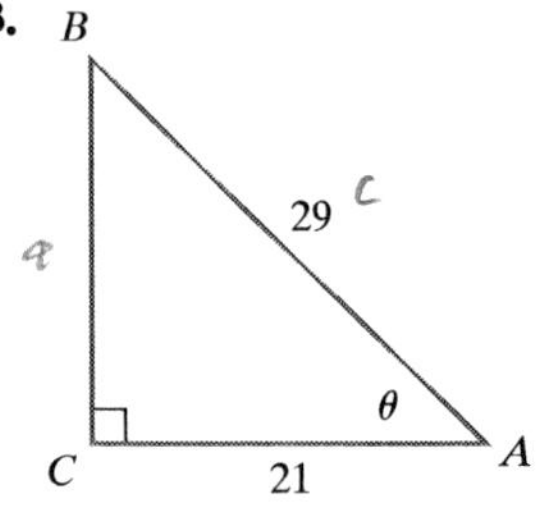

4.

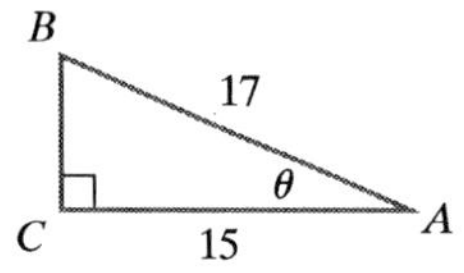

5.

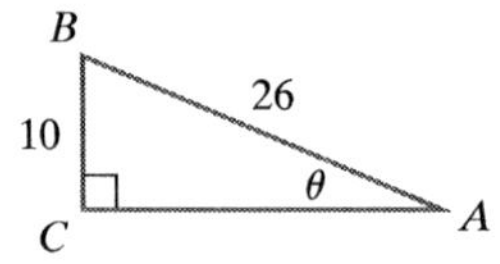

6.

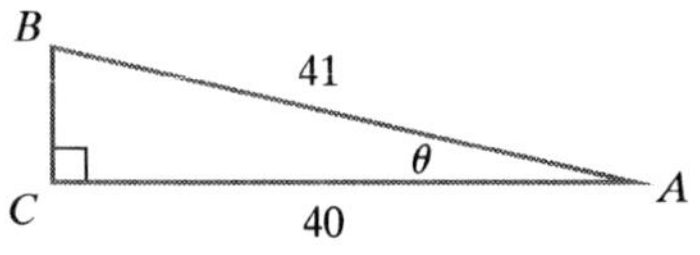

7.

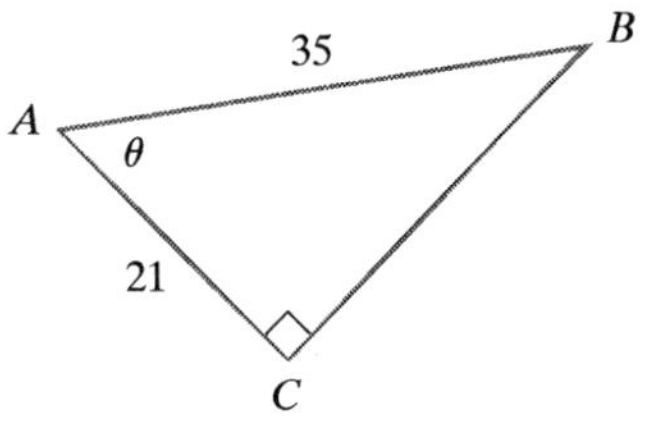

8.

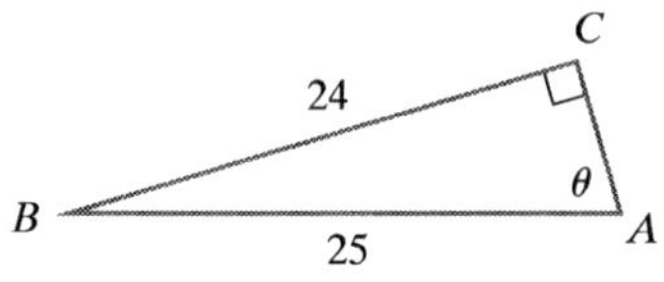

In Exercises 9–16, use the given triangles to evaluate each expression. If necessary, express the value without a square root in the denominator by rationalizing the denominator.

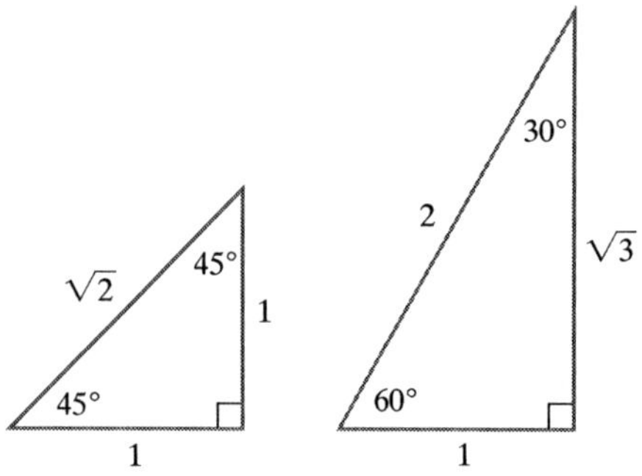

9. $\cos 30°$

10. $\tan 30°$

11. $\sec 45°$

12. $\csc 45°$

13. $\tan \frac{\pi}{3}$

14. $\cot \frac{\pi}{3}$

15. $\sin \frac{\pi}{4} - \cos \frac{\pi}{4}$

16. $\tan \frac{\pi}{4} + \csc \frac{\pi}{6}$

In Exercises 17–20, θ is an acute angle and $\sin\theta$ and $\cos\theta$ are given. Use identities to find $\tan\theta$, $\csc\theta$, $\sec\theta$, and $\cot\theta$. Where necessary, rationalize denominators.

17. $\sin\theta = \frac{8}{17}, \quad \cos\theta = \frac{15}{17}$

18. $\sin\theta = \frac{3}{5}, \quad \cos\theta = \frac{4}{5}$

19. $\sin\theta = \frac{1}{3}, \quad \cos\theta = \frac{2\sqrt{2}}{3}$

20. $\sin\theta = \frac{2}{3}, \quad \cos\theta = \frac{\sqrt{5}}{3}$

In Exercises 21–24, θ is an acute angle and $\sin\theta$ is given. Use the Pythagorean identity $\sin^2\theta + \cos^2\theta = 1$ to find $\cos\theta$.

21. $\sin\theta = \frac{6}{7}$

22. $\sin\theta = \frac{7}{8}$

23. $\sin\theta = \frac{\sqrt{39}}{8}$

24. $\sin\theta = \frac{\sqrt{21}}{5}$

In Exercises 25–30, use an identity to find the value of each expression. Do not use a calculator.

25. $\sin 37° \csc 37°$

26. $\cos 53° \sec 53°$

27. $\sin^2 \frac{\pi}{9} + \cos^2 \frac{\pi}{9}$

28. $\sin^2 \frac{\pi}{10} + \cos^2 \frac{\pi}{10}$

29. $\sec^2 23° - \tan^2 23°$

30. $\csc^2 63° - \cot^2 63°$

In Exercises 31–38, find a cofunction with the same value as the given expression.

31. $\sin 7°$

32. $\sin 19°$

33. $\csc 25°$

34. $\csc 35°$

35. $\tan \frac{\pi}{9}$

36. $\tan \frac{\pi}{7}$

37. $\cos \frac{2\pi}{5}$

38. $\cos \frac{3\pi}{8}$

In Exercises 39–48, use a calculator to find the value of the trigonometric function to four decimal places.

39. $\sin 38°$

40. $\cos 21°$

41. $\tan 32.7°$

42. $\tan 52.6°$

43. $\csc 17°$

44. $\sec 55°$

45. $\cos \frac{\pi}{10}$

46. $\sin \frac{3\pi}{10}$

47. $\cot \frac{\pi}{12}$

48. $\cot \frac{\pi}{18}$

In Exercises 49–54, find the measure of the side of the right triangle whose length is designated by a lowercase letter. Round answers to the nearest whole number.

49.

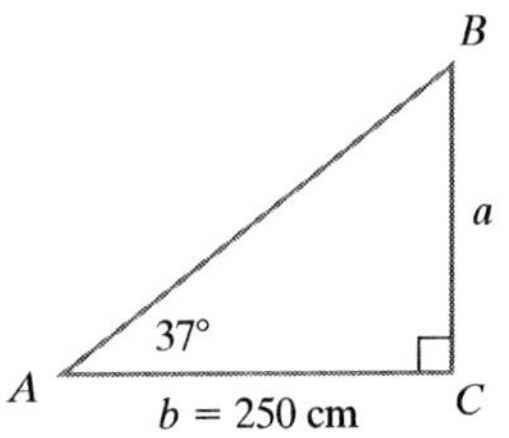

50.

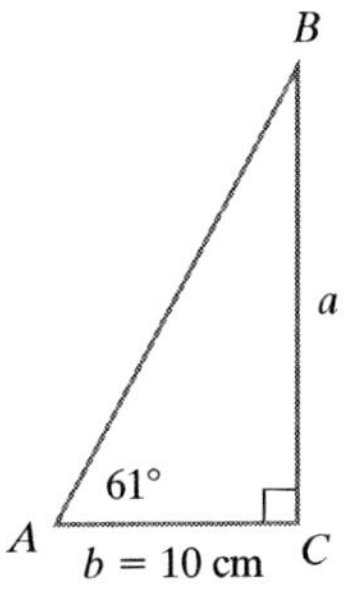

51.

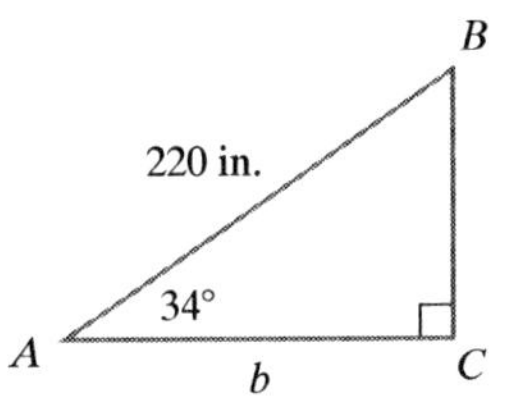

52.

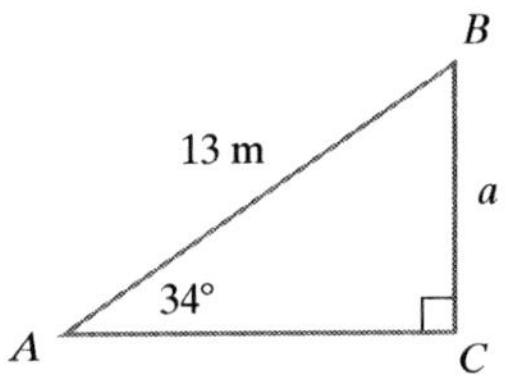

53.

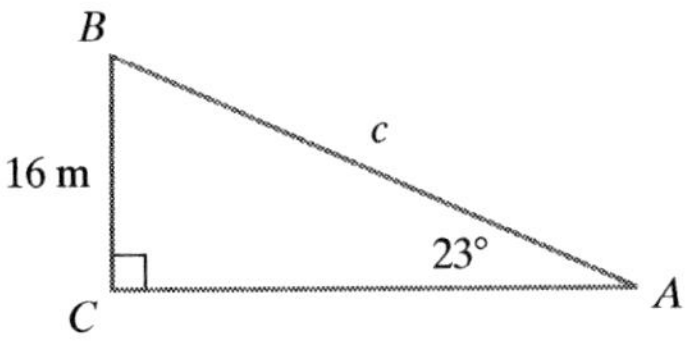

54.

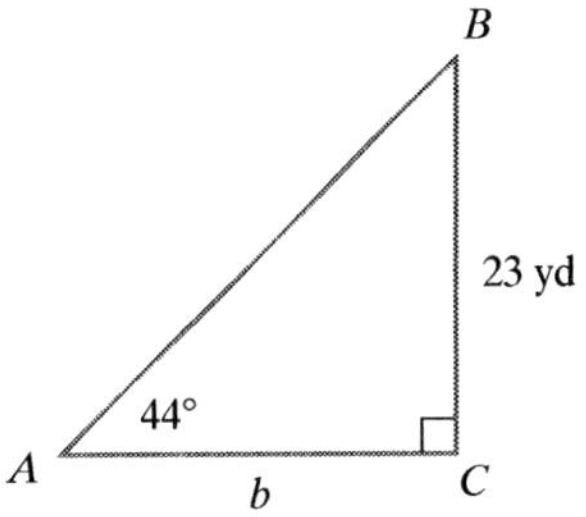

In Exercises 55–58, use a calculator to find the value of the acute angle θ to the nearest degree.

55. $\sin \theta = 0.2974$

56. $\cos \theta = 0.8771$

57. $\tan \theta = 4.6252$

58. $\tan \theta = 26.0307$

In Exercises 59–62, use a calculator to find the value of the acute angle θ in radians, rounded to three decimal places.

59. $\cos \theta = 0.4112$

60. $\sin \theta = 0.9499$

61. $\tan \theta = 0.4169$

62. $\tan \theta = 0.5117$

Application Exercises

63. To find the distance across a lake, a surveyor took the measurements in the figure shown. Use these measurements to determine how far it is across the lake. Round to the nearest yard.

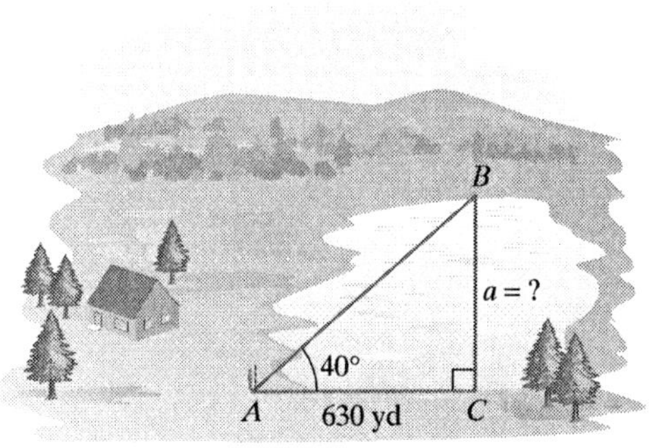

64. At a certain time of day, the angle of elevation of the sun is 40°. To the nearest foot, find the height of a tree whose shadow is 35 feet long.

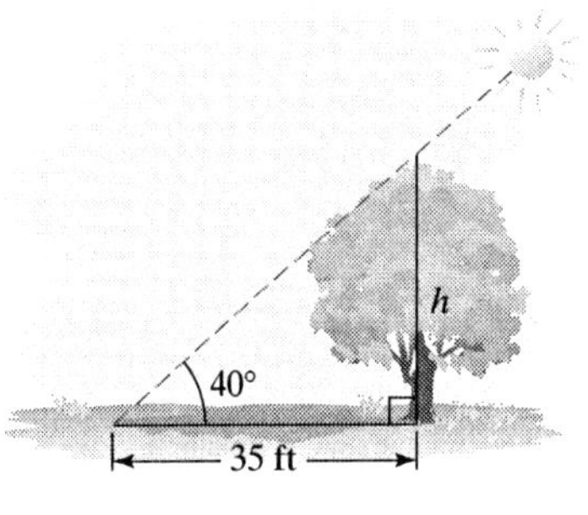

65. A tower that is 125 feet tall casts a shadow 172 feet long. Find the angle of elevation of the sun to the nearest degree.

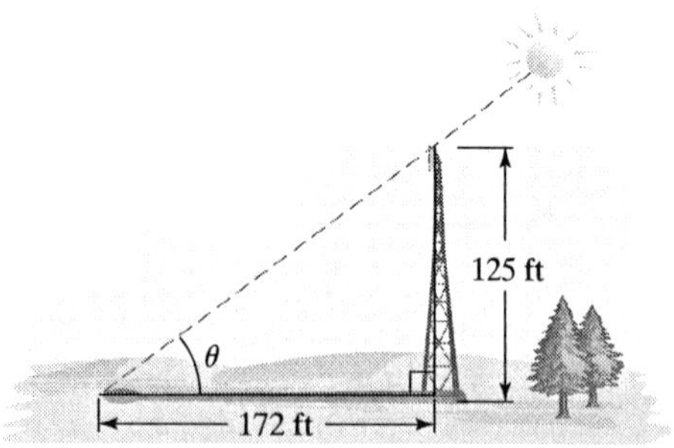

66. The Washington Monument is 555 feet high. If you stand one quarter of a mile, or 1320 feet, from the base of the monument and look to the top, find the angle of elevation to the nearest degree.

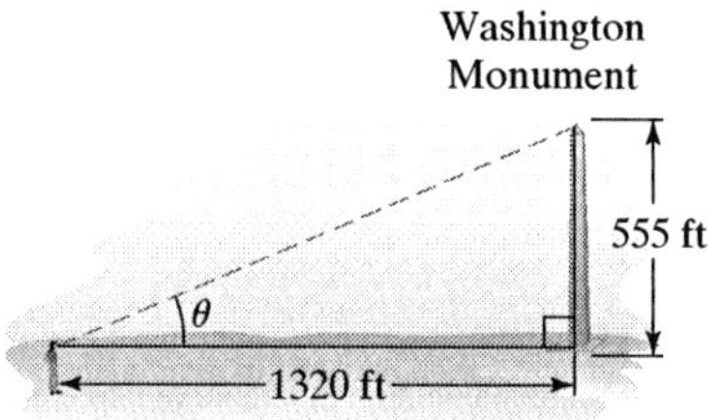

67. A plane rises from take-off and flies at an angle of 10° with the horizontal runway. When it has gained 500 feet, find the distance, to the nearest foot, the plane has flown.

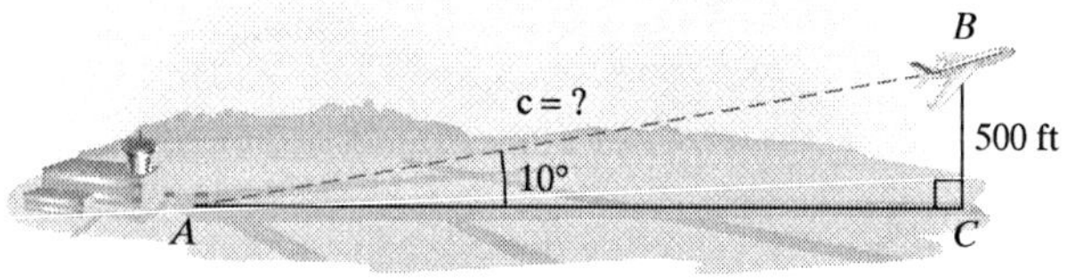

68. A road is inclined at an angle of 5°. After driving 5000 feet along this road, find the driver's increase in altitude. Round to the nearest foot.

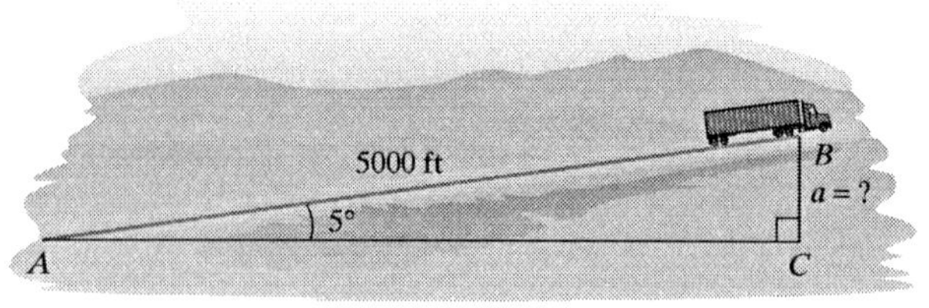

69. A telephone pole, shown at the top of the next column, is 60 feet tall. A guy wire 75 feet long is attached from the ground to the top of the pole. Find the angle between the wire and the pole to the nearest degree.

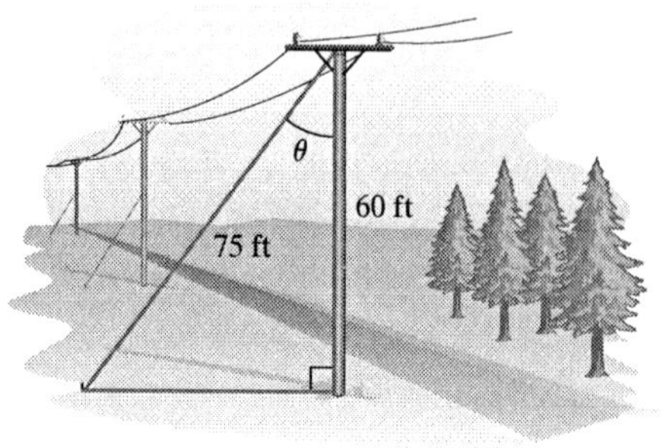

70. A telephone pole is 55 feet tall. A guy wire 80 feet long is attached from the ground to the top of the pole. Find the angle between the wire and the pole to the nearest degree.

Writing in Mathematics

71. If you are given the lengths of the sides of a right triangle, describe how to find the sine of either acute angle.

72. Describe one similarity and one difference between the definitions of $\sin\theta$ and $\cos\theta$, where θ is an acute angle of a right triangle.

73. Describe the triangle used to find the trigonometric functions of 45°.

74. Describe the triangle used to find the trigonometric functions of 30° and 60°.

75. What is a trigonometric identity?

76. Use words (not an equation) to describe one of the reciprocal identities.

77. Use words (not an equation) to describe one of the quotient identities.

78. Use words (not an equation) to describe one of the Pythagorean identities.

79. Describe a relationship among trigonometric functions that is based on angles that are complements.

80. Describe what is meant by an angle of elevation and an angle of depression.

81. Stonehenge, the famous "stone circle" in England, was built between 2750 B.C. and 1300 B.C. using solid stone blocks weighing over 99,000 pounds each. It required 550 people to pull a single stone up a ramp inclined at a 9° angle. Describe how right triangle trigonometry can be used to determine the distance the 550 workers had to drag a stone in order to raise it to a height of 30 feet.

Technology Exercises

82. Use a calculator in the radian mode to fill in the values in the following table. Then draw a conclusion about $\frac{\sin\theta}{\theta}$ as θ approaches 0.

θ	0.4	0.3	0.2	0.1	0.01	0.001	0.0001	0.00001
$\sin\theta$								
$\frac{\sin\theta}{\theta}$								

83. Use a calculator in the radian mode to fill in the values in the following table. Then draw a conclusion about $\frac{\cos\theta - 1}{\theta}$ as θ approaches 0.

θ	0.4	0.3	0.2	0.1	0.01	0.001	0.0001	0.00001
$\cos\theta$								
$\frac{\cos\theta - 1}{\theta}$								

Critical Thinking Exercises

84. Which one of the following is true?

a. $\frac{\tan 45°}{\tan 15°} = \tan 3°$

b. $\tan^2 15° - \sec^2 15° = -1$

c. $\sin 45° + \cos 45° = 1$

d. $\tan^2 5° = \tan 25°$

85. Explain why the sine or cosine of an acute angle cannot be greater than or equal to 1.

86. Describe what happens to the tangent of an acute angle as the angle gets close to 90°. What happens at 90°?

87. From the top of a 250-foot lighthouse, a plane is sighted overhead and a ship is observed directly below the plane. The angle of elevation of the plane is 22° and the angle of depression of the ship is 35°. Find **a.** the distance of the ship from the lighthouse; **b.** the plane's height above the water. Round to the nearest foot.

SECTION 5.3 *Trigonometric Functions of Any Angle*

Objectives

1. Use the definitions of trigonometric functions of any angle.
2. Use the signs of the trigonometric functions.
3. Find reference angles.
4. Use reference angles to evaluate trigonometric functions.

There is something comforting in the repetition of some of nature's patterns. The ocean level at a beach varies between high and low tide approximately every 12 hours. The number of hours of daylight oscillates from a maximum on the summer solstice, June 21, to a minimum on the winter solstice, December 21. Then it increases to the same maximum the following June 21. Some believe that cycles, called biorhythms, represent physical, emotional, and intellectual aspects of our lives. Throughout the remainder of this chapter, we will see how the trigonometric functions are used to model phenomena that occur again and again. To do this, we need to move beyond right triangles.

1 Use the definitions of trigonometric functions of any angle.

Trigonometric Functions of Any Angle

In the last section, we evaluated trigonometric functions of acute angles, such as that shown in Figure 5.26(a). Note that this angle is in standard position. The point $P = (x, y)$ is a point r units from the origin on the terminal side of θ. A right triangle is formed by drawing a perpendicular from $P = (x, y)$ to the x-axis. Note that y is the length of the side opposite θ and x is the length of the side adjacent to θ.

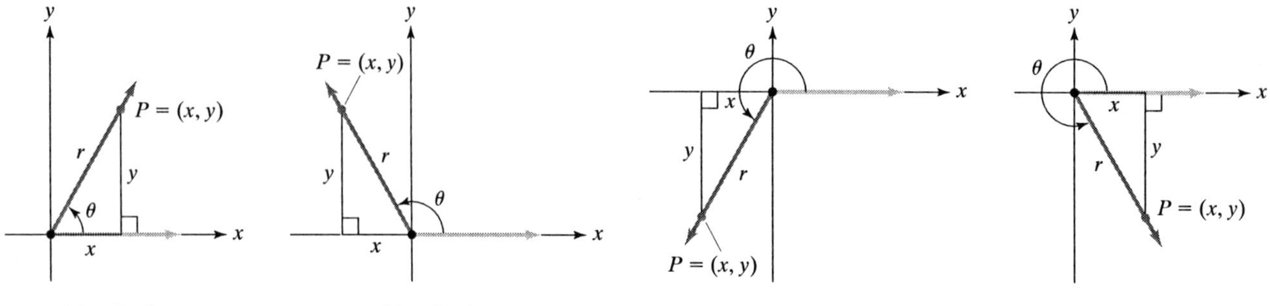

Figure 5.26

Figures 5.26(b), (c), and (d) show angles in standard position, but they are not acute. We can extend our definitions of the six trigonometric functions to include such angles, as well as quadrantal angles. (Recall that a quadrantal angle has its terminal side on the x-axis or y-axis; such angles are *not* shown in Figure 5.26.) The point $P = (x, y)$ may be any point on the terminal side of the angle θ other than the origin, (0, 0).

Study Tip

If θ is acute, we have the right triangle shown in Figure 5.26(a). In this situation, the definitions in the box are the right triangle definitions of the trigonometric functions. This should make it easier for you to remember the six definitions.

Definitions of Trigonometric Functions of Any Angle

Let θ be any angle in standard position, and let $P = (x, y)$ be a point on the terminal side of θ. If $r = \sqrt{x^2 + y^2}$ is the distance from (0, 0) to (x, y), as shown in Figure 5.26, the **six trigonometric functions of θ** are defined by the following ratios:

$$\sin\theta = \frac{y}{r} \qquad \cos\theta = \frac{x}{r} \qquad \tan\theta = \frac{y}{x},\ x \neq 0$$

$$\csc\theta = \frac{r}{y},\ y \neq 0 \qquad \sec\theta = \frac{r}{x},\ x \neq 0 \qquad \cot\theta = \frac{x}{y},\ y \neq 0.$$

Because the point $P = (x, y)$ is any point on the terminal side of θ other than the origin, $(0, 0), r = \sqrt{x^2 + y^2}$ cannot be zero. Examine the six trigonometric functions defined previously. Note that the denominator of the sine and cosine functions is r. Because $r \neq 0$, the sine and cosine functions are defined for any real value of the angle θ. This is not true for the other four trigonometric functions. Note that the denominator of the tangent and secant functions is x. These functions are not defined if $x = 0$. If the point $P = (x, y)$ is on the y-axis, then $x = 0$. Thus, the tangent and secant functions are undefined for all quadrantal angles with terminal sides on the positive or negative y-axis. Likewise, if $P = (x, y)$ is on the x-axis, then $y = 0$, and the cotangent and cosecant functions are undefined. The cotangent and cosecant functions are undefined for all quadrantal angles with terminal sides on the positive or negative x-axis.

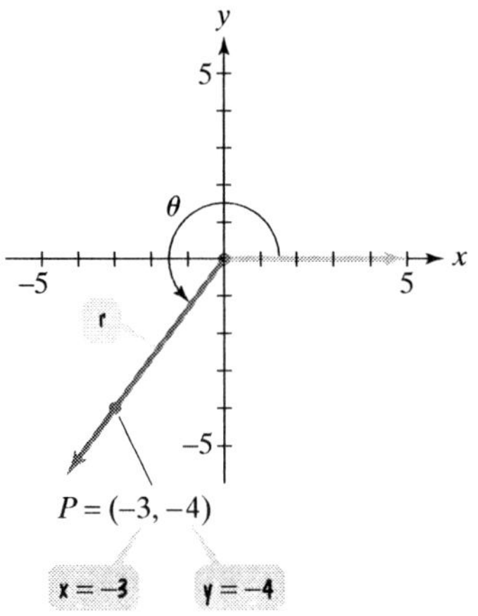

Figure 5.27

EXAMPLE 1 Evaluating Trigonometric Functions

Let $P = (-3, -4)$ be a point on the terminal side of θ. Find each of the six trigonometric functions of θ.

Solution The situation is shown in Figure 5.27. We need values for x, y, and r to evaluate all six trigonometric functions. We are given the values of x and y. Because $P = (-3, -4)$ is a point on the terminal side of θ, $x = -3$ and $y = -4$. Furthermore,

$$r = \sqrt{x^2 + y^2} = \sqrt{(-3)^2 + (-4)^2} = \sqrt{9 + 16} = \sqrt{25} = 5.$$

Now that we know x, y, and r, we can find the six trigonometric functions of θ.

$$\sin\theta = \frac{y}{r} = \frac{-4}{5} = -\frac{4}{5}, \quad \cos\theta = \frac{x}{r} = \frac{-3}{5} = -\frac{3}{5}, \quad \tan\theta = \frac{y}{x} = \frac{-4}{-3} = \frac{4}{3}$$

$$\csc\theta = \frac{r}{y} = \frac{5}{-4} = -\frac{5}{4}, \quad \sec\theta = \frac{r}{x} = \frac{5}{-3} = -\frac{5}{3}, \quad \cot\theta = \frac{x}{y} = \frac{-3}{-4} = \frac{3}{4}$$

These ratios are the reciprocals of those shown directly above.

Check Point 1 Let $P = (4, -3)$ be a point on the terminal side of θ. Find each of the six trigonometric functions of θ.

How do we find the values of the trigonometric functions for a quadrantal angle? First, draw the angle in standard position. Second, choose a point P on the angle's terminal side. The trigonometric function values of θ depend only on the size of θ and not on the distance of point P from the origin. Thus, we choose a point that is 1 unit from the origin. Finally, apply the definition of the appropriate trigonometric function.

EXAMPLE 2 Trigonometric Functions of Quadrantal Angles

Evaluate, if possible, the sine function and the tangent function at the following four quadrantal angles:

a. $\theta = 0° = 0$ **b.** $\theta = 90° = \frac{\pi}{2}$ **c.** $\theta = 180° = \pi$ **d.** $\theta = 270° = \frac{3\pi}{2}$.

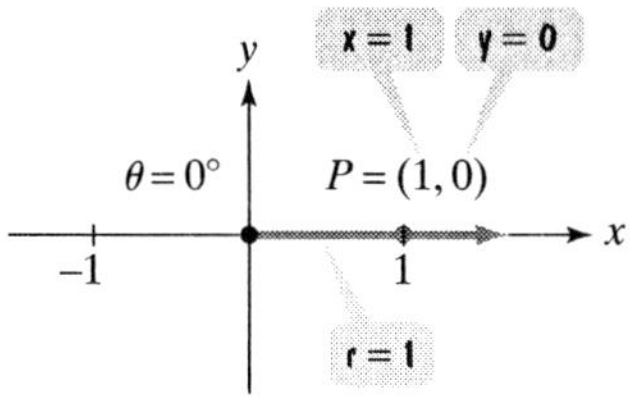

Solution

a. If $\theta = 0° = 0$ radians, then the terminal side of the angle is on the positive x-axis. Let us select the point $P = (1, 0)$ with $x = 1$ and $y = 0$. This point is 1 unit from the origin, so $r = 1$. Now that we know x, y, and r, we can apply the definitions of the sine and tangent functions. (The figure on the left is repeated at the top of the next page.)

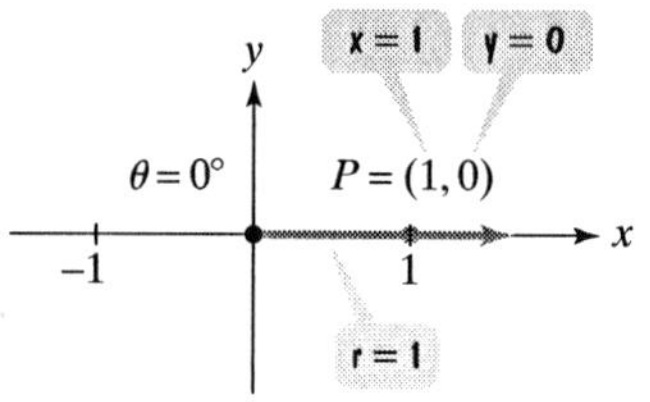

$$\sin 0^\circ = \sin 0 = \frac{y}{r} = \frac{0}{1} = 0$$

$$\tan 0^\circ = \tan 0 = \frac{y}{x} = \frac{0}{1} = 0$$

b. If $\theta = 90^\circ = \frac{\pi}{2}$ radians, then the terminal side of the angle is on the positive y-axis. Let us select the point $P = (0, 1)$ with $x = 0$ and $y = 1$. This point is 1 unit from the origin, so $r = 1$. Now that we know x, y, and r, we can apply the definitions of the sine and tangent functions.

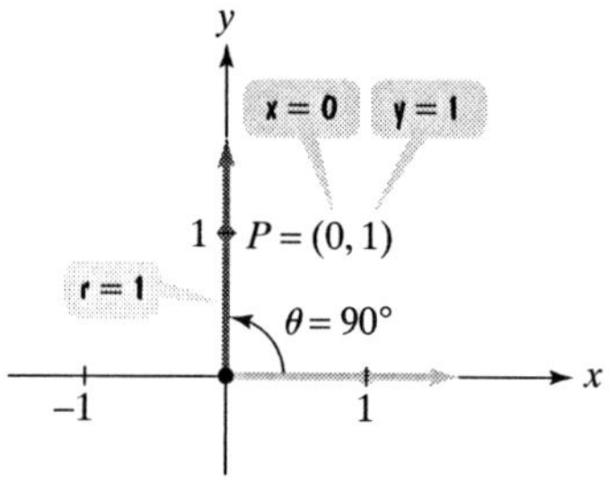

$$\sin 90^\circ = \sin \frac{\pi}{2} = \frac{y}{r} = \frac{1}{1} = 1$$

$$\tan 90^\circ = \tan \frac{\pi}{2} = \frac{y}{x} = \frac{1}{0}$$

Because division by 0 is undefined, tan 90° is undefined.

c. If $\theta = 180^\circ = \pi$ radians, then the terminal side of the angle is on the negative x-axis. Let us select the point $P = (-1, 0)$ with $x = -1$ and $y = 0$. This point is 1 unit from the origin, so $r = 1$. Now that we know x, y, and r, we can apply the definitions of the sine and tangent functions.

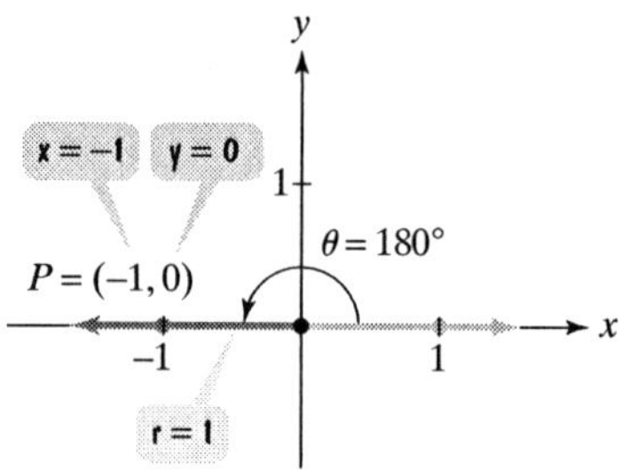

$$\sin 180^\circ = \sin \pi = \frac{y}{r} = \frac{0}{1} = 0$$

$$\tan 180^\circ = \tan \pi = \frac{y}{x} = \frac{0}{-1} = 0$$

Discovery

Try finding tan 90° and tan 270° with your calculator. Describe what occurs.

d. If $\theta = 270° = \frac{3\pi}{2}$ radians, then the terminal side of the angle is on the negative y-axis. Let us select the point $P = (0, -1)$ with $x = 0$ and $y = -1$. This point is 1 unit from the origin, so $r = 1$. Now that we know x, y, and r, we can apply the definitions of the sine and tangent functions.

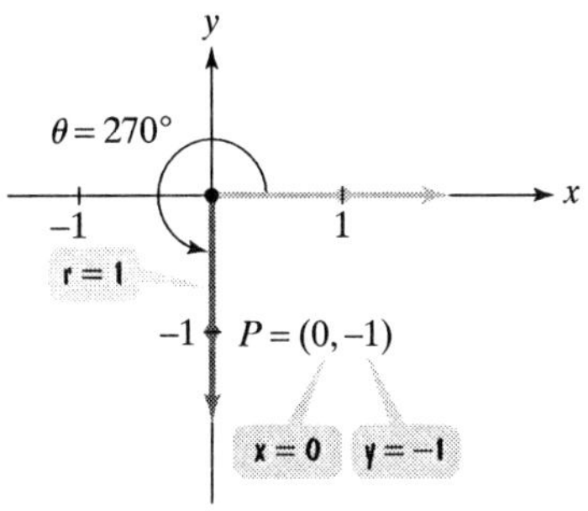

$$\sin 270° = \sin\frac{3\pi}{2} = \frac{y}{r} = \frac{-1}{1} = -1$$

$$\tan 270° = \tan\frac{3\pi}{2} = \frac{y}{x} = \frac{-1}{0}$$

Because division by 0 is undefined, tan 270° is undefined.

Check Point 2 Evaluate, if possible, the cosine function and the cosecant function at the following four quadrantal angles:

a. $\theta = 0° = 0$ **b.** $\theta = 90° = \frac{\pi}{2}$

c. $\theta = 180° = \pi$ **d.** $\theta = 270° = \frac{3\pi}{2}$.

2 Use the signs of the trigonometric functions.

The Signs of the Trigonometric Functions

In Example 2, we evaluated trigonometric functions of quadrantal angles. However, we will now return to the trigonometric functions of nonquadrantal angles. **If θ is not a quadrantal angle, the sign of a trigonometric function depends on the quadrant in which θ lies.** In all four quadrants, r is positive. However, x and y can be positive or negative. For example, if θ lies in quadrant II, x is negative and y is positive. Thus, the only positive ratios in this quadrant are $\frac{y}{r}$ and its reciprocal, $\frac{r}{y}$. These ratios are the function values for the sine and cosecant, respectively. In short, if θ lies in quadrant II, $\sin\theta$ and $\csc\theta$ are positive. The other four trigonometric functions are negative.

Figure 5.28 summarizes the signs of the trigonometric functions. If θ lies in quadrant I, all six functions are positive. If θ lies in quadrant II, only $\sin\theta$ and $\csc\theta$ are positive. If θ lies in quadrant III, only $\tan\theta$ and $\cot\theta$ are positive. Finally, if θ lies in quadrant IV, only $\cos\theta$ and $\sec\theta$ are positive. Observe that the positive functions in each quadrant occur in reciprocal pairs.

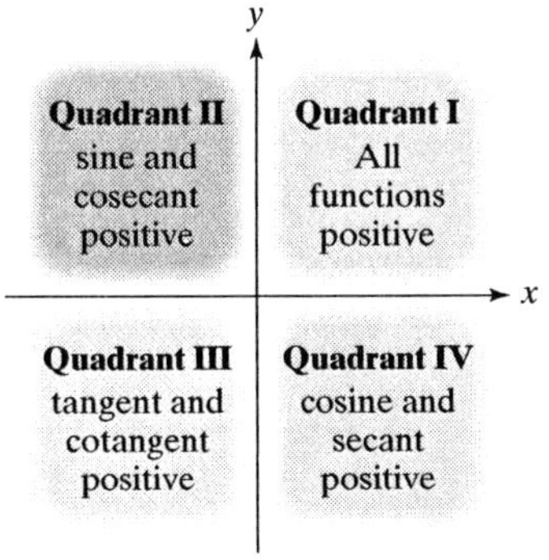

Figure 5.28 The signs of the trigonometric functions

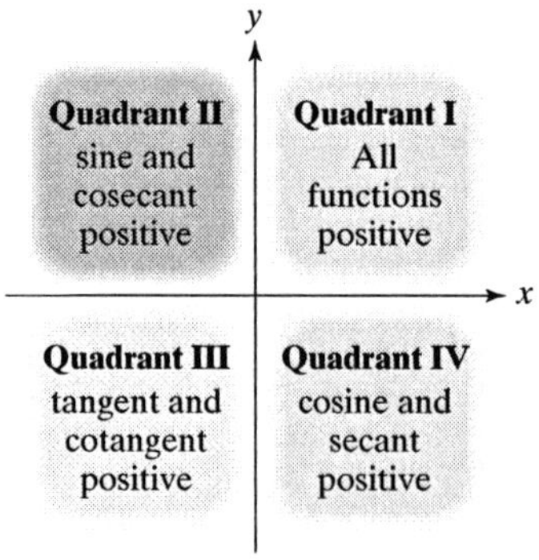

Figure 5.28 The signs of the trigonometric functions, repeated

EXAMPLE 3 Finding the Quadrant in Which an Angle Lies

If $\tan\theta < 0$ and $\cos\theta > 0$, name the quadrant in which angle θ lies.

Solution Because $\tan\theta < 0$, θ cannot lie in quadrant I; all the functions are positive in quadrant I. Furthermore, θ cannot lie in quadrant III; $\tan\theta$ is positive in quadrant III. Thus, with $\tan\theta < 0$, θ lies in quadrant II or quadrant IV. We are also given that $\cos\theta > 0$. Because quadrant IV is the only quadrant in which the cosine is positive and the tangent is negative, we conclude that θ lies in quadrant IV.

Check Point 3 If $\sin\theta < 0$ and $\cos\theta < 0$, name the quadrant in which angle θ lies.

EXAMPLE 4 Evaluating Trigonometric Functions

Given $\tan\theta = -\frac{2}{3}$ and $\cos\theta > 0$, find $\cos\theta$ and $\csc\theta$.

Solution Because the tangent is negative and the cosine is positive, θ lies in quadrant IV. This will help us to determine whether the negative sign in $\tan\theta = -\frac{2}{3}$ should be associated with the numerator or the denominator. Keep in mind that in quadrant IV, x is positive and y is negative. Thus,

In quadrant IV, y is negative.

$$\tan\theta = -\frac{2}{3} = \frac{y}{x} = \frac{-2}{3}.$$

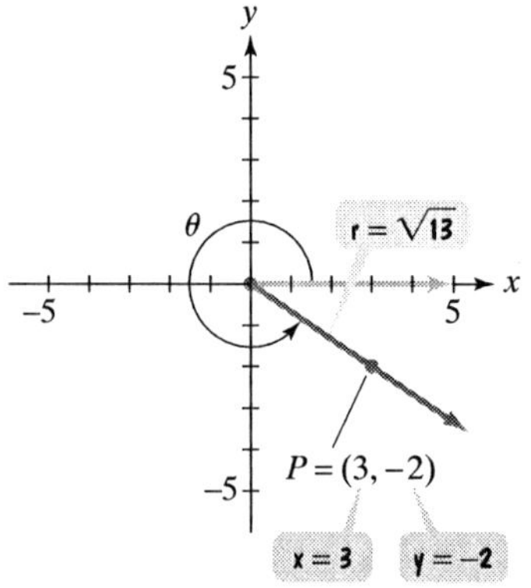

Figure 5.29 $\tan\theta = -\frac{2}{3}$ and $\cos\theta > 0$

(See Figure 5.29.) Thus, $x = 3$ and $y = -2$. Furthermore,

$$r = \sqrt{x^2 + y^2} = \sqrt{3^2 + (-2)^2} = \sqrt{9 + 4} = \sqrt{13}.$$

Now that we know x, y, and r, we can find $\cos\theta$ and $\csc\theta$.

$$\cos\theta = \frac{x}{r} = \frac{3}{\sqrt{13}} = \frac{3}{\sqrt{13}} \cdot \frac{\sqrt{13}}{\sqrt{13}} = \frac{3\sqrt{13}}{13} \qquad \csc\theta = \frac{r}{y} = \frac{\sqrt{13}}{-2} = -\frac{\sqrt{13}}{2}$$

Check Point 4 Given $\tan\theta = -\frac{1}{3}$ and $\cos\theta < 0$, find $\sin\theta$ and $\sec\theta$.

3 Find reference angles.

Reference Angles

We will often evaluate trigonometric functions of positive angles greater than 90° and all negative angles by making use of a positive acute angle. This positive acute angle is called a *reference angle*.

Definition of a Reference Angle

Let θ be a nonacute angle in standard position that lies in a quadrant. Its **reference angle** is the positive acute angle θ' formed by the terminal side of θ and the x-axis.

Figure 5.30 shows the reference angle for θ lying in quadrants II, III, and IV. Notice that the formula used to find θ', the reference angle, varies according to the quadrant in which θ lies. You may find it easier to find the reference angle for a given angle by making a figure that shows the angle in standard position. The acute angle formed by the terminal side of this angle and the x-axis is the reference angle.

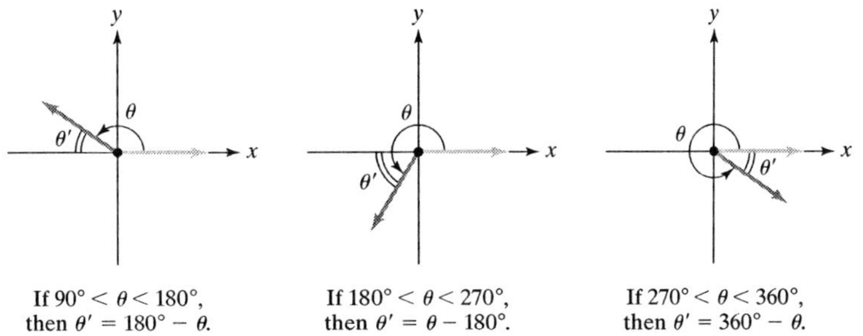

Figure 5.30 Reference angles, θ', for positive angles, θ, in quadrants II, III, and IV

EXAMPLE 5 Finding Reference Angles

Find the reference angle, θ', for each of the following angles:

a. $\theta = 345°$ **b.** $\theta = \frac{5\pi}{6}$ **c.** $\theta = -135°$ **d.** $\theta = 2.5$.

Solution

a. A 345° angle in standard position is shown in Figure 5.31. Because 345° lies in quadrant IV, the reference angle is

$$\theta' = 360° - 345° = 15°.$$

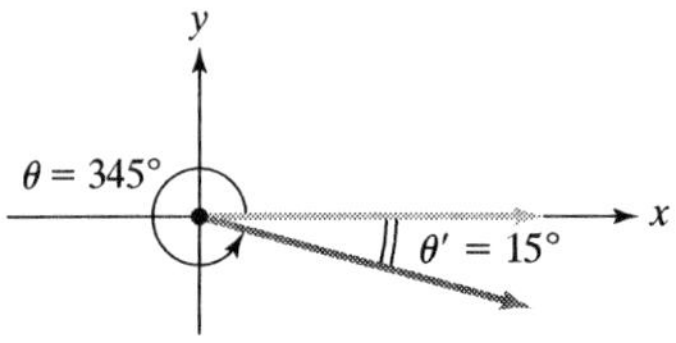

Figure 5.31

b. Because $\frac{5\pi}{6}$ lies between $\frac{\pi}{2} = \frac{3\pi}{6}$ and $\pi = \frac{6\pi}{6}$, $\theta = \frac{5\pi}{6}$ lies in quadrant II. The angle is shown in Figure 5.32. The reference angle is

$$\theta' = \pi - \frac{5\pi}{6} = \frac{6\pi}{6} - \frac{5\pi}{6} = \frac{\pi}{6}.$$

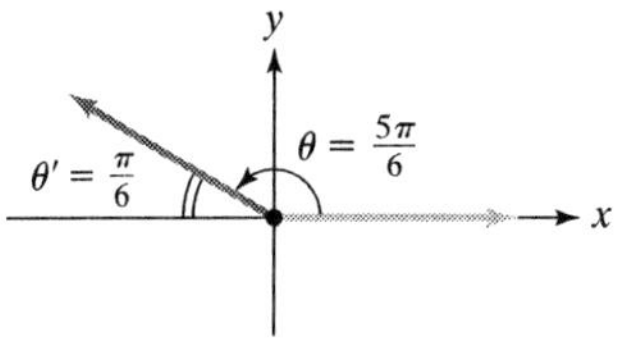

Figure 5.32

c. A −135° angle in standard position is shown in Figure 5.33. The figure indicates that the positive acute angle formed by the terminal side of θ and the x-axis is 45°. The reference angle is

$$\theta' = 45°.$$

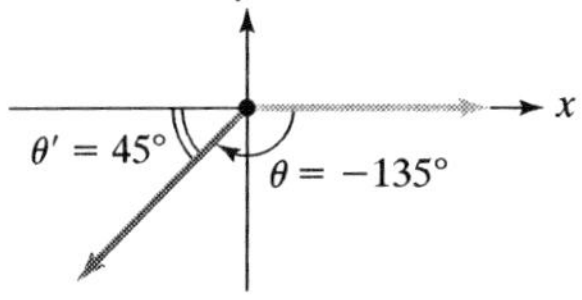

Figure 5.33

d. The angle $\theta = 2.5$ lies between $\frac{\pi}{2} \approx 1.57$ and $\pi \approx 3.14$. This means that $\theta = 2.5$ is in quadrant II, shown in Figure 5.34. The reference angle is

$$\theta' = \pi - 2.5 \approx 0.64.$$

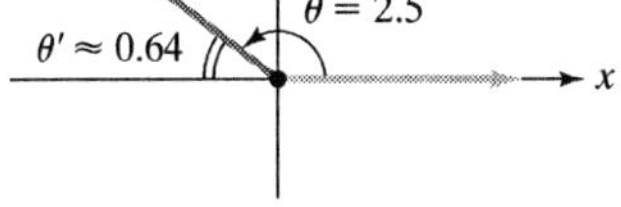

Figure 5.34

Check Point 5 Find the reference angle, θ', for each of the following angles:

a. $\theta = 210°$ **b.** $\theta = \frac{7\pi}{4}$ **c.** $\theta = -240°$ **d.** $\theta = 3.6$

The way that reference angles are defined makes them useful in evaluating trigonometric functions.

4 Use reference angles to evaluate trigonometric functions.

Using Reference Angles to Evaluate Trigonometric Functions

The values of the trigonometric functions of a given angle, θ, are the same as the values of the trigonometric functions of the reference angle, θ', except possibly for the sign. A function value of the acute reference angle, θ', is always positive. However, the same function value for θ may be positive or negative.

For example, we can use a reference angle, θ', to obtain an exact value for $\tan 120°$. The reference angle for $\theta = 120°$ is $\theta' = 180° - 120° = 60°$. We know the exact value of the tangent function of the reference angle: $\tan 60° = \sqrt{3}$. We also know that the value of a trigonometric function of a given angle, θ, is the same as that of its reference angle, θ', except possibly for the sign. Thus, we can conclude that $\tan 120°$ equals $-\sqrt{3}$ or $\sqrt{3}$.

What sign should we attach to $\sqrt{3}$? A 120° angle lies in quadrant II, where only the sine and cosecant are positive. Thus, the tangent function is negative for a 120° angle. Therefore,

Prefix by a negative sign to show tangent is negative in quadrant II.

$$\tan 120° = -\tan 60° = -\sqrt{3}.$$

The reference angle for 120° is 60°.

In the previous section, we used two right triangles to find exact trigonometric values of 30°, 45°, and 60°. Using a procedure similar to finding $\tan 120°$, we can now find the function values of all angles for which 30°, 45°, or 60° are reference angles.

A Procedure for Using Reference Angles to Evaluate Trigonometric Functions

The value of a trigonometric function of any angle θ is found as follows:

1. Find the associated reference angle, θ', and the function value for θ'.
2. Use the quadrant in which θ lies to prefix the appropriate sign to the function value in step 1.

Discovery

Draw the two right triangles involving 30°, 45°, and 60°. Indicate the length of each side. Use these lengths to verify the function values for the reference angles in the solution to Example 6.

EXAMPLE 6 Using Reference Angles to Evaluate Trigonometric Functions

Use reference angles to find the exact value of each of the following trigonometric functions:

a. $\sin 135°$ **b.** $\cos \frac{4\pi}{3}$ **c.** $\cot\left(-\frac{\pi}{3}\right)$.

Solution

a. We use our two-step procedure to find $\sin 135°$.

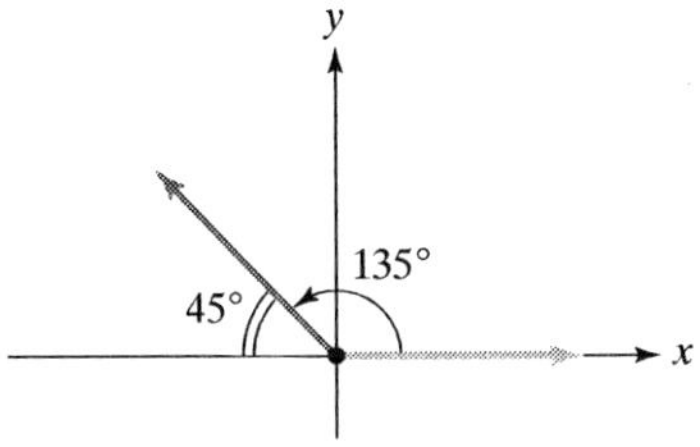

Figure 5.35 Reference angle for 135°

Step 1 Find the reference angle, θ', and $\sin\theta'$. Figure 5.35 shows 135° lies in quadrant II. The reference angle is

$$\theta' = 180° - 135° = 45°.$$

The function value for the reference angle is $\sin 45° = \dfrac{\sqrt{2}}{2}$.

Step 2 Use the quadrant in which θ lies to prefix the appropriate sign to the function value in step 1. The angle $\theta = 135°$ lies in quadrant II. Because the sine is positive in quadrant II, we put a + sign before the function value of the reference angle. Thus,

The sine is positive in quadrant II.

$$\sin 135° = +\sin 45° = \frac{\sqrt{2}}{2}.$$

The reference angle for 135° is 45°.

b. We use our two-step procedure to find $\cos\dfrac{4\pi}{3}$.

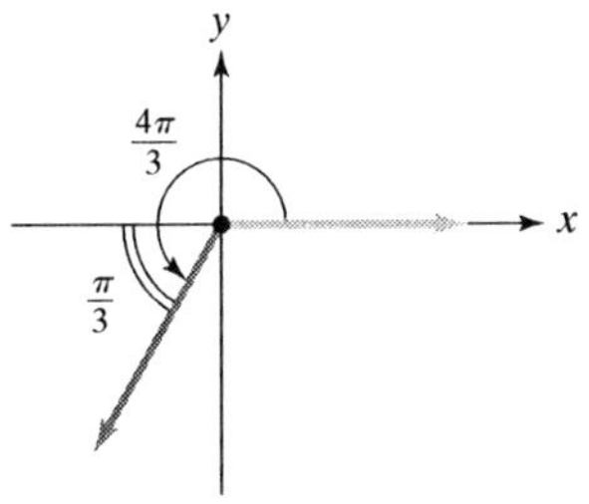

Figure 5.36 Reference angle for $\dfrac{4\pi}{3}$

Step 1 Find the reference angle, θ', and $\cos\theta'$. Figure 5.36 shows that $\theta = \dfrac{4\pi}{3}$ lies in quadrant III. The reference angle is

$$\theta' = \frac{4\pi}{3} - \pi = \frac{4\pi}{3} - \frac{3\pi}{3} = \frac{\pi}{3}.$$

The function value for the reference angle is

$$\cos\frac{\pi}{3} = \frac{1}{2}.$$

Step 2 Use the quadrant in which θ lies to prefix the appropriate sign to the function value in step 1. The angle $\theta = \dfrac{4\pi}{3}$ lies in quadrant III. Because only the tangent and cotangent are positive in quadrant III, the cosine is negative in this quadrant. We put a – sign before the function value of the reference angle. Thus,

The cosine is negative in quadrant III.

$$\cos\frac{4\pi}{3} = -\cos\frac{\pi}{3} = -\frac{1}{2}.$$

The reference angle for $\frac{4\pi}{3}$ is $\frac{\pi}{3}$.

c. We use our two-step procedure to find $\cot\left(-\frac{\pi}{3}\right)$.

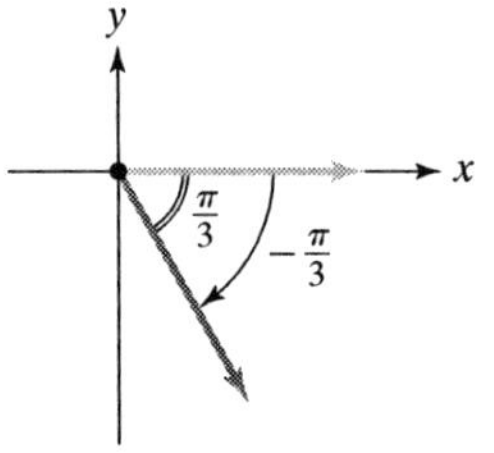

Figure 5.37 Reference angle for $-\frac{\pi}{3}$

Step 1 Find the reference angle, θ', and $\cot\theta'$. Figure 5.37 shows that $\theta = -\frac{\pi}{3}$ lies in quadrant IV. The reference angle is $\theta' = \frac{\pi}{3}$. The function value for the reference angle is $\cot\frac{\pi}{3} = \frac{\sqrt{3}}{3}$.

Step 2 Use the quadrant in which θ lies to prefix the appropriate sign to the function value in step 1. The angle $\theta = -\frac{\pi}{3}$ lies in quadrant IV. Because only the cosine and secant are positive in quadrant IV, the cotangent is negative in this quadrant. We put a − sign before the function value of the reference angle. Thus,

The cotangent is negative in quadrant IV.

$$\cot\left(-\frac{\pi}{3}\right) = -\cot\frac{\pi}{3} = -\frac{\sqrt{3}}{3}.$$

The reference angle for $-\frac{\pi}{3}$ is $\frac{\pi}{3}$.

Check Point 6 Use reference angles to find the exact value of the following trigonometric functions:

a. $\sin 300°$ **b.** $\tan\frac{5\pi}{4}$ **c.** $\sec\left(-\frac{\pi}{6}\right)$.

EXERCISE SET 5.3

Practice Exercises

In Exercises 1–8, a point on the terminal side of angle θ is given. Find the exact value of each of the six trigonometric functions of θ.

1. $(-4, 3)$
2. $(-12, 5)$
3. $(2, 3)$
4. $(3, 7)$
5. $(3, -3)$
6. $(5, -5)$
7. $(-2, -5)$
8. $(-1, -3)$

In Exercises 9–16, evaluate the trigonometric function at the quadrantal angle, or state that the expression is undefined.

9. $\cos\pi$
10. $\tan\pi$
11. $\sec\pi$
12. $\csc\pi$
13. $\tan\frac{3\pi}{2}$
14. $\cos\frac{3\pi}{2}$
15. $\cot\frac{\pi}{2}$
16. $\tan\frac{\pi}{2}$

In Exercises 17–22, let θ be an angle in standard position. Name the quadrant in which θ lies.

17. $\sin\theta > 0, \quad \cos\theta > 0$
18. $\sin\theta < 0, \quad \cos\theta > 0$
19. $\sin\theta < 0, \quad \cos\theta < 0$
20. $\tan\theta < 0, \quad \sin\theta < 0$
21. $\tan\theta < 0, \quad \cos\theta < 0$
22. $\cot\theta > 0, \quad \sec\theta < 0$

In Exercises 23–34, find the exact value of each of the remaining trigonometric functions of θ.

23. $\cos\theta = -\frac{3}{5}$, θ in quadrant III
24. $\sin\theta = -\frac{12}{13}$, θ in quadrant III
25. $\sin\theta = \frac{5}{13}$, θ in quadrant II
26. $\cos\theta = \frac{4}{5}$, θ in quadrant IV
27. $\cos\theta = \frac{8}{17}$, $270° < \theta < 360°$
28. $\cos\theta = \frac{1}{3}$, $270° < \theta < 360°$
29. $\tan\theta = -\frac{2}{3}$, $\sin\theta > 0$
30. $\tan\theta = -\frac{1}{3}$, $\sin\theta > 0$
31. $\tan\theta = \frac{4}{3}$, $\cos\theta < 0$
32. $\tan\theta = \frac{5}{12}$, $\cos\theta < 0$
33. $\sec\theta = -3$, $\tan\theta > 0$
34. $\csc\theta = -4$, $\tan\theta > 0$

In Exercises 35–50, find the reference angle for each angle.

35. $160°$ **36.** $170°$

37. $205°$ **38.** $210°$

39. $355°$ **40.** $351°$

41. $\frac{7\pi}{4}$ **42.** $\frac{5\pi}{4}$

43. $\frac{5\pi}{6}$ **44.** $\frac{5\pi}{7}$

45. $-150°$ **46.** $-250°$

47. $-335°$ **48.** $-359°$

49. 4.7 **50.** 5.5

In Exercises 51–66, use reference angles to find the exact value of each expression. Do not use a calculator.

51. $\cos 225°$ **52.** $\sin 300°$

53. $\tan 210°$ **54.** $\sec 240°$

55. $\tan 420°$ **56.** $\tan 405°$

57. $\sin \frac{2\pi}{3}$ **58.** $\cos \frac{3\pi}{4}$

59. $\csc \frac{7\pi}{6}$ **60.** $\cot \frac{7\pi}{4}$

61. $\tan \frac{9\pi}{4}$ **62.** $\tan \frac{9\pi}{2}$

63. $\sin(-240°)$ **64.** $\sin(-225°)$

65. $\tan\left(-\frac{\pi}{4}\right)$ **66.** $\tan\left(-\frac{\pi}{6}\right)$

Writing in Mathematics

67. If you are given a point on the terminal side of angle θ, explain how to find $\sin\theta$.

68. Explain why $\tan 90°$ is undefined.

69. If $\cos\theta > 0$ and $\tan\theta < 0$, explain how to find the quadrant in which θ lies.

70. What is a reference angle? Give an example with your description.

71. Explain how reference angles are used to evaluate trigonometric functions. Give an example with your description.

SECTION 5.4 *Trigonometric Functions of Real Numbers; Periodic Functions*

Objectives

1. Use a unit circle to define trigonometric functions of real numbers.
2. Recognize the domain and range of sine and cosine functions.
3. Use even and odd trigonometric functions.
4. Use periodic properties.

Cycles govern many aspects of life—heartbeats, sleep patterns, seasons, and tides all follow regular, predictable cycles. In this section, we will see why trigonometric functions are used to model phenomena that occur in cycles. To do this, we need to move beyond angles and consider trigonometric functions of real numbers.

1 Use a unit circle to define trigonometric functions of real numbers.

Trigonometric Functions of Real Numbers

Thus far, we have considered trigonometric functions of angles measured in degrees or radians. To define trigonometric functions of real numbers, rather than angles, we use a unit circle. A **unit circle** is a circle of radius 1, with its center at the origin of a rectangular coordinate system. The equation of this unit circle is $x^2 + y^2 = 1$. Figure 5.38 shows a unit circle in which the central angle measures t radians. We can use the formula for the length of a circular arc, $s = r\theta$, to find the length of the intercepted arc.

$$s = r\theta = 1 \cdot t = t$$

The radius of a unit circle is 1.

The radian measure of the central angle is t.

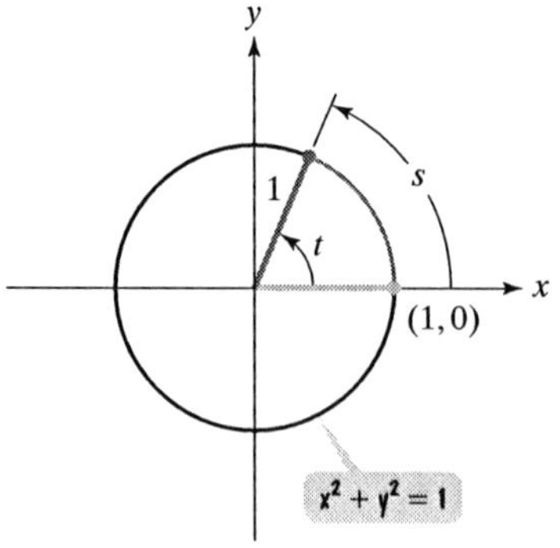

Figure 5.38 Unit circle with a central angle measuring t radians

Thus, the length of the intercepted arc is t. This is also the radian measure of the central angle. Thus, **in a unit circle, the radian measure of the angle is equal to the measure of the intercepted arc.** Both are given by the same *real number t*.

In Figure 5.39, the radian measure of the angle and the length of the intercepted arc are both shown by t. Let $P = (x, y)$ denote the point on the unit circle that has arc length t from $(1, 0)$. Figure 5.39(a) shows that if t is positive, point P is reached by moving counterclockwise along the unit circle from $(1, 0)$. Figure 5.39(b) shows that if t is negative, point P is reached by moving clockwise along the unit circle from $(1, 0)$. For each real number t, there corresponds a point $P = (x, y)$ on the unit circle.

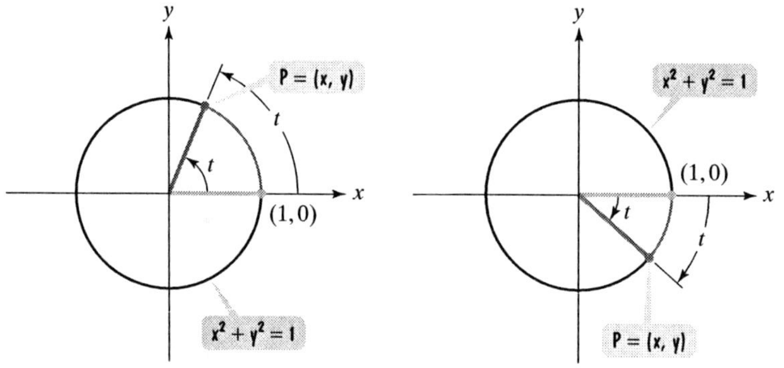

Figure 5.39 (a) t is positive. (b) t is negative.

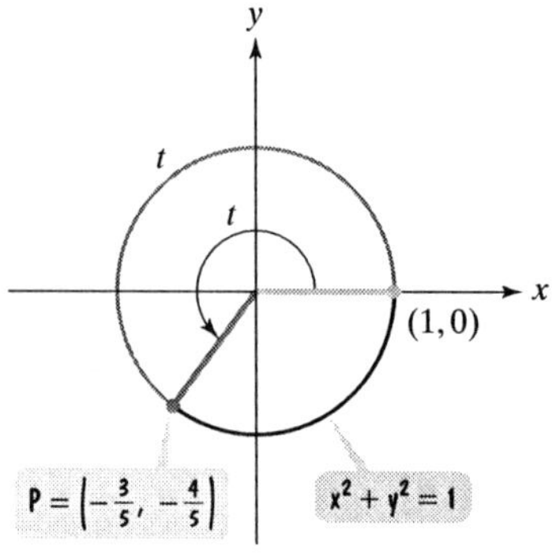

Figure 5.40

Using Figure 5.39, we define the cosine function at t as the x-coordinate of P and the sine function at t as the y-coordinate of P. Thus,

$$x = \cos t \quad \text{and} \quad y = \sin t.$$

For example, a point $P = (x, y)$ on the unit circle corresponding to a real number t is shown in Figure 5.40 for $\pi < t < \frac{3\pi}{2}$. We see that the coordinates of $P = (x, y)$ are $x = -\frac{3}{5}$ and $y = -\frac{4}{5}$. Because the cosine function is the x-coordinate of P and the sine function is the y-coordinate of P, the values of these trigonometric functions at the real number t are

$$\cos t = -\frac{3}{5} \quad \text{and} \quad \sin t = -\frac{4}{5}.$$

Definitions of the Trigonometric Functions in Terms of a Unit Circle

If t is a real number and $P = (x, y)$ is a point on the unit circle that corresponds to t, then

$$\sin t = y \qquad \cos t = x \qquad \tan t = \frac{y}{x}, x \neq 0$$

$$\csc t = \frac{1}{y}, y \neq 0 \qquad \sec t = \frac{1}{x}, x \neq 0 \qquad \cot t = \frac{x}{y}, y \neq 0.$$

Because this definition expresses function values in terms of coordinates of a point on a unit circle, the trigonometric functions are sometimes called the **circular functions.**

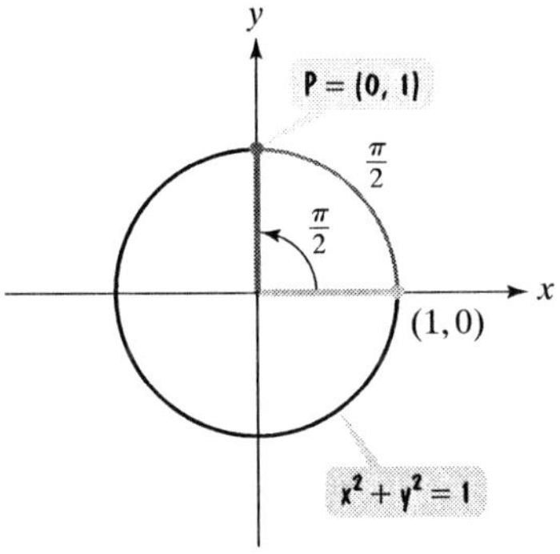

Figure 5.41

EXAMPLE 1 Finding Values of the Trigonometric Functions

Use Figure 5.41 to find the values of the trigonometric functions at $t = \frac{\pi}{2}$.

Solution The point P on the unit circle that corresponds to $t = \frac{\pi}{2}$ has coordinates $(0, 1)$. We use $x = 0$ and $y = 1$ to find the values of the trigonometric functions.

$$\sin\frac{\pi}{2} = y = 1 \qquad \cos\frac{\pi}{2} = x = 0$$

$$\csc\frac{\pi}{2} = \frac{1}{y} = \frac{1}{1} = 1 \qquad \cot\frac{\pi}{2} = \frac{x}{y} = \frac{0}{1} = 0$$

By definition, $\tan t = \frac{y}{x}$ and $\sec t = \frac{1}{x}$. Because $x = 0$, $\tan\frac{\pi}{2}$ and $\sec\frac{\pi}{2}$, are undefined.

Check Point 1 Use the figure on the right to find the values of the trigonometric functions at $t = \pi$.

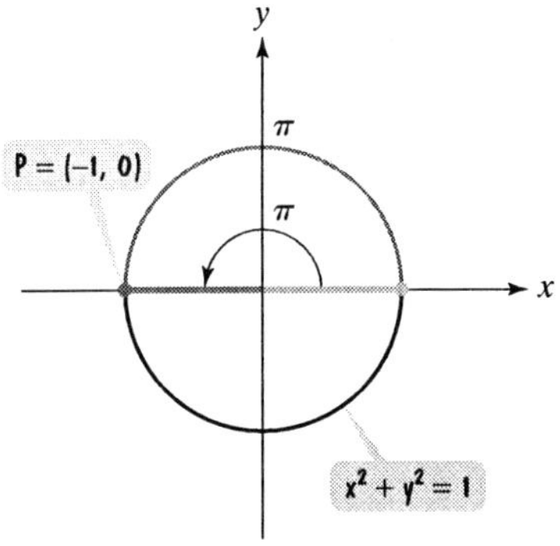

2 Recognize the domain and range of sine and cosine functions.

Domain and Range of Sine and Cosine Functions

The value of a trigonometric function at the real number t is its value at an angle of t radians. However, using real number domains, we can observe properties of trigonometric functions that are not as apparent using the angle approach. For example, the domain and range of each trigonometric function can be found from the unit circle definition. At this point, let's look only at the sine and cosine functions,

$$\sin t = y \quad \text{and} \quad \cos t = x.$$

Because t can be the radian measure of any angle or, equivalently, the measure of any intercepted arc, the domain of the sine function and the cosine function is the set of all real numbers. Because the radius of the unit circle is 1, we have

$$-1 \le x \le 1 \quad \text{and} \quad -1 \le y \le 1.$$

Therefore, with $x = \cos t$ and $y = \sin t$, we obtain

$$-1 \le \cos t \le 1 \quad \text{and} \quad -1 \le \sin t \le 1.$$

The range of the cosine and sine functions is $[-1, 1]$.

The Domain and Range of the Sine and Cosine Functions

The domain of the sine function and the cosine function is the set of all real numbers. The range of these functions is the set of all real numbers from −1 to 1, inclusive.

3 Use even and odd trigonometric functions.

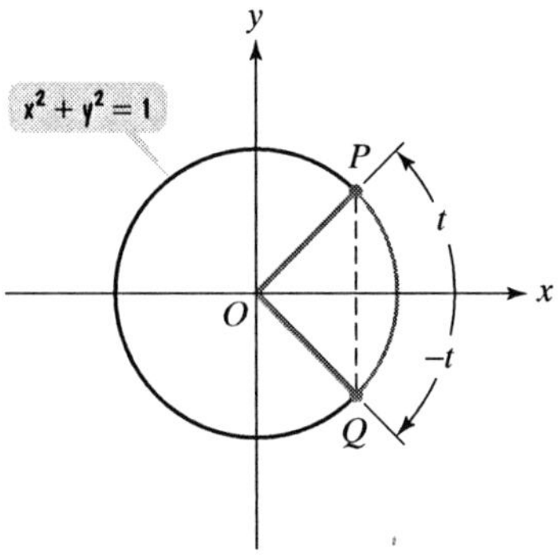

Figure 5.42

Even and Odd Trigonometric Functions

In Chapter 2, we saw that a function is even if $f(-t) = f(t)$ and odd if $f(-t) = -f(t)$. We can use Figure 5.42 to show that the cosine function is an even function and the sine function is an odd function. By definition, the coordinates of the points P and Q in Figure 5.42 are as follows:

$$P: (\cos t, \sin t)$$

$$Q: (\cos(-t), \sin(-t)).$$

In Figure 5.42, the x-coordinates of P and Q are the same. Thus,

$$\cos(-t) = \cos t.$$

This shows that the cosine function is an even function. By contrast, the y-coordinates of P and Q are negatives of each other. Thus,

$$\sin(-t) = -\sin t.$$

This shows that the sine function is an odd function.

This argument is valid regardless of the length of t. Thus, the arc may terminate in any of the four quadrants or on any axis. Using the unit circle definition of the trigonometric functions, we obtain the following results:

Even and Odd Trigonometric Functions

The cosine and secant functions are *even*.

$$\cos(-t) = \cos t \qquad \sec(-t) = \sec t$$

The sine, cosecant, tangent, and cotangent functions are *odd*.

$$\sin(-t) = -\sin t \qquad \csc(-t) = -\csc t$$

$$\tan(-t) = -\tan t \qquad \cot(-t) = -\cot t$$

EXAMPLE 2 Using Even and Odd Functions to Find Exact Values

Find the exact value of:

a. $\cos(-45°)$ **b.** $\tan\left(-\frac{\pi}{3}\right)$.

Solution

a. $\cos(-45°) = \cos 45° = \frac{\sqrt{2}}{2}$ **b.** $\tan\left(-\frac{\pi}{3}\right) = -\tan\frac{\pi}{3} = -\sqrt{3}$

Check Point 2 Find the exact value of:

a. $\cos(-60°)$ **b.** $\tan\left(-\frac{\pi}{6}\right)$.

4 Use periodic properties.

Periodic Functions

Certain patterns in nature repeat again and again. For example, the ocean level at a beach varies from low tide to high tide and then back to low tide approximately every 12 hours. If low tide occurs at noon, then high tide will be around 6 P.M. and low tide will occur again around midnight, and so on infinitely. If $f(t)$ represents the ocean level at the beach at any time t, then the level is the same 12 hours later. Thus,

$$f(t + 12) = f(t).$$

The word *periodic* means that this tidal behavior repeats infinitely. The *period*, 12 hours, is the time it takes to complete one full cycle.

Definition of a Periodic Function

A function f is **periodic** if there exists a positive number p such that

$$f(t + p) = f(t)$$

for all t in the domain of f. The smallest number p for which f is periodic is called the **period** of f.

The trigonometric functions are used to model periodic phenomena. Why? If we begin at any point P on the unit circle and travel a distance of 2π units along the perimeter, we will return to the same point P. Because the trigonometric functions are defined in terms of the coordinates of that point P, we obtain the following results:

Periodic Properties of the Sine and Cosine Functions

$$\sin(t + 2\pi) = \sin t \quad \text{and} \quad \cos(t + 2\pi) = \cos t$$

The sine and cosine functions are periodic functions and have period 2π.

EXAMPLE 3 Using Periodic Properties to Find Exact Values

Find the exact value of: **a.** $\tan 420°$ **b.** $\sin \frac{9\pi}{4}$.

Solution

a. $\tan 420° = \tan(60° + 360°) = \tan 60° = \sqrt{3}$

b. $\sin \frac{9\pi}{4} = \sin\left(\frac{\pi}{4} + 2\pi\right) = \sin \frac{\pi}{4} = \frac{\sqrt{2}}{2}$

Check Point 3 Find the exact value of:

a. $\cos 405°$ **b.** $\tan \frac{7\pi}{3}$.

Like the sine and cosine functions, the secant and cosecant functions have period 2π. However, the tangent and cotangent functions have a smaller period. If we begin at any point $P(x, y)$ on the unit circle and travel a distance of π units along the perimeter, we arrive at the point $(-x, -y)$. The tangent function, defined in terms of the coordinates of a point, is the same at (x, y) and $(-x, -y)$.

Tangent function at (x, y) $\quad \frac{y}{x} = \frac{-y}{-x} \quad$ Tangent function π radians later

We see that $\tan(t + \pi) = \tan t$. The same observations apply to the cotangent function.

Periodic Properties of the Tangent and Cotangent Functions

$$\tan(t + \pi) = \tan t \quad \text{and} \quad \cot(t + \pi) = \cot t$$

The tangent and cotangent functions are periodic functions and have period π.

Why do the trigonometric functions model phenomena that repeat *indefinitely*? By starting at point P on the unit circle and traveling a distance of 2π units, 4π units, 6π units, and so on, we return to the starting point P. Because the trigonometric functions are defined in terms of the coordinates of that point P, if we add (or subtract) multiples of 2π, the trigonometric values do not change. Furthermore, the trigonometric values for the tangent and cotangent functions do not change if we add (or subtract) multiples of π.

Repetitive Behavior of the Sine, Cosine, and Tangent Functions

For any integer n and real number t,

$$\sin(t + 2\pi n) = \sin t, \quad \cos(t + 2\pi n) = \cos t, \quad \text{and} \quad \tan(t + \pi n) = \tan t.$$

EXERCISE SET 5.4

Practice Exercises

In Exercises 1–4, a point $P(x, y)$ is shown on the unit circle corresponding to a real number t. Find the values of the trigonometric functions at t.

1.

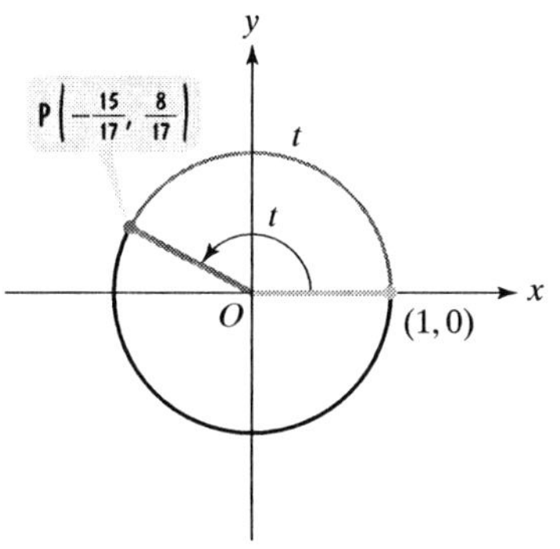

2.

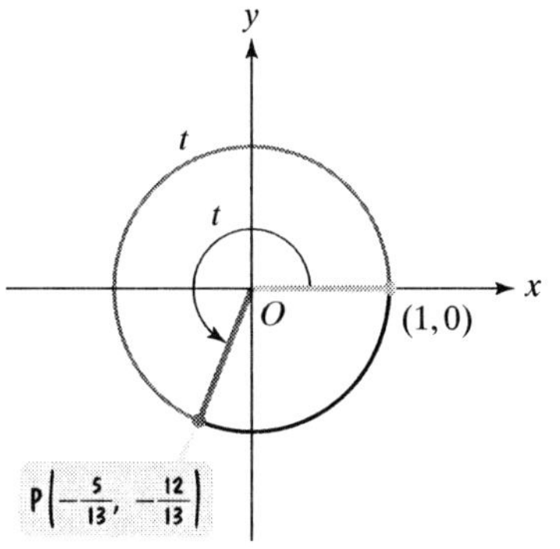

3.

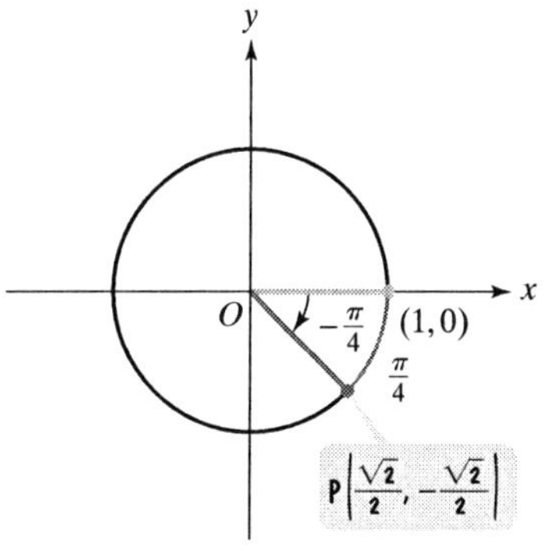

4.

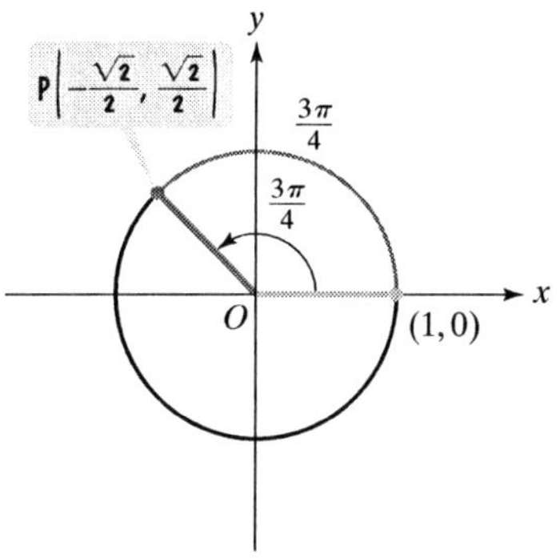

In Exercises 5–8, use even and odd properties of the trigonometric functions to find the exact value of each expression.

5. $\sin(-45°)$

6. $\tan(-45°)$

7. $\sec\left(-\frac{\pi}{3}\right)$

8. $\sec\left(-\frac{\pi}{6}\right)$

In Exercises 9–12, use periodic properties to find the exact value of each expression.

9. $\cos 585°$

10. $\cos 570°$

11. $\cot \frac{7\pi}{3}$

12. $\cot \frac{9\pi}{4}$

Application Exercises

13. The number of hours of daylight, H, on day t of any given year (on January 1, $t = 1$) in Fairbanks, Alaska, can be modeled by the function

$$H(t) = 12 + 8.3 \sin\left[\frac{2\pi}{365}(t - 80)\right].$$

a. March 21, the 80th day of the year, is the spring equinox. Find the number of hours of daylight in Fairbanks on this day.

b. June 21, the 172nd day of the year, is the summer solstice, the day with the maximum number of hours of daylight. To the nearest tenth of an hour, find the number of hours of daylight in Fairbanks on this day.

c. December 21, the 355th day of the year, is the winter solstice, the day with the minimum number of hours of daylight. Find, to the nearest tenth of an hour, the number of hours of daylight in Fairbanks on this day.

14. The number of hours of daylight, H, on day t of any given year (on January 1, $t = 1$) in San Diego, California, can be modeled by the function

$$H(t) = 12 + 2.4 \sin\left[\frac{2\pi}{365}(t - 80)\right].$$

a. March 21, the 80th day of the year, is the spring equinox. Find the number of hours of daylight in San Diego on this day.

b. June 21, the 172nd day of the year, is the summer solstice, the day with the maximum number of hours of daylight. Find, to the nearest tenth of an hour, the number of hours of daylight in San Diego on this day.

c. December 21, the 355th day of the year, is the winter solstice, the day with the minimum number of hours of daylight. To the nearest tenth of an hour, find the number of hours of daylight in San Diego on this day.

15. People who believe in biorhythms claim that there are three cycles that rule our behavior—the physical, emotional, and mental. Each is a sine function of a certain period. The function for our emotional fluctuations is

$$E = \sin \frac{\pi}{14} t$$

where t is measured in days starting at birth. Emotional fluctuations, E, are measured from -1 to 1, inclusive, with 1 representing peak emotional well-being, -1 representing the low for emotional well-being, and 0 representing feeling neither emotionally high nor low.

a. Find E corresponding to $t = 7$, 14, 21, 28, and 35. Describe what you observe.

b. What is the period of the emotional cycle?

16. The height of the water, H, in feet, at a boat dock t hours after 6 A.M. is given by

$$H = 10 + 4 \sin \frac{\pi}{6} t.$$

a. Find the height of the water at the dock at 6 A.M., 9 A.M., noon, 6 P.M., midnight, and 3 A.M.

b. When is low tide and when is high tide?

c. What is the period of this function and what does this mean about the tides?

Writing in Mathematics

17. Why are the trigonometric functions sometimes called circular functions?

18. What is the range of the sine function? Use the unit circle to explain where this range comes from.

19. What do we mean by even trigonometric functions? Which of the six functions fall into this category?

20. What is a periodic function? Why are the sine and cosine functions periodic?

21. Explain how you can use the function for emotional fluctuations in Exercise 15 to determine good days for having dinner with your moody boss.

22. Describe a phenomenon that repeats infinitely. What is its period?

Critical Thinking Exercises

23. Find the exact value of $\cos 0° + \cos 1° + \cos 2° + \cos 3° + \cdots + \cos 179° + \cos 180°$.

24. If $f(x) = \sin x$ and $f(a) = \frac{1}{4}$, find the value of

$$f(a) + f(a + 2\pi) + f(a + 4\pi) + f(a + 6\pi).$$

25. If $f(x) = \sin x$ and $f(a) = \frac{1}{4}$, find the value of $f(a) + 2f(-a)$.

26. The seats of a ferris wheel are 40 feet from the wheel's center. When you get on the ride, your seat is 5 feet above the ground. How far above the ground are you after rotating through an angle of 765°? Round to the nearest foot.

SECTION 5.5 *Graphs of Sine and Cosine Functions*

Objectives

1. Understand the graph of $y = \sin x$.
2. Graph variations of $y = \sin x$.
3. Understand the graph of $y = \cos x$.
4. Graph variations of $y = \cos x$.
5. Use vertical shifts of sine and cosine curves.
6. Model periodic behavior.

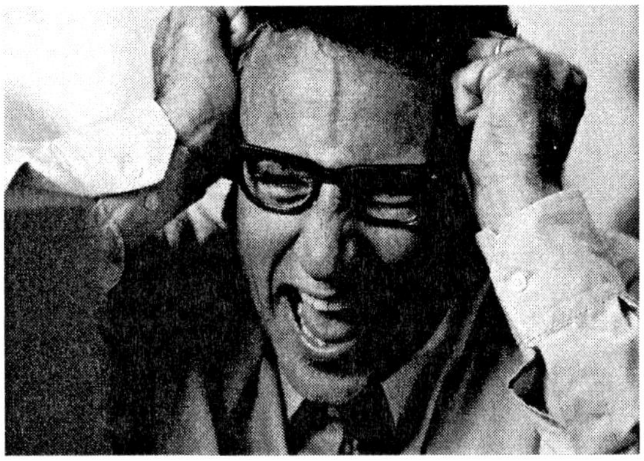

Take a deep breath and relax. Many relaxation exercises involve slowing down our breathing. Some people suggest that the way we breathe affects every part of our lives. Did you know that graphs of trigonometric functions can be used to analyze the breathing cycle, which is our closest link to both life and death?

In this section, we use graphs of sine and cosine functions to visualize their properties. We use the traditional symbol x, rather than θ or t, to represent the independent variable. We use the symbol y for the dependent variable, or the function's value at x. Thus, we will be graphing $y = \sin x$ and $y = \cos x$ in rectangular coordinates. In all graphs of trigonometric functions, the independent variable, x, is measured in radians.

1 Understand the graph of $y = \sin x$.

The Graph of $y = \sin x$

The trigonometric functions can be graphed in a rectangular coordinate system by plotting points whose coordinates satisfy the function. Thus, we graph $y = \sin x$ by listing some points on the graph. Because the period of the sine function is 2π, we will graph the function on the interval $[0, 2\pi]$. The rest of the graph is made up of repetitions of this portion.

Table 5.1 lists some values of (x, y) on the graph of $y = \sin x$, $0 \le x \le 2\pi$.

Table 5.1 Values of (x, y) on $y = \sin x$

x	0	$\frac{\pi}{6}$	$\frac{\pi}{3}$	$\frac{\pi}{2}$	$\frac{2\pi}{3}$	$\frac{5\pi}{6}$	π	$\frac{7\pi}{6}$	$\frac{4\pi}{3}$	$\frac{3\pi}{2}$	$\frac{5\pi}{3}$	$\frac{11\pi}{6}$	2π
$y = \sin x$	0	$\frac{1}{2}$	$\frac{\sqrt{3}}{2}$	1	$\frac{\sqrt{3}}{2}$	$\frac{1}{2}$	0	$-\frac{1}{2}$	$-\frac{\sqrt{3}}{2}$	-1	$-\frac{\sqrt{3}}{2}$	$-\frac{1}{2}$	0

As x increases from 0 to $\frac{\pi}{2}$, y increases from 0 to 1.

As x increases from $\frac{\pi}{2}$ to π, y decreases from 1 to 0.

As x increases from π to $\frac{3\pi}{2}$, y decreases from 0 to −1.

As x increases from $\frac{3\pi}{2}$ to 2π, y increases from −1 to 0.

In plotting the points obtained in Table 5.1, we will use the approximation $\frac{\sqrt{3}}{2} \approx 0.87$. Rather than approximating π, we will mark off units on the x-axis in terms of π. If we connect these points with a smooth curve, we obtain the graph shown in Figure 5.43 on the next page. The figure shows one period of the graph of $y = \sin x$.

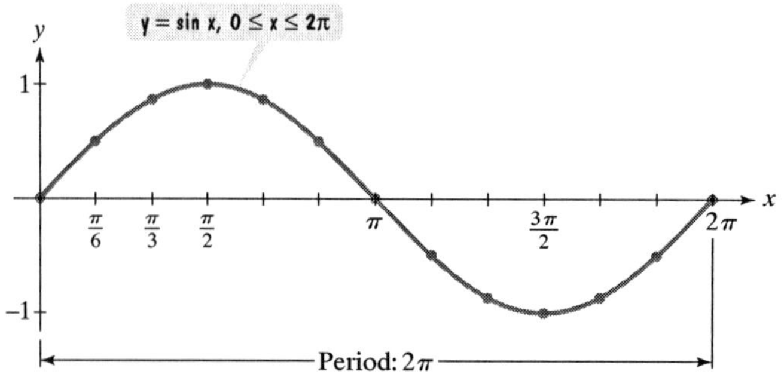

Figure 5.43 One period of the graph of $y = \sin x$

We can obtain a more complete graph of $y = \sin x$ by continuing the portion shown in Figure 5.43 to the left and to the right. The graph of the sine function, called a **sine curve,** is shown in Figure 5.44. Any part of the graph that corresponds to one period (2π) is one cycle of the graph of $y = \sin x$.

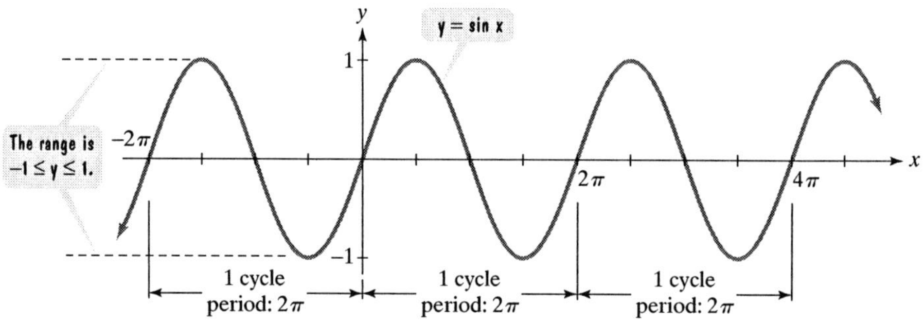

Figure 5.44 The graph of $y = \sin x$

The graph of $y = \sin x$ allows us to visualize some of the properties of the sine function.

- The domain is the set of all real numbers. The graph extends indefinitely to the left and to the right with no gaps or holes.
- The range consists of all numbers between -1 and 1, inclusive. The graph never rises above 1 or falls below -1.
- The period is 2π. The graph's pattern repeats in every interval of length 2π.
- The function is an odd function: $\sin(-x) = -\sin x$. This can be seen by observing that the graph is symmetric with respect to the origin.

2 Graph variations of $y = \sin x$.

Graphing Variations of $y = \sin x$

To graph variations of $y = \sin x$ by hand, it is helpful to find x-intercepts, maximum points, and minimum points. One complete cycle of the sine curve includes three x-intercepts, one maximum point, and one minimum point. The graph of $y = \sin x$ has x-intercepts at the beginning, middle, and end of its full period, shown in Figure 5.45. The curve reaches its maximum point $\frac{1}{4}$ of the way through the period. It reaches its minimum point $\frac{3}{4}$ of the way through the period. Thus, key points in graphing sine functions are

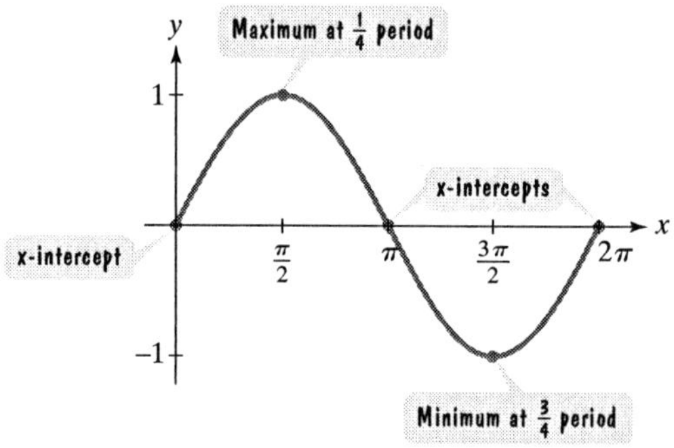

Figure 5.45 Key points in graphing the sine function

obtained by dividing the period into four equal parts. The x-coordinates of the five key points are as follows:

$$x_1 = \text{value of } x \text{ where the cycle begins}$$

$$x_2 = x_1 + \frac{\text{period}}{4}$$

$$x_3 = x_2 + \frac{\text{period}}{4}$$

$$x_4 = x_3 + \frac{\text{period}}{4}$$

$$x_5 = x_4 + \frac{\text{period}}{4}.$$

Add "quarter-periods" to find successive value of x.

The y-coordinates of the five key points are obtained by evaluating the given function at each of these values of x.

The graph of $y = \sin x$ forms the basis for graphing functions of the form

$$y = A \sin x.$$

For example, consider $y = 2\sin x$, in which $A = 2$. We can obtain the graph of $y = 2\sin x$ from that of $y = \sin x$ if we multiply each y-coordinate on the graph of $y = \sin x$ by 2. Figure 5.46 shows the graphs. The basic sine curve is *stretched* and ranges between -2 and 2, rather than between -1 and 1. However, both $y = \sin x$ and $y = 2\sin x$ have a period of 2π.

In general, the graph of $y = A\sin x$ ranges between $-|A|$ and $|A|$. Thus, the range of the function is $-|A| \le y \le |A|$. If $|A| > 1$, the basic sine curve is *stretched*, as in Figure 5.46. If $|A| < 1$, the basic sine curve is *shrunk*. We call $|A|$ the **amplitude** of $y = A\sin x$. The maximum value of y on the graph of $y = A\sin x$ is $|A|$, the amplitude.

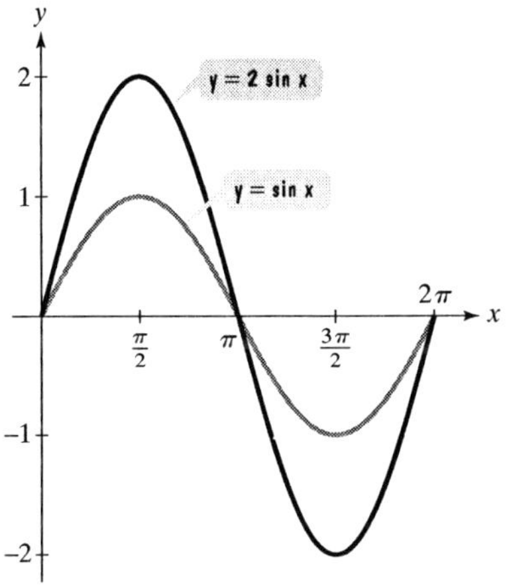

Figure 5.46 Comparing the graphs of $y = \sin x$ and $y = 2\sin x$

Graphing Variations of $y = \sin x$

1. Identify the amplitude and the period.
2. Find the values of x for the five key points—the three x-intercepts, the maximum point, and the minimum point. Start with the value of x where the cycle begins and add quarter-periods—that is, $\frac{\text{period}}{4}$—to find successive values of x.
3. Find the values of y for the five key points by evaluating the function at each value of x from step 2.
4. Connect the five key points with a smooth curve and graph one complete cycle of the given function.
5. Extend the graph in step 4 to the left or right as desired.

EXAMPLE 1 Graphing a Variation of $y = \sin x$

Determine the amplitude of $y = \frac{1}{2}\sin x$. Then graph $y = \sin x$ and $y = \frac{1}{2}\sin x$ for $0 \le x \le 2\pi$.

Solution

Step 1 Identify the amplitude and the period. The equation $y = \frac{1}{2}\sin x$ is of the form $y = A\sin x$ with $A = \frac{1}{2}$. Thus, the amplitude is $|A| = \frac{1}{2}$. This means that the maximum value of y is $\frac{1}{2}$ and the minimum value of y is $-\frac{1}{2}$. The period for both $y = \frac{1}{2}\sin x$ and $y = \sin x$ is 2π.

Step 2 **Find the values of x for the five key points.** We need to find the three x-intercepts, the maximum point, and the minimum point on the interval $[0, 2\pi]$. To do so, we begin by dividing the period, 2π, by 4.

$$\frac{\text{period}}{4} = \frac{2\pi}{4} = \frac{\pi}{2}$$

We start with the value of x where the cycle begins: $x = 0$. Now we add quarter-periods, $\frac{\pi}{2}$, to generate x-values for each of the key points. The five x-values are

$$x = 0, \quad x = 0 + \frac{\pi}{2} = \frac{\pi}{2}, \quad x = \frac{\pi}{2} + \frac{\pi}{2} = \pi,$$
$$x = \pi + \frac{\pi}{2} = \frac{3\pi}{2}, \quad x = \frac{3\pi}{2} + \frac{\pi}{2} = 2\pi.$$

Step 3 **Find the values of y for the five key points.** We evaluate the function at each value of x from step 2.

Value of x	Value of y: $y = \frac{1}{2}\sin x$	Coordinates of key point	
0	$y = \frac{1}{2}\sin 0 = \frac{1}{2} \cdot 0 = 0$	$(0, 0)$	
$\frac{\pi}{2}$	$y = \frac{1}{2}\sin\frac{\pi}{2} = \frac{1}{2} \cdot 1 = \frac{1}{2}$	$\left(\frac{\pi}{2}, \frac{1}{2}\right)$	maximum point
π	$y = \frac{1}{2}\sin\pi = \frac{1}{2} \cdot 0 = 0$	$(\pi, 0)$	
$\frac{3\pi}{2}$	$y = \frac{1}{2}\sin\frac{3\pi}{2} = \frac{1}{2}(-1) = -\frac{1}{2}$	$\left(\frac{3\pi}{2}, -\frac{1}{2}\right)$	minimum point
2π	$y = \frac{1}{2}\sin 2\pi = \frac{1}{2} \cdot 0 = 0$	$(2\pi, 0)$	

There are x-intercepts at 0, π, and 2π. The maximum and minimum points are indicated by the voice balloons.

Step 4 **Connect the five key points with a smooth curve and graph one complete cycle of the given function.** The five key points for $y = \frac{1}{2}\sin x$ are shown in Figure 5.47. By connecting the points with a smooth curve, the figure shows one complete cycle of $y = \frac{1}{2}\sin x$. Also shown is the graph of $y = \sin x$. The graph of $y = \frac{1}{2}\sin x$ shrinks the graph of $y = \sin x$.

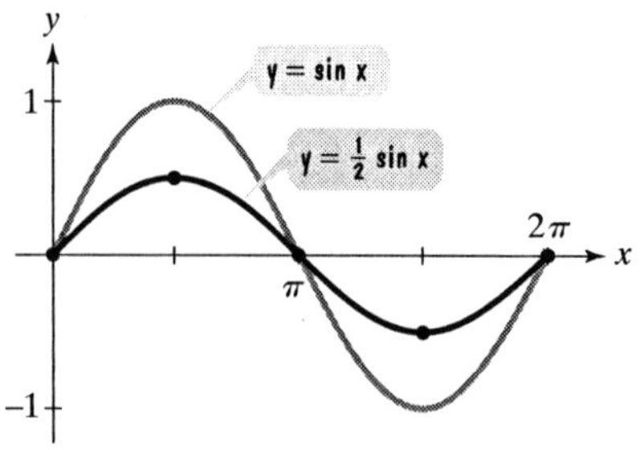

Figure 5.47 The graphs of $y = \sin x$ and $y = \frac{1}{2}\sin x$, $0 \le x \le 2\pi$

Check Point 1 Determine the amplitude of $y = 3 \sin x$. Then graph $y = \sin x$ and $y = 3 \sin x$ for $0 \le x \le 2\pi$.

EXAMPLE 2 Graphing a Variation of $y = \sin x$

Determine the amplitude of $y = -2 \sin x$. Then graph $y = \sin x$ and $y = -2 \sin x$ for $-\pi \le x \le 3\pi$.

Solution

Step 1 Identify the amplitude and the period. The equation $y = -2 \sin x$ is of the form $y = A \sin x$ with $A = -2$. Thus, the amplitude is $|A| = |-2| = 2$. This means that the maximum value of y is 2 and the minimum value of y is -2. Both $y = \sin x$ and $y = -2 \sin x$ have a period of 2π.

Step 2 Find the x-values for the five key points. Begin by dividing the period, 2π, by 4.

$$\frac{\text{period}}{4} = \frac{2\pi}{4} = \frac{\pi}{2}$$

Start with the value of x where the cycle begins: $x = 0$. Adding quarter-periods, $\frac{\pi}{2}$, the five x-values for the key points are

$$x = 0, \quad x = 0 + \frac{\pi}{2} = \frac{\pi}{2}, \quad x = \frac{\pi}{2} + \frac{\pi}{2} = \pi,$$

$$x = \pi + \frac{\pi}{2} = \frac{3\pi}{2}, \quad x = \frac{3\pi}{2} + \frac{\pi}{2} = 2\pi.$$

Step 3 Find the values of y for the five key points. We evaluate the function at each value of x from step 2.

Value of x	Value of y: $y = -2 \sin x$	Coordinates of key point
0	$y = -2 \sin 0 = -2 \cdot 0 = 0$	$(0, 0)$
$\frac{\pi}{2}$	$y = -2 \sin \frac{\pi}{2} = -2 \cdot 1 = -2$	$\left(\frac{\pi}{2}, -2\right)$ minimum point
π	$y = -2 \sin \pi = -2 \cdot 0 = 0$	$(\pi, 0)$
$\frac{3\pi}{2}$	$y = -2 \sin \frac{3\pi}{2} = -2(-1) = 2$	$\left(\frac{3\pi}{2}, 2\right)$ maximum point
2π	$y = -2 \sin 2\pi = -2 \cdot 0 = 0$	$(2\pi, 0)$

There are x-intercepts at 0, π, and 2π. The minimum and maximum points are indicated by the voice balloons.

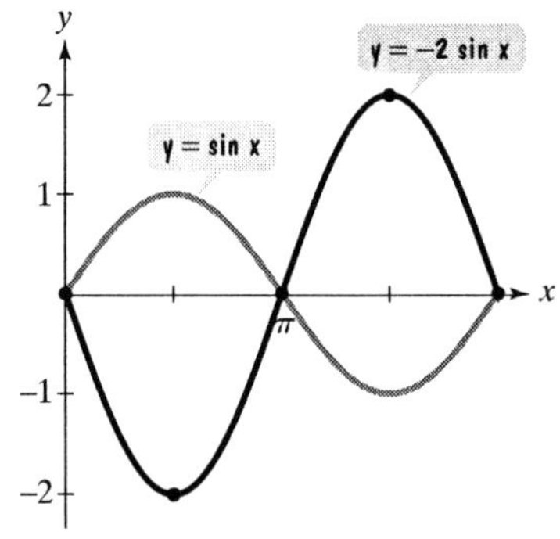

Figure 5.48 The graphs of $y = \sin x$ and $y = -2 \sin x$, $0 \le x \le 2\pi$

Step 4 Connect the five key points with a smooth curve and graph one complete cycle of the given function. The five key points for $y = -2 \sin x$ are shown in Figure 5.48. By connecting the points with a smooth curve, the dark red portion shows one complete cycle of $y = -2 \sin x$. Also shown in dark blue is one complete cycle of the graph of $y = \sin x$.

Step 5 Extend the graph in step 4 to the left or right as desired. The dark red and dark blue portions of the graphs in Figure 5.48 are from 0 to 2π. In order to graph for $-\pi \le x \le 3\pi$, continue the pattern of each graph to the left and to the right. These extensions are shown by the lighter colors in Figure 5.48.

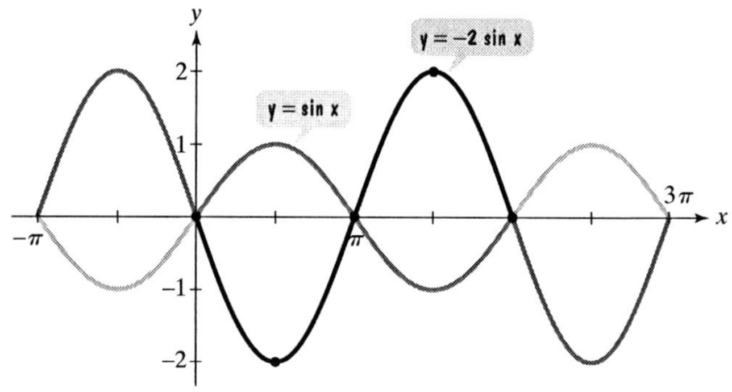

Figure 5.48, extended
The graphs of $y = \sin x$ and $y = -2\sin x, -\pi \le x \le 3\pi$

Check Point 2 Determine the amplitude of $y = -\frac{1}{2}\sin x$. Then graph $y = \sin x$ and $y = -\frac{1}{2}\sin x$ for $-\pi \le x \le 3\pi$.

Now let us examine the graphs of functions of the form $y = A\sin Bx$, where B is the coefficient of x. How do such graphs compare to those of functions of the form $y = A\sin x$? We know that $y = A\sin x$ completes one cycle from $x = 0$ to $x = 2\pi$. Thus, $y = A\sin Bx$ completes one cycle from $Bx = 0$ to $Bx = 2\pi$. Solve each of these equations for x.

$$Bx = 0 \qquad Bx = 2\pi$$

$$x = 0 \qquad x = \frac{2\pi}{B} \qquad \text{Divide both sides of each equation by } B.$$

This means that $y = A\sin Bx$ completes one cycle from 0 to $\frac{2\pi}{B}$. The period is $\frac{2\pi}{B}$.

Amplitudes and Periods

The graph of $y = A\sin Bx$ has

$$\text{amplitude} = |A|$$

$$\text{period} = \frac{2\pi}{B}.$$

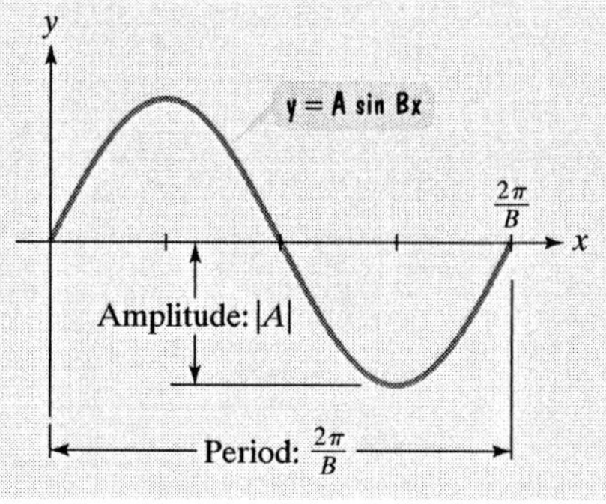

EXAMPLE 3 Graphing a Function of the Form $y = A\sin Bx$

Determine the amplitude and period of $y = 3\sin 2x$. Then graph the function for $0 \le x \le 2\pi$.

Solution

Step 1 Identify the amplitude and the period. The equation $y = 3 \sin 2x$ is of the form $y = A \sin Bx$ with $A = 3$ and $B = 2$.

$$\text{amplitude:} \quad |A| = |3| = 3$$

$$\text{period:} \quad \frac{2\pi}{B} = \frac{2\pi}{2} = \pi$$

The amplitude, 3, tells us that the maximum value of y is 3 and the minimum value of y is -3.

Step 2 Find the x-values for the five key points. Begin by dividing the period, π, by 4.

$$\frac{\text{period}}{4} = \frac{\pi}{4}$$

Start with the value of x where the cycle begins: $x = 0$. Adding quarter-periods, $\frac{\pi}{4}$, the five x-values for the key points are

$$x = 0, \quad x = 0 + \frac{\pi}{4} = \frac{\pi}{4}, \quad x = \frac{\pi}{4} + \frac{\pi}{4} = \frac{\pi}{2},$$

$$x = \frac{\pi}{2} + \frac{\pi}{4} = \frac{3\pi}{4}, \quad x = \frac{3\pi}{4} + \frac{\pi}{4} = \pi.$$

Step 3 Find the values of y for the five key points. We evaluate the function at each value of x from step 2.

Value of x	Value of y: $y = 3 \sin 2x$	Coordinates of key point	
0	$y = 3 \sin 2 \cdot 0$ $= 3 \sin 0 = 3 \cdot 0 = 0$	$(0, 0)$	
$\frac{\pi}{4}$	$y = 3 \sin 2 \cdot \frac{\pi}{4}$ $= 3 \sin \frac{\pi}{2} = 3 \cdot 1 = 3$	$\left(\frac{\pi}{4}, 3\right)$	maximum point
$\frac{\pi}{2}$	$y = 3 \sin 2 \cdot \frac{\pi}{2}$ $= 3 \sin \pi = 3 \cdot 0 = 0$	$\left(\frac{\pi}{2}, 0\right)$	
$\frac{3\pi}{4}$	$y = 3 \sin 2 \cdot \frac{3\pi}{4}$ $= 3 \sin \frac{3\pi}{2} = 3(-1) = -3$	$\left(\frac{3\pi}{4}, -3\right)$	minimum point
π	$y = 3 \sin 2 \cdot \pi$ $= 3 \sin 2\pi = 3 \cdot 0 = 0$	$(\pi, 0)$	

In the interval $[0, \pi]$, there are x-intercepts at 0, $\frac{\pi}{2}$, and π. The maximum and minimum points are indicated by the voice balloons.

Step 4 Connect the five key points with a smooth curve and graph one complete cycle of the given function. The five key points for $y = 3 \sin 2x$ are shown in Figure 5.49. By connecting the points with a smooth curve, the blue portion shows one complete cycle of $y = 3 \sin 2x$ from 0 to π.

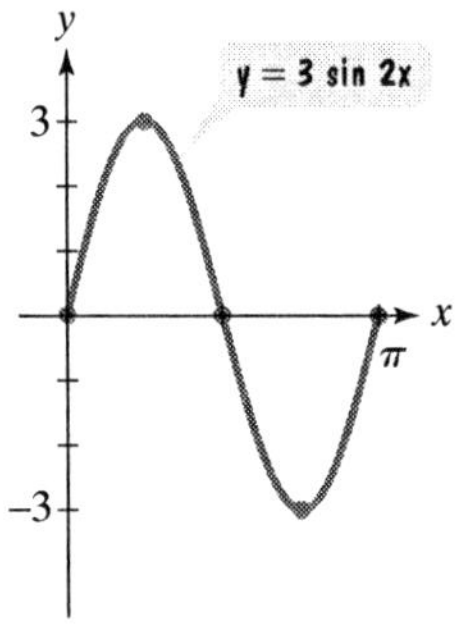

Figure 5.49 The graph of $y = 3 \sin 2x, 0 \le x \le \pi$

Technology

The graph of $y = 3\sin 2x$ in a $\left[0, 2\pi, \frac{\pi}{2}\right]$ by $[-4, 4, 1]$ viewing rectangle verifies our hand-drawn graph in Figure 5.49.

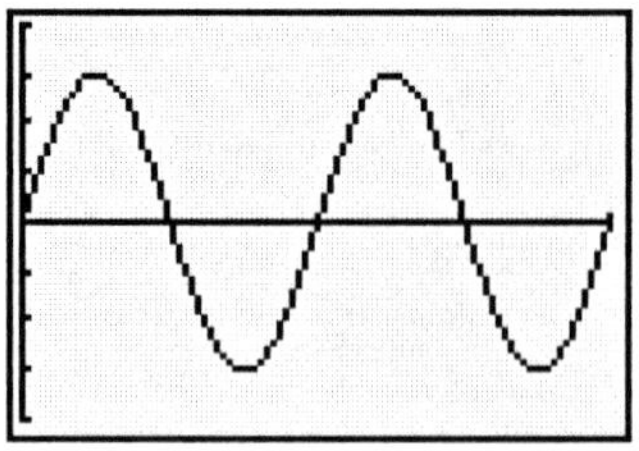

Step 5 Extend the graph in step 4 to the left or right as desired. The blue portion of the graph in Figure 5.49 is from 0 to π. In order to graph for $0 \le x \le 2\pi$, we continue this portion and extend the graph another full period to the right. This extension is shown in black in Figure 5.49.

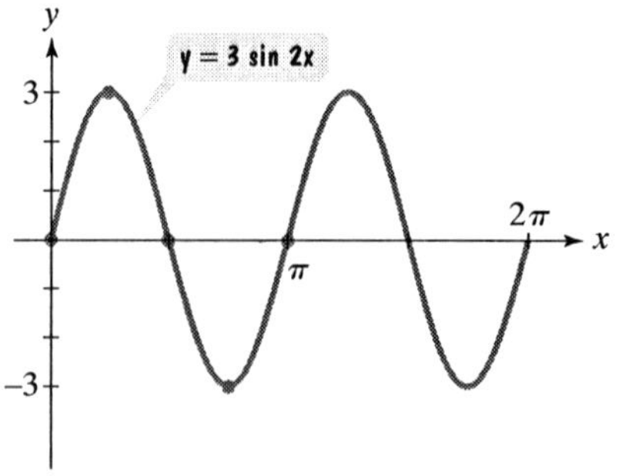

Figure 5.49, extended

Check Point 3 Determine the amplitude and period of $y = 2\sin\frac{1}{2}x$. Then graph the function for $0 \le x \le 8\pi$.

Now let us examine the graphs of functions of the form $y = A\sin(Bx - C)$. How do such graphs compare to those of functions of the form $y = A\sin Bx$?

In both cases, the amplitude is $|A|$ and the period is $\frac{2\pi}{B}$. One complete cycle occurs if $Bx - C$ increases from 0 to 2π. This means that we can find an interval containing one cycle by solving the equations

$Bx - C = 0$ and $Bx - C = 2\pi$.

$$Bx = C \qquad Bx = C + 2\pi \qquad \text{Add } C \text{ to both sides in each equation.}$$

$$x = \frac{C}{B} \qquad x = \frac{C}{B} + \frac{2\pi}{B}. \qquad \text{Divide both sides by } B \text{ in each equation.}$$

This is the x-coordinate on the left where the cycle begins.

This is the x-coordinate on the right where the cycle ends. $\frac{2\pi}{B}$ is the period.

The voice balloon on the left indicates that $y = A\sin(Bx - C)$ shifts the graph of $y = A\sin Bx$ horizontally by $\frac{C}{B}$. Thus, the number $\frac{C}{B}$ is the **phase shift** associated with the graph.

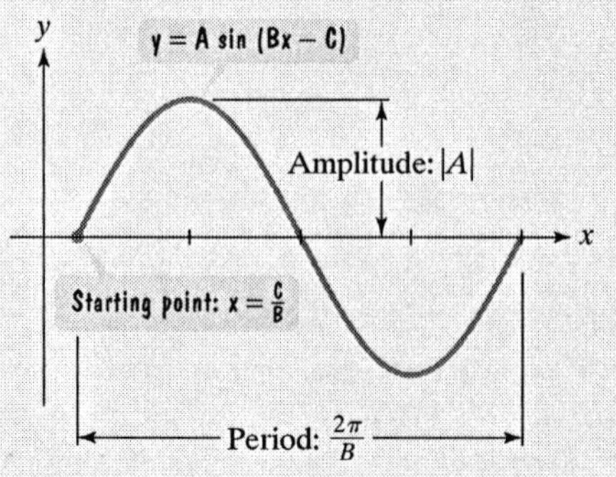

The Graph of $y = A\sin(Bx - C)$

The graph of $y = A\sin(Bx - C)$ is obtained by horizontally shifting the graph of $y = A\sin Bx$ so that the starting point of the cycle is shifted from $x = 0$ to $x = \frac{C}{B}$. If $\frac{C}{B} > 0$, the shift is to the right. If $\frac{C}{B} < 0$, the shift is to the left. The number $\frac{C}{B}$ is called the **phase shift.**

$$\text{amplitude} = |A|$$

$$\text{period} = \frac{2\pi}{B}$$

EXAMPLE 4 Graphing a Function of the Form $y = A\sin(Bx - C)$

Determine the amplitude, period, and phase shift of $y = 4\sin\left(2x - \frac{2\pi}{3}\right)$. Then graph one period of the function.

Solution

Step 1 Identify the amplitude, the period, and the phase shift. We must first identify values for A, B, and C.

This equation is of the form $y = A\sin(Bx - C)$.

$$y = 4\sin\left(2x - \frac{2\pi}{3}\right)$$

Using the voice balloon, we see that $A = 4$, $B = 2$, and $C = \frac{2\pi}{3}$.

amplitude: $|A| = |4| = 4$ — The maximum y is 4 and the minimum is -4.

period: $\frac{2\pi}{B} = \frac{2\pi}{2} = \pi$ — Each cycle is completed in π radians.

phase shift: $\frac{C}{B} = \frac{\frac{2\pi}{3}}{2} = \frac{2\pi}{3} \cdot \frac{1}{2} = \frac{\pi}{3}$ — A cycle starts at $x = \frac{\pi}{3}$.

Step 2 Find the x-values for the five key points. Begin by dividing the period, π, by 4.

$$\frac{\text{period}}{4} = \frac{\pi}{4}$$

Start with the value of x where the cycle begins: $x = \frac{\pi}{3}$. Adding quarter-periods, $\frac{\pi}{4}$, the five x-values for the key points are

$$x = \frac{\pi}{3}, \quad x = \frac{\pi}{3} + \frac{\pi}{4} = \frac{4\pi}{12} + \frac{3\pi}{12} = \frac{7\pi}{12},$$

$$x = \frac{7\pi}{12} + \frac{\pi}{4} = \frac{7\pi}{12} + \frac{3\pi}{12} = \frac{10\pi}{12} = \frac{5\pi}{6},$$

$$x = \frac{5\pi}{6} + \frac{\pi}{4} = \frac{10\pi}{12} + \frac{3\pi}{12} = \frac{13\pi}{12},$$

$$x = \frac{13\pi}{12} + \frac{\pi}{4} = \frac{13\pi}{12} + \frac{3\pi}{12} = \frac{16\pi}{12} = \frac{4\pi}{3}.$$

Study Tip

You can speed up the additions on the right by first writing the starting point, $\frac{\pi}{3}$, and the quarter-period, $\frac{\pi}{4}$, with a common denominator, 12.

starting point $= \frac{\pi}{3} = \frac{4\pi}{12}$

quarter-period $= \frac{\pi}{4} = \frac{3\pi}{12}$

Step 3 Find the values of y for the five key points. We evaluate the function at each value of x from step 2.

Value of x	Value of y: $y = 4\sin\left(2x - \frac{2\pi}{3}\right)$	Coordinates of key point	
$\frac{\pi}{3}$	$y = 4\sin\left(2 \cdot \frac{\pi}{3} - \frac{2\pi}{3}\right)$ $= 4\sin 0 = 4 \cdot 0 = 0$	$\left(\frac{\pi}{3}, 0\right)$	
$\frac{7\pi}{12}$	$y = 4\sin\left(2 \cdot \frac{7\pi}{12} - \frac{2\pi}{3}\right)$ $= 4\sin\left(\frac{7\pi}{6} - \frac{2\pi}{3}\right)$ $= 4\sin\frac{3\pi}{6} = 4\sin\frac{\pi}{2} = 4 \cdot 1 = 4$	$\left(\frac{7\pi}{12}, 4\right)$	maximum point
$\frac{5\pi}{6}$	$y = 4\sin\left(2 \cdot \frac{5\pi}{6} - \frac{2\pi}{3}\right)$ $= 4\sin\left(\frac{5\pi}{3} - \frac{2\pi}{3}\right)$ $= 4\sin\frac{3\pi}{3} = 4\sin\pi = 4 \cdot 0 = 0$	$\left(\frac{5\pi}{6}, 0\right)$	
$\frac{13\pi}{12}$	$y = 4\sin\left(2 \cdot \frac{13\pi}{12} - \frac{2\pi}{3}\right)$ $= 4\sin\left(\frac{13\pi}{6} - \frac{4\pi}{6}\right)$ $= 4\sin\frac{9\pi}{6} = 4\sin\frac{3\pi}{2} = 4(-1) = -4$	$\left(\frac{13\pi}{12}, -4\right)$	minimum point
$\frac{4\pi}{3}$	$y = 4\sin\left(2 \cdot \frac{4\pi}{3} - \frac{2\pi}{3}\right)$ $= 4\sin\frac{6\pi}{3} = 4\sin 2\pi = 4 \cdot 0 = 0$	$\left(\frac{4\pi}{3}, 0\right)$	

In the interval $\left[\frac{\pi}{3}, \frac{4\pi}{3}\right]$, there are x-intercepts at $\frac{\pi}{3}$, $\frac{5\pi}{6}$, and $\frac{4\pi}{3}$. The maximum and minimum points are indicated by the voice balloons.

Step 4 Connect the five key points with a smooth curve and graph one complete cycle of the given function. The key points and the graph of $y = 4\sin\left(2x - \frac{2\pi}{3}\right)$ are shown in Figure 5.50.

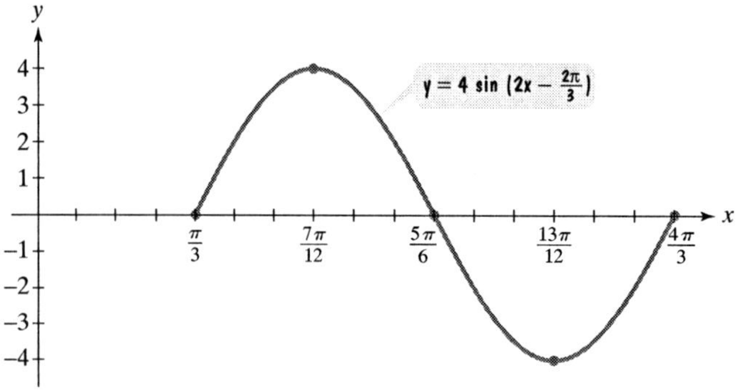

Figure 5.50

Check Point 4 Determine the amplitude, period, and phase shift of $y = 3\sin\left(2x - \frac{\pi}{3}\right)$. Then graph one period of the function.

3 Understand the graph of $y = \cos x$.

The Graph of $y = \cos x$

We graph $y = \cos x$ by listing some points on the graph. Because the period of the cosine function is 2π, we will concentrate on the graph of the basic cosine curve on the interval $[0, 2\pi]$. The rest of the graph is made up of repetitions of this portion. Table 5.2 lists some values of (x, y) on the graph of $y = \cos x$.

Table 5.2 Values of (x, y) on $y = \cos x$

x	0	$\frac{\pi}{6}$	$\frac{\pi}{3}$	$\frac{\pi}{2}$	$\frac{2\pi}{3}$	$\frac{5\pi}{6}$	π	$\frac{7\pi}{6}$	$\frac{4\pi}{3}$	$\frac{3\pi}{2}$	$\frac{5\pi}{3}$	$\frac{11\pi}{6}$	2π
$y = \cos x$	1	$\frac{\sqrt{3}}{2}$	$\frac{1}{2}$	0	$-\frac{1}{2}$	$-\frac{\sqrt{3}}{2}$	-1	$-\frac{\sqrt{3}}{2}$	$-\frac{1}{2}$	0	$\frac{1}{2}$	$\frac{\sqrt{3}}{2}$	1

As x increases from 0 to $\frac{\pi}{2}$, y decreases from 1 to 0.

As x increases from $\frac{\pi}{2}$ to π, y decreases from 0 to -1.

As x increases from π to $\frac{3\pi}{2}$, y increases from -1 to 0.

As x increases from $\frac{3\pi}{2}$ to 2π, y increases from 0 to 1.

Plotting the points in Table 5.2 and connecting them with a smooth curve, we obtain the graph shown in Figure 5.51. The portion of the graph in dark blue shows one complete period. We can obtain a more complete graph of $y = \cos x$ by extending this dark blue portion to the left and to the right.

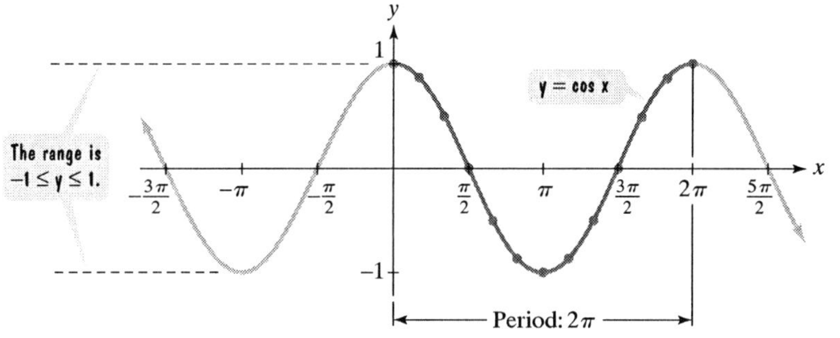

Figure 5.51 The graph of $y = \cos x$

The graph of $y = \cos x$ allows us to visualize some of the properties of the cosine function.

- The domain is the set of all real numbers. The graph extends indefinitely to the left and to the right with no gaps or holes.
- The range consists of all numbers between -1 and 1, inclusive. The graph never rises above 1 or falls below -1.
- The period is 2π. The graph's pattern repeats in every interval of length 2π.
- The function is an even function: $\cos(-x) = \cos x$. This can be seen by observing that the graph is symmetric with respect to the y-axis.

Turn back the page and take a second look at Figure 5.51. Can you see that the graph of $y = \cos x$ is the graph of $y = \sin x$ with a phase shift of $-\frac{\pi}{2}$ radians? If you trace along the curve from $x = -\frac{\pi}{2}$ to $x = \frac{3\pi}{2}$, you are tracing one complete cycle of the sine curve. This can be expressed as an identity:

$$\cos x = \sin\left(x + \frac{\pi}{2}\right).$$

Because of this similarity, the graphs of sine functions and cosine functions are called **sinusoidal graphs.**

4 Graph variations of $y = \cos x$.

Graphing Variations of $y = \cos x$

We use the same steps to graph variations of $y = \cos x$ as we did for graphing variations of $y = \sin x$. We will continue finding key points by dividing the period into four equal parts. Amplitudes, periods, and phase shifts play an important role when graphing by hand.

The Graph of $y = A \cos Bx$

The graph of $y = A \cos Bx$ has

$$\text{amplitude} = |A|$$

$$\text{period} = \frac{2\pi}{B}.$$

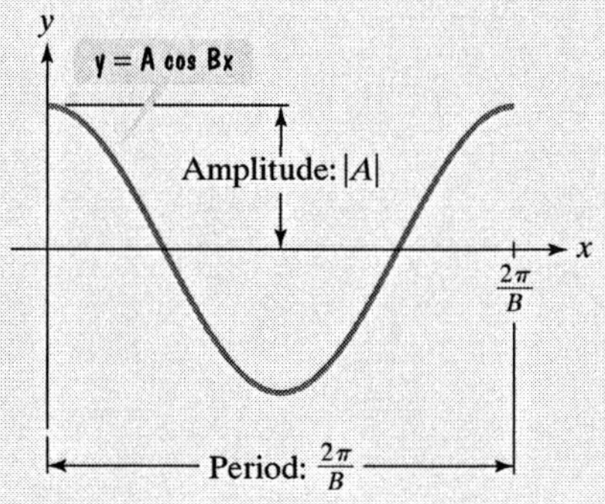

EXAMPLE 5 Graphing a Function of the Form $y = A \cos Bx$

Determine the amplitude and period of $y = -3 \cos \frac{\pi}{2} x$. Then graph the function for $-4 \le x \le 4$.

Solution

Step 1 Identify the amplitude and the period. The equation $y = -3 \cos \frac{\pi}{2} x$ is of the form $y = A \cos Bx$ with $A = -3$ and $B = \frac{\pi}{2}$.

amplitude: $|A| = |-3| = 3$ — The maximum y is 3 and the minimum is −3.

period: $\frac{2\pi}{B} = \frac{2\pi}{\frac{\pi}{2}} = 2\pi \cdot \frac{2}{\pi} = 4$ — Each cycle is completed in 4 radians.

Step 2 Find the x-values for the five key points. Begin by dividing the period, 4, by 4.

$$\frac{\text{period}}{4} = \frac{4}{4} = 1$$

Start with the value of x where the cycle begins: $x = 0$. Adding quarter-periods, 1, the five x-values for the key points are

$$x = 0, \quad x = 0 + 1 = 1, \quad x = 1 + 1 = 2, \quad x = 2 + 1 = 3, \quad x = 3 + 1 = 4$$

Step 3 **Find the values of y for the five key points.** We evaluate the function at each value of x from step 2.

Value of x	Value of y: $y = -3\cos\frac{\pi}{2}x$	Coordinates of key point
0	$y = -3\cos\frac{\pi}{2}\cdot 0$ $= -3\cos 0 = -3\cdot 1 = -3$	$(0, -3)$ minimum point
1	$y = -3\cos\frac{\pi}{2}\cdot 1$ $= -3\cos\frac{\pi}{2} = -3\cdot 0 = 0$	$(1, 0)$
2	$y = -3\cos\frac{\pi}{2}\cdot 2$ $= -3\cos\pi = -3(-1) = 3$	$(2, 3)$ maximum point
3	$y = -3\cos\frac{\pi}{2}\cdot 3$ $= -3\cos\frac{3\pi}{2} = -3(0) = 0$	$(3, 0)$
4	$y = -3\cos\frac{\pi}{2}\cdot 4$ $= -3\cos 2\pi = -3(1) = -3$	$(4, -3)$ minimum point

In the interval $[0, 4]$, there are x-intercepts at 1 and 3. The minimum and maximum points are indicated by the voice balloons.

Step 4 **Connect the five key points with a smooth curve and graph one complete cycle of the given function.** The five key points for $y = -3\cos\frac{\pi}{2}x$ are shown in Figure 5.52. By connecting the points with a smooth curve, the blue portion shows one complete cycle of $y = -3\cos\frac{\pi}{2}x$ from 0 to 4.

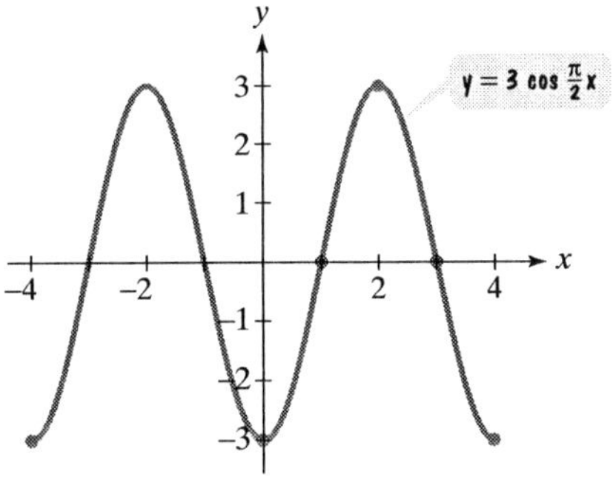

Figure 5.52

Step 5 **Extend the graph in step 4 to the left or right as desired.** The blue portion of the graph in Figure 5.52 is for x from 0 to 4. In order to graph for $-4 \le x \le 4$, we continue this portion and extend the graph another full period to the left. This extension is shown in black in Figure 5.52.

Technology

The graph of $y = -3\cos\frac{\pi}{2}x$ in a $[-4, 4, 1]$ by $[-4, 4, 1]$ viewing rectangle verifies our hand-drawn graph in Figure 5.52.

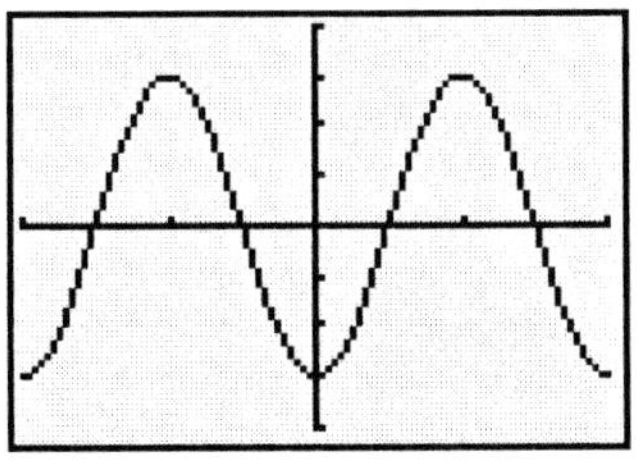

Check Point 5 Determine the amplitude and period of $y = -4\cos \pi x$. Then graph the function for $-2 \le x \le 2$.

Finally, let us examine the graphs of functions of the form $y = A\cos(Bx - C)$.

Graphs of these functions shift the graph of $y = A\cos Bx$ horizontally by $\frac{C}{B}$.

The Graph of $y = A\cos(Bx - C)$

The graph of $y = A\cos(Bx - C)$ is obtained by horizontally shifting the graph of $y = A\cos Bx$ so that the starting point of the cycle is shifted from $x = 0$ to $x = \frac{C}{B}$. If $\frac{C}{B} > 0$, the shift is to the right. If $\frac{C}{B} < 0$, the shift is to the left. The number $\frac{C}{B}$ is called the **phase shift.**

$$\text{amplitude} = |A|$$

$$\text{period} = \frac{2\pi}{B}$$

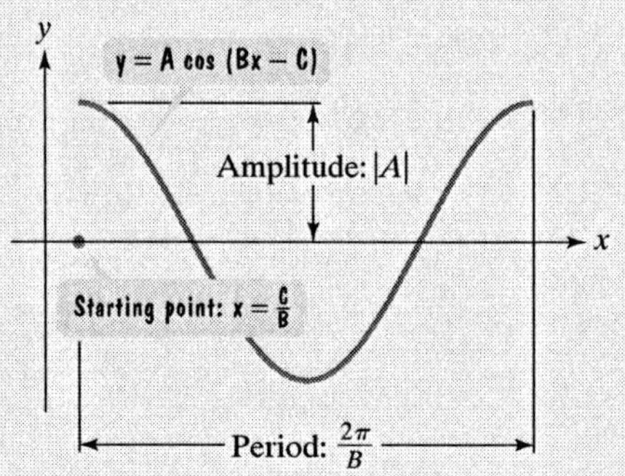

EXAMPLE 6 Graphing a Function of the Form $y = A\cos(Bx - C)$

Determine the amplitude, period, and phase shift of $y = \frac{1}{2}\cos(4x + \pi)$. Then graph one period of the function.

Solution

Step 1 Identify the amplitude, the period, and the phase shift. We must first identify values for A, B, and C. To do this, we need to express the equation in the form $y = A\cos(Bx - C)$. Thus, we write $y = \frac{1}{2}\cos(4x + \pi)$ as $y = \frac{1}{2}\cos[4x - (-\pi)]$. Now we can identify values for A, B, and C.

This equation is of the form $y = A\cos(Bx - C)$.

$$y = \frac{1}{2}\cos[4x - (-\pi)]$$

Using the voice balloon, we see that $A = \frac{1}{2}$, $B = 4$, and $C = -\pi$.

amplitude: $|A| = \left|\frac{1}{2}\right| = \frac{1}{2}$ — The maximum y is $\frac{1}{2}$ and the minimum is $-\frac{1}{2}$.

period: $\frac{2\pi}{B} = \frac{2\pi}{4} = \frac{\pi}{2}$ — Each cycle is completed in $\frac{\pi}{2}$ radians.

phase shift: $\frac{C}{B} = -\frac{\pi}{4}$ — A cycle starts at $x = -\frac{\pi}{4}$.

Step 2 Find the x-values for the five key points. Begin by dividing the period, $\frac{\pi}{2}$, by 4.

$$\frac{\text{period}}{4} = \frac{\frac{\pi}{2}}{4} = \frac{\pi}{8}$$

Start with the value of x where the cycle begins: $x = -\frac{\pi}{4}$. Adding quarter-periods, $\frac{\pi}{8}$, the five x-values for the key points are

$$x = -\frac{\pi}{4}, \quad x = -\frac{\pi}{4} + \frac{\pi}{8} = -\frac{2\pi}{8} + \frac{\pi}{8} = -\frac{\pi}{8}, \quad x = -\frac{\pi}{8} + \frac{\pi}{8} = 0,$$

$$x = 0 + \frac{\pi}{8} = \frac{\pi}{8}, \quad x = \frac{\pi}{8} + \frac{\pi}{8} = \frac{2\pi}{8} = \frac{\pi}{4}.$$

Step 3 Find the values of y for the five key points. Take a few minutes and use your calculator to evaluate the function at each value of x from step 2. Show that the key points are

$$\left(-\frac{\pi}{4}, \frac{1}{2}\right), \quad \left(-\frac{\pi}{8}, 0\right), \quad \left(0, -\frac{1}{2}\right), \quad \left(\frac{\pi}{8}, 0\right), \quad \text{and} \quad \left(\frac{\pi}{4}, \frac{1}{2}\right).$$

maximum point | x-intercept at $-\frac{\pi}{8}$ | minimum point | x-intercept at $\frac{\pi}{8}$ | maximum point

Step 4 Connect the five key points with a smooth curve and graph one complete cycle of the given function. The key points and the graph of $y = \frac{1}{2}\cos(4x + \pi)$ are shown in Figure 5.53.

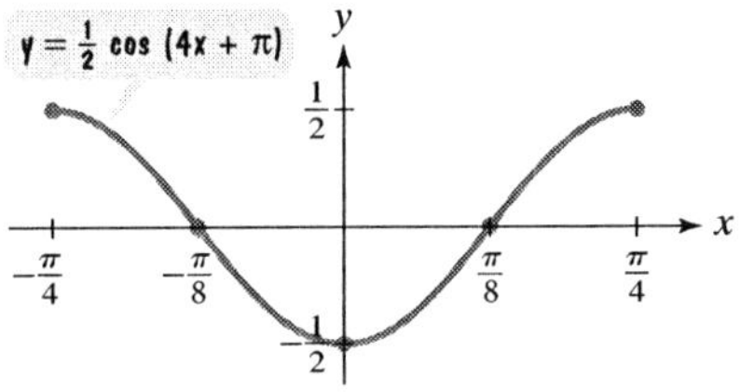

Figure 5.53

Technology

The graph of

$$y = \frac{1}{2}\cos(4x + \pi)$$

in a $\left[-\frac{\pi}{4}, \frac{\pi}{4}, \frac{\pi}{8}\right]$ by $[-1, 1, 1]$ viewing rectangle verifies our hand-drawn graph in Figure 5.53.

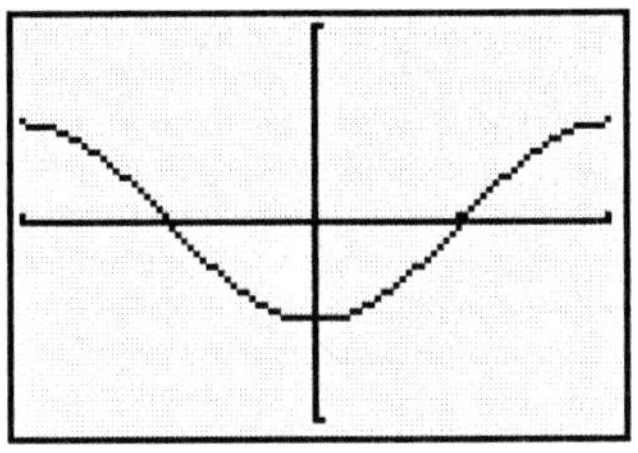

Check Point 6 Determine the amplitude, period, and phase shift of $y = \frac{3}{2}\cos(2x + \pi)$. Then graph one period of the function.

5 Use vertical shifts of sine and cosine curves.

Vertical Shifts of Sinusoidal Graphs

We now look at sinusoidal graphs of

$$y = A\sin(Bx - C) + D \quad \text{and} \quad y = A\cos(Bx - C) + D.$$

The constant D causes vertical shifts in the graphs of $y = A\sin(Bx - C)$ and $y = A\cos(Bx - C)$. If D is positive, the shift is D units upward. If D is negative, the shift is D units downward. These vertical shifts result in sinusoidal graphs oscillating about the horizontal line $y = D$ rather than about the x-axis. Thus, the maximum y is $D + |A|$ and the minimum y is $D - |A|$.

EXAMPLE 7 A Vertical Shift

Graph one period of the function $y = \frac{1}{2}\cos x - 1$.

Solution The graph of $y = \frac{1}{2}\cos x - 1$ is the graph of $y = \frac{1}{2}\cos x$ shifted one unit downward. The period of $y = \frac{1}{2}\cos x$ is 2π, which is also the period for the vertically shifted graph. The key points on the interval $[0, 2\pi]$ for $y = \frac{1}{2}\cos x - 1$ are found by first determining their x-coordinates. The quarter-period is $\frac{2\pi}{4}$, or $\frac{\pi}{2}$. The cycle begins at $x = 0$. As always, we add quarter-periods to generate x-values for each of the key points. The five x-values are

$$x = 0, \quad x = 0 + \frac{\pi}{2} = \frac{\pi}{2}, \quad x = \frac{\pi}{2} + \frac{\pi}{2} = \pi,$$

$$x = \pi + \frac{\pi}{2} = \frac{3\pi}{2}, \quad x = \frac{3\pi}{2} + \frac{\pi}{2} = 2\pi.$$

The values of y for the five key points and their coordinates are determined as follows.

Value of x	Value of y: $y = \frac{1}{2}\cos x - 1$	Coordinates of key point
0	$y = \frac{1}{2}\cos 0 - 1$ $= \frac{1}{2} \cdot 1 - 1 = -\frac{1}{2}$	$\left(0, -\frac{1}{2}\right)$
$\frac{\pi}{2}$	$y = \frac{1}{2}\cos\frac{\pi}{2} - 1$ $= \frac{1}{2} \cdot 0 - 1 = -1$	$\left(\frac{\pi}{2}, -1\right)$
π	$y = \frac{1}{2}\cos\pi - 1$ $= \frac{1}{2}(-1) - 1 = -\frac{3}{2}$	$\left(\pi, -\frac{3}{2}\right)$
$\frac{3\pi}{2}$	$y = \frac{1}{2}\cos\frac{3\pi}{2} - 1$ $= \frac{1}{2} \cdot 0 - 1 = -1$	$\left(\frac{3\pi}{2}, -1\right)$
2π	$y = \frac{1}{2}\cos 2\pi - 1$ $= \frac{1}{2} \cdot 1 - 1 = -\frac{1}{2}$	$\left(2\pi, -\frac{1}{2}\right)$

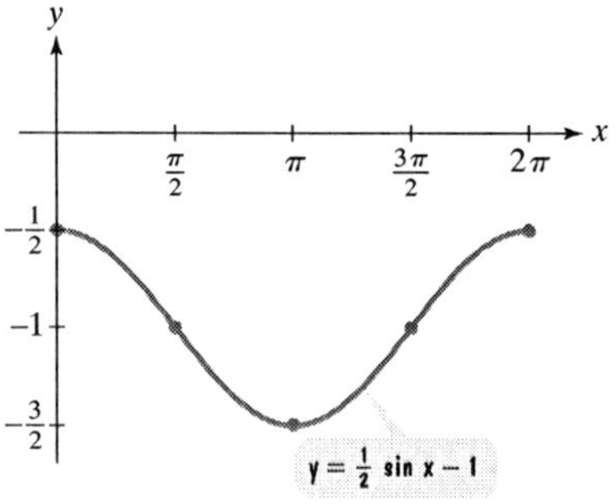

Figure 5.54

The five key points for $y = \frac{1}{2}\cos x - 1$ are shown in Figure 5.54. By connecting the points with a smooth curve, we obtain one period of the graph.

Check Point 7 Graph one period of the function $y = 2\cos x + 1$.

6 Model periodic behavior.

Modeling Periodic Behavior

Our breathing consists of alternating periods of inhaling and exhaling. Each complete pumping cycle of the human heart can be described using a sine function. Our brain waves during deep sleep are sinusoidal. Viewed in this way, trigonometry becomes an intimate experience.

Some graphing utilities have a SINe REGression feature. This feature gives the sine function in the form $y = A\sin(Bx + C) + D$ of best fit for wavelike data. At least four data points must be used. However, it is not always necessary to use technology. In our next example, we use our understanding of sinusoidal graphs to model the process of breathing.

EXAMPLE 8 A Trigonometric Breath of Life

The graph in Figure 5.55 shows one complete normal breathing cycle. The cycle consists of inhaling and exhaling. It takes place every 5 seconds. Velocity of air flow is positive when we inhale and negative when we exhale. It is measured in liters per second. If y represents velocity of air flow after x seconds, find a function of the form $y = A\sin Bx$ that models air flow in a normal breathing cycle.

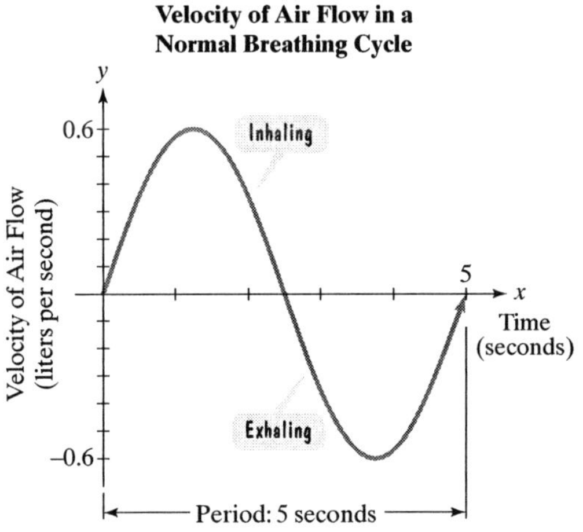

Figure 5.55

Solution We need to determine values for A and B in the equation $y = A\sin Bx$. A, the amplitude, is the maximum value of y. Figure 5.55 shows that this maximum value is 0.6. Thus, $A = 0.6$.

The value of B in $y = A\sin Bx$ can be found using the formula for the period: period $= \frac{2\pi}{B}$. The period of our breathing cycle is 5 seconds. Thus,

$$5 = \frac{2\pi}{B} \quad \text{Our goal is to solve this equation for } B.$$

$$5B = 2\pi \quad \text{Multiply both sides of the equation by } B.$$

$$B = \frac{2\pi}{5}. \quad \text{Divide both sides of the equation by 5.}$$

We see that $A = 0.6$ and $B = \frac{2\pi}{5}$. Substitute these values into $y = A\sin Bx$. The breathing cycle is modeled by

$$y = 0.6\sin\frac{2\pi}{5}x.$$

Check Point 8 Find an equation of the form $y = A \sin Bx$ that produces the graph shown in the figure on the right.

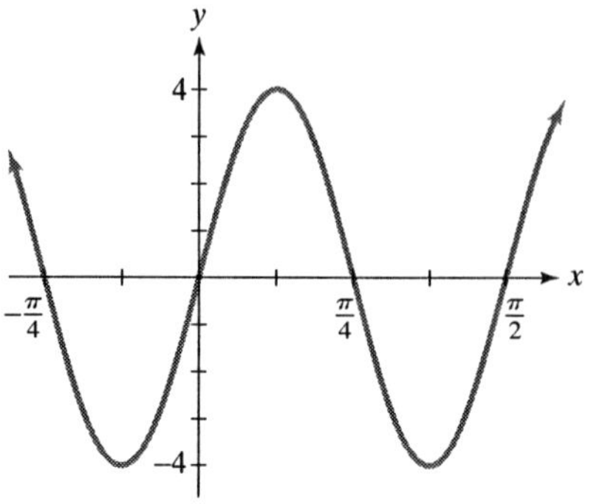

EXAMPLE 9 Modeling a Tidal Cycle

Figure 5.56 shows that the depth of water at a boat dock varies with the tides. The depth is 5 feet at low tide and 13 feet at high tide. On a certain day, low tide occurs at 4 A.M. and high tide at 10 A.M. If y represents the depth of the water, in feet, x hours after midnight, use a sine function of the form $y = A\sin(Bx - C) + D$ to model the water's depth.

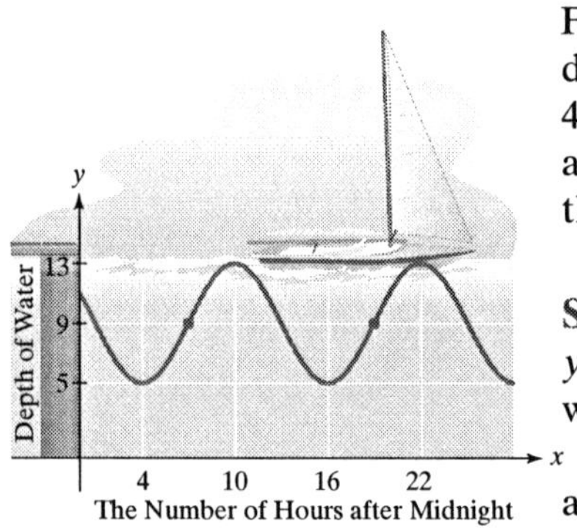

Figure 5.56

Solution We need to determine values for A, B, C, and D in the equation $y = A\sin(Bx - C) + D$. We can find these values using Figure 5.56. We begin with D.

To find D, we use the vertical shift. Because the water's depth ranges from a minimum of 5 feet to a maximum of 13 feet, the curve oscillates about the middle value, 9 feet. Thus, $D = 9$, which is the vertical shift.

At maximum depth, the water is 4 feet above 9 feet. Thus, A, the amplitude, is 4: $A = 4$.

To find B, we use the period. The blue portion of the graph shows that one complete tidal cycle occurs in $19 - 7$, or 12 hours. The period is 12. Thus,

$$12 = \frac{2\pi}{B} \quad \text{Our goal is to solve this equation for } B.$$

$$12B = 2\pi \quad \text{Multiply both sides by } B.$$

$$B = \frac{2\pi}{12} = \frac{\pi}{6}. \quad \text{Divide both sides by 12.}$$

To find C, we use the phase shift. The blue portion of the graph shows that the starting point of the cycle is shifted from 0 to 7. The phase shift, $\frac{C}{B}$, is 7.

$$7 = \frac{C}{B} \quad \text{The phase shift of } y = A\sin(Bx - C) \text{ is } \frac{C}{B}.$$

$$7 = \frac{C}{\frac{\pi}{6}} \quad \text{From above, we have } B = \frac{\pi}{6}.$$

$$\frac{7\pi}{6} = C \quad \text{Multiply both sides of the equation by } \frac{\pi}{6}.$$

Technology

We can use a graphing utility to verify that the model in Example 9

$$y = 4\sin\left(\frac{\pi}{6}x - \frac{7\pi}{6}\right) + 9$$

is correct. The graph of the function is shown in a $[0, 28, 4]$ by $[0, 15, 5]$ viewing rectangle.

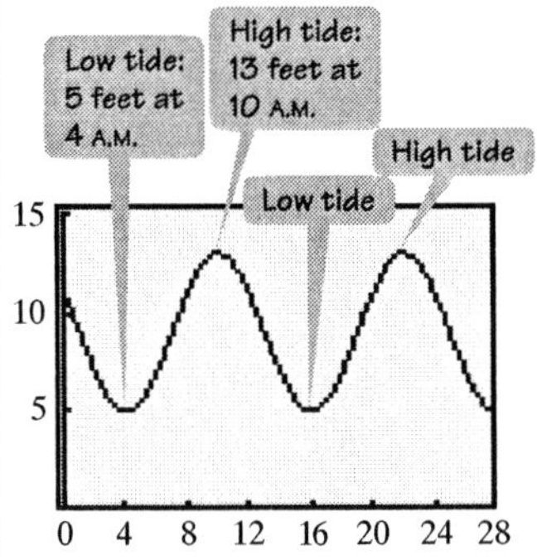

We see that $A = 4$, $B = \frac{\pi}{6}$, $C = \frac{7\pi}{6}$, and $D = 9$. Substitute these values into $y = A\sin(Bx - C) + D$. The water's depth, in feet, x hours after midnight is modeled by

$$y = 4\sin\left(\frac{\pi}{6}x - \frac{7\pi}{6}\right) + 9.$$

Check Point 9 The figure shows the number of hours of daylight for a region that is 30° north of the equator. Hours of daylight are at a minimum of 10 hours in December. Hours of daylight are at a maximum of 14 hours in June. Let x represent the month of the year, with 1 for January, 2 for February, 3 for March, and 12 for December. If y represents the number of hours of daylight in month x, use a sine function of the form $y = A\sin(Bx - C) + D$ to model the hours of daylight.

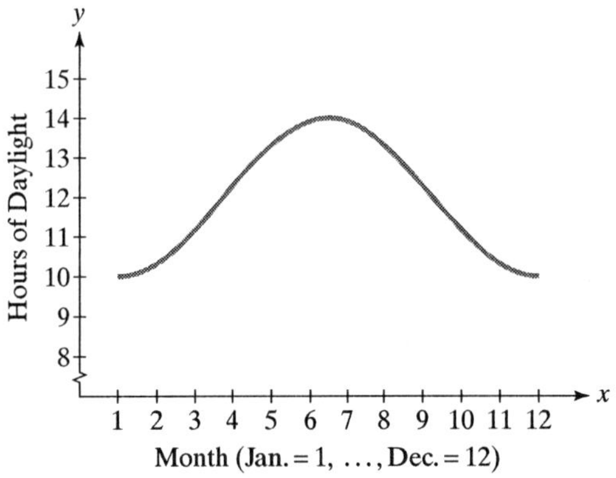

EXERCISE SET 5.5

Practice Exercises

In Exercises 1–6, determine the amplitude of each function. Then graph the function and $y = \sin x$ in the same rectangular coordinate system for $0 \le x \le 2\pi$.

1. $y = 4\sin x$

2. $y = 5\sin x$

3. $y = \frac{1}{3}\sin x$

4. $y = \frac{1}{4}\sin x$

5. $y = -3\sin x$

6. $y = -4\sin x$

In Exercises 7–16, determine the amplitude and period of each function. Then graph one period of the function.

7. $y = \sin 2x$

8. $y = \sin 4x$

9. $y = 3\sin\frac{1}{2}x$

10. $y = 2\sin\frac{1}{4}x$

11. $y = 4\sin\pi x$

12. $y = 3\sin 2\pi x$

13. $y = -3\sin 2\pi x$

14. $y = -2\sin\pi x$

15. $y = -\sin\frac{2}{3}x$

16. $y = -\sin\frac{4}{3}x$

In Exercises 17–30, determine the amplitude, period, and phase shift of each function. Then graph one period of the function.

17. $y = \sin(x - \pi)$

18. $y = \sin\left(x - \frac{\pi}{2}\right)$

19. $y = \sin(2x - \pi)$

20. $y = \sin\left(2x - \frac{\pi}{2}\right)$

21. $y = 3\sin(2x - \pi)$
22. $y = 3\sin\left(2x - \frac{\pi}{2}\right)$
23. $y = \frac{1}{2}\sin\left(x + \frac{\pi}{2}\right)$
24. $y = \frac{1}{2}\sin(x + \pi)$
25. $y = -2\sin\left(2x + \frac{\pi}{2}\right)$
26. $y = -3\sin\left(2x + \frac{\pi}{2}\right)$
27. $y = 3\sin(\pi x + 2)$
28. $y = 3\sin(2\pi x + 4)$
29. $y = -2\sin(2\pi x + 4\pi)$
30. $y = -3\sin(2\pi x + 4\pi)$

In Exercises 31–34, determine the amplitude of each function. Then graph the function and $y = \cos x$ in the same rectangular coordinate system for $0 \le x \le 2\pi$.

31. $y = 2\cos x$
32. $y = 3\cos x$
33. $y = -2\cos x$
34. $y = -3\cos x$

In Exercises 35–42, determine the amplitude and period of each function. Then graph one period of the function.

35. $y = \cos 2x$
36. $y = \cos 4x$
37. $y = 4\cos 2\pi x$
38. $y = 5\cos 2\pi x$
39. $y = -4\cos\frac{1}{2}x$
40. $y = -3\cos\frac{1}{3}x$
41. $y = -\frac{1}{2}\cos\frac{\pi}{3}x$
42. $y = -\frac{1}{2}\cos\frac{\pi}{4}x$

In Exercises 43–50, determine the amplitude, period, and phase shift of each function. Then graph one period of the function.

43. $y = 3\cos(2x - \pi)$
44. $y = 4\cos(2x - \pi)$
45. $y = \frac{1}{2}\cos\left(3x + \frac{\pi}{2}\right)$
46. $y = \frac{1}{2}\cos(2x + \pi)$
47. $y = -3\cos\left(2x - \frac{\pi}{2}\right)$
48. $y = -4\cos\left(2x - \frac{\pi}{2}\right)$
49. $y = 2\cos(2\pi x + 8\pi)$
50. $y = 3\cos(2\pi x + 4\pi)$

In Exercises 51–58, use a vertical shift to graph one period of the function.

51. $y = \sin x + 2$
52. $y = \sin x - 2$
53. $y = \cos x - 3$
54. $y = \cos x + 3$
55. $y = 2\sin\frac{1}{2}x + 1$
56. $y = 2\cos\frac{1}{2}x + 1$
57. $y = -3\cos 2\pi x + 2$
58. $y = -3\sin 2\pi x + 2$

Application Exercises

In the theory of biorhythms, sine functions are used to measure a person's potential. You can obtain your biorhythm chart online by simply entering your date of birth, the date you want your biorhythm chart to begin, and the number of months you wish to be included in the plot. At the top of the next column is your author's chart, beginning January 25, 2003, when he was 21,093 days old. We all have cycles with the same amplitudes and periods as those shown here. Each of our three basic cycles begins at birth. Use the biorhythm chart shown to solve Exercises 59–66. The longer tick marks correspond to the dates shown.

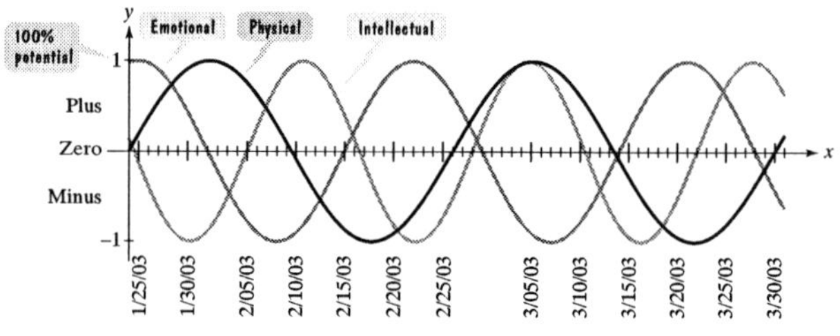

59. What is the period of the physical cycle?

60. What is the period of the emotional cycle?

61. What is the period of the intellectual cycle?

62. For the period shown, what is the worst day in February for your author to run in a marathon?

63. For the period shown, what is the best day in March for your author to meet an online friend for the first time?

64. For the period shown, what is the best day in February for your author to begin writing this trigonometry chapter?

65. If you extend these sinusoidal graphs to the end of the year, is there a day when your author should not even bother getting out of bed?

66. If you extend these sinusoidal graphs to the end of the year, are there any days where your author is at near-peak physical, emotional, and intellectual potential?

67. Rounded to the nearest hour, Los Angeles averages 14 hours of daylight in June, 10 hours in December, and 12 hours in March and September. Let x represent the number of months after June and y represent the number of hours of daylight in month x. Make a graph that displays the information from June of one year to June of the following year.

68. A clock with an hour hand that is 15 inches long is hanging on a wall. At noon, the distance between the tip of the hour hand and the ceiling is 23 inches. At 3 P.M., the distance is 38 inches; at 6 P.M., 53 inches; at 9 P.M., 38 inches; and at midnight the distance is again 23 inches. If y represents the distance between the tip of the hour hand and the ceiling x hours after noon, make a graph that displays the information for $0 \le x \le 24$.

69. The number of hours of daylight in Boston is given by

$$y = 3\sin\frac{2\pi}{365}(x - 79) + 12$$

where x is the number of days after January 1.

a. What is the amplitude of this function?
b. What is the period of this function?

c. How many hours of daylight are there on the longest day of the year?

d. How many hours of daylight are there on the shortest day of the year?

e. Graph the function for one period, starting on January 1.

70. The average monthly temperature, y, in degrees Fahrenheit, for Juneau, Alaska, can be modeled by $y = 16 \sin\left(\frac{\pi}{6}x - \frac{2\pi}{3}\right) + 40$, where x is the month of the year (January $= 1$, February $= 2$, ... December $= 12$). Graph the function for $1 \le x \le 12$. What is the highest average monthly temperature? In which month does this occur?

71. The figure shows the depth of water at the end of a boat dock. The depth is 6 feet at low tide and 12 feet at high tide. On a certain day, low tide occurs at 6 A.M. and high tide at noon. If y represents the depth of the water x hours after midnight, use a cosine function of the form $y = A \cos Bx + D$ to model the water's depth.

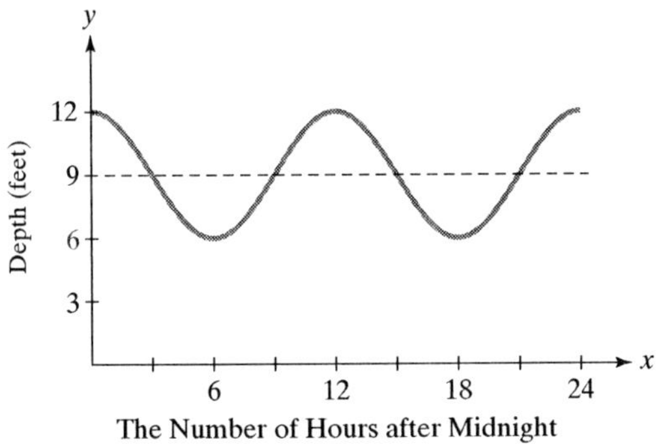

72. The figure shows the depth of water at the end of a boat dock. The depth is 5 feet at high tide and 3 feet at low tide. On a certain day, high tide occurs at noon and low tide at 6 P.M. If y represents the depth of the water x hours after noon, use a cosine function of the form $y = A \cos Bx + D$ to model the water's depth.

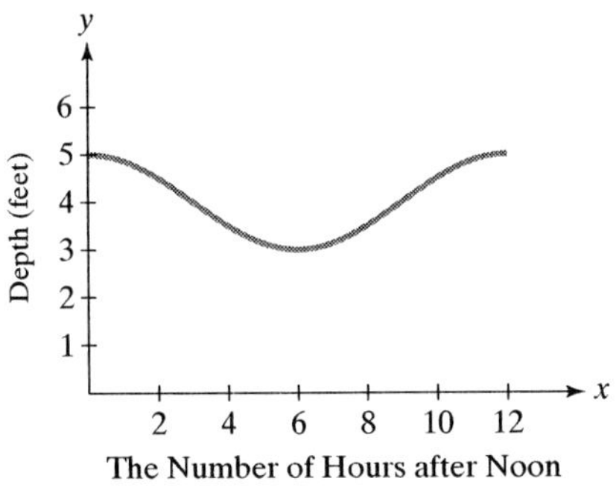

Writing in Mathematics

73. Without drawing a graph, describe the behavior of the basic sine curve.

74. What is the amplitude of the sine function? What does this tell you about the graph?

75. If you are given the equation of a sine function, how do you determine the period?

76. What does a phase shift indicate about the graph of a sine function? How do you determine the phase shift from the function's equation?

77. Describe a general procedure for obtaining the graph of $y = A \sin(Bx - C)$.

78. Without drawing a graph, describe the behavior of the basic cosine curve.

79. Describe a relationship between the graphs of $y = \sin x$ and $y = \cos x$.

80. Describe the relationship between the graphs of $y = A\cos(Bx - C)$ and $y = A\cos(Bx - C) + D$.

81. Biorhythm cycles provide interesting applications of sinusoidal graphs. But do you believe in the validity of biorhythms? Write a few sentences explaining why or why not.

Technology Exercises

82. Use a graphing utility to verify any five of the sine curves that you drew by hand in Exercises 7–30. The amplitude, period, and phase shift should help you to determine appropriate range settings.

83. Use a graphing utility to verify any five of the cosine curves that you drew by hand in Exercises 31–50.

84. Use a graphing utility to verify any two of the sinusoidal curves with vertical shifts that you drew in Exercises 51–58.

In Exercises 85–88, use a graphing utility to graph two periods of the function.

85. $y = 3\sin(2x + \pi)$

86. $y = -2\cos\left(2\pi x - \frac{\pi}{2}\right)$

87. $y = 0.2\sin\left(\frac{\pi}{10}x + \pi\right)$

88. $y = 3\sin(2x - \pi) + 5$

89. Use a graphing utility to graph $y = \sin x$ and $y = x - \frac{x^3}{6} + \frac{x^5}{120}$ in a $\left[-\pi, \pi, \frac{\pi}{2}\right]$ by $[-2, 2, 1]$ viewing rectangle. How do the graphs compare?

90. Use a graphing utility to graph $y = \cos x$ and $y = 1 - \frac{x^2}{2} + \frac{x^4}{24}$ in a $\left[-\pi, \pi, \frac{\pi}{2}\right]$ by $[-2, 2, 1]$ viewing rectangle. How do the graphs compare?

91. Use a graphing utility to graph

$$y = \sin x + \frac{\sin 2x}{2} + \frac{\sin 3x}{3} + \frac{\sin 4x}{4}$$

in a $\left[-2\pi, 2\pi, \frac{\pi}{2}\right]$ by $[-2, 2, 1]$ viewing rectangle. How do these waves compare to the smooth rolling waves of the basic sine curve?

92. Use a graphing utility to graph

$$y = \sin x - \frac{\sin 3x}{9} + \frac{\sin 5x}{25}$$

in a $\left[-2\pi, 2\pi, \frac{\pi}{2}\right]$ by $[-2, 2, 1]$ viewing rectangle. How do these waves compare to the smooth rolling waves of the basic sine curve?

93. The data show the average monthly temperatures for Washington, D.C.

a. Use your graphing utility to draw a scatter plot of the data from $x = 1$ through $x = 12$.

b. Use the SINe REGression feature to find the sinusoidal function of the form $y = A\sin(Bx + C) + D$ that best fits the data.

c. Use your graphing utility to draw the sinusoidal function of best fit on the scatter plot.

x Month		Average Monthly Temperature, °F
1	(January)	34.6
2	(February)	37.5
3	(March)	47.2
4	(April)	56.5
5	(May)	66.4
6	(June)	75.6
7	(July)	80.0
8	(August)	78.5
9	(September)	71.3
10	(October)	59.7
11	(November)	49.8
12	(December)	39.4

Source: U.S. National Oceanic and Atmospheric Administration

94. Repeat Exercise 93 for data of your choice. The data can involve the average monthly temperatures for the region where you live or any data whose scatter plot takes the form of a sinusoidal function.

Critical Thinking Exercises

Graph each function in Exercises 95–96 by hand.

95. $y = \sin x + \cos x$ for $0 \le x \le 2\pi$

96. $y = x + \cos x$ for $0 \le x \le \frac{5\pi}{2}$

97. Use the cosine function to find an equation of the graph in the figure shown.

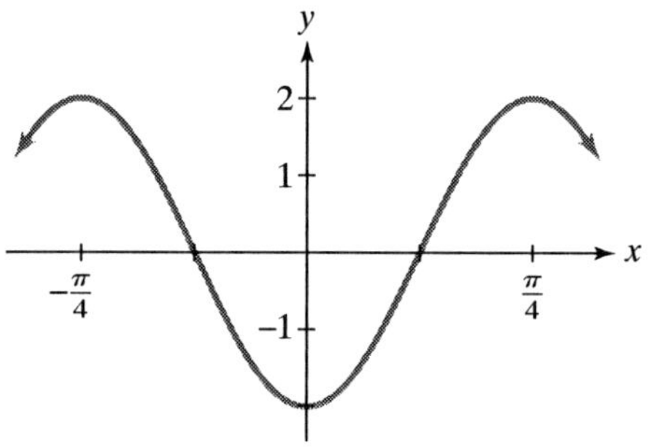

Group Exercise

98. This exercise is intended to provide some fun with biorhythms, regardless of whether you believe they have any validity. We will use each member's chart to determine biorhythmic compatibility. Before meeting, each group member should go online and obtain his or her biorhythm chart. The date of the group meeting is the date on which your chart should begin. Include 12 months in the plot. At the meeting, compare differences and similarities among the intellectual sinusoidal curves. Using these comparisons, each person should find the one other person with whom he or she would be most intellectually compatible.

SECTION 5.6 Graphs of Other Trigonometric Functions

Objectives

1. Understand the graph of $y = \tan x$.
2. Graph variations of $y = \tan x$.
3. Understand the graph of $y = \cot x$.
4. Graph variations of $y = \cot x$.
5. Understand the graphs of $y = \csc x$ and $y = \sec x$.
6. Graph variations of $y = \csc x$ and $y = \sec x$.

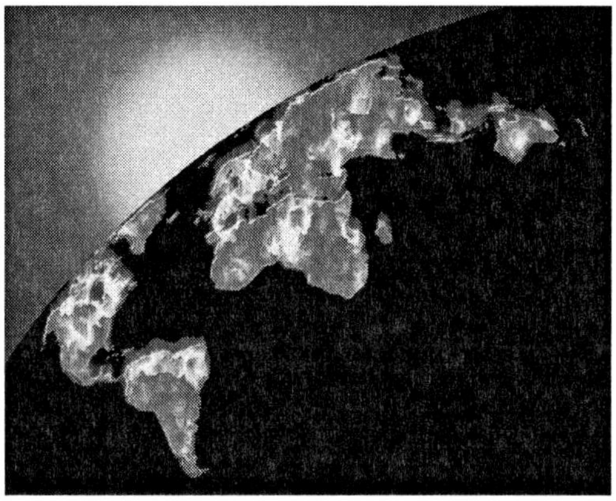

Recent advances in our understanding of climate have changed global warming from a subject for a disaster movie (the Statue of Liberty up to its chin in water) to a serious but manageable scientific and policy issue. Global warming is related to the burning of fossil fuels, which adds carbon dioxide to the atmosphere. In the new millennium, we will see whether our use of fossil fuels will add enough carbon dioxide to the atmosphere to change it (and our climate) in significant ways. In this section's exercise set, you will see how trigonometric graphs reveal interesting patterns in carbon dioxide concentration from 1990 through 2005. In the section itself, trigonometric graphs will reveal patterns involving the tangent, cotangent, secant, and cosecant functions.

1 Understand the graph of $y = \tan x$.

The Graph of $y = \tan x$

The properties of the tangent function discussed in Section 5.4 will help us determine its graph. Because the tangent function has properties that are different from sinusoidal functions, its graph differs significantly from those of sine and cosine. Properties of the tangent function include the following:

- The period is π. It is only necessary to graph $y = \tan x$ over an interval of length π. The remainder of the graph consists of repetitions of that graph at intervals of π.
- The tangent function is an odd function: $\tan(-x) = -\tan x$. The graph is symmetric with respect to the origin.
- The tangent function is undefined at $\frac{\pi}{2}$. The graph of $y = \tan x$ has a vertical asymptote at $x = \frac{\pi}{2}$.

We obtain the graph of $y = \tan x$ using some points on the graph and origin symmetry. Table 5.3 lists some values of (x, y) on the graph of $y = \tan x$ on the interval $\left[0, \frac{\pi}{2}\right)$.

Table 5.3 Values of (x, y) on $y = \tan x$

x	0	$\frac{\pi}{6}$	$\frac{\pi}{4}$	$\frac{\pi}{3}$	$\frac{5\pi}{12}$ (75°)	$\frac{17\pi}{36}$ (85°)	$\frac{89\pi}{180}$ (89°)	1.57	$\frac{\pi}{2}$
$y = \tan x$	0	$\frac{\sqrt{3}}{3} \approx 0.6$	1	$\sqrt{3} \approx 1.7$	3.7	11.4	57.3	1255.8	undefined

As x increases from 0 to $\frac{\pi}{2}$, y increases slowly at first, then more and more rapidly.

The graph in Figure 5.57(a) is based on our observation that as x increases from 0 to $\frac{\pi}{2}$, y increases slowly at first, then more and more rapidly. Notice that y increases without bound as x approaches $\frac{\pi}{2}$. As the figure shows, the graph of $y = \tan x$ has a vertical asymptote at $x = \frac{\pi}{2}$.

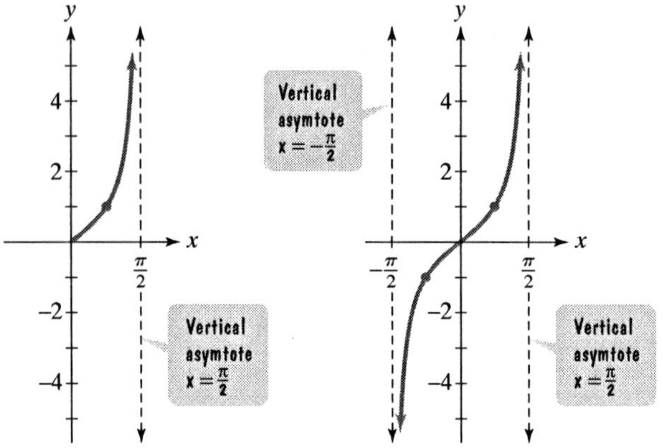

Figure 5.57 Graphing the tangent function

(a) $y = \tan x, 0 \le x < \frac{\pi}{2}$ (b) $y = \tan x, -\frac{\pi}{2} < x < \frac{\pi}{2}$

The graph of $y = \tan x$ can be completed on the interval $\left(-\frac{\pi}{2}, \frac{\pi}{2}\right)$ by using origin symmetry. Figure 5.57(b) shows the result of reflecting the graph in Figure 5.57(a) about the origin. The graph of $y = \tan x$ has another vertical asymptote at $x = -\frac{\pi}{2}$. Notice that y decreases without bound as x approaches $-\frac{\pi}{2}$.

Because the period of the tangent function is π radians, the graph in Figure 5.57(b) shows one complete period of $y = \tan x$. We obtain the complete graph of $y = \tan x$ by repeating the graph in Figure 5.57(b) to the left and right over intervals of π. The resulting graph and its main characteristics are shown in the following box:

The Tangent Curve: The Graph of $y = \tan x$ and Its Characteristics

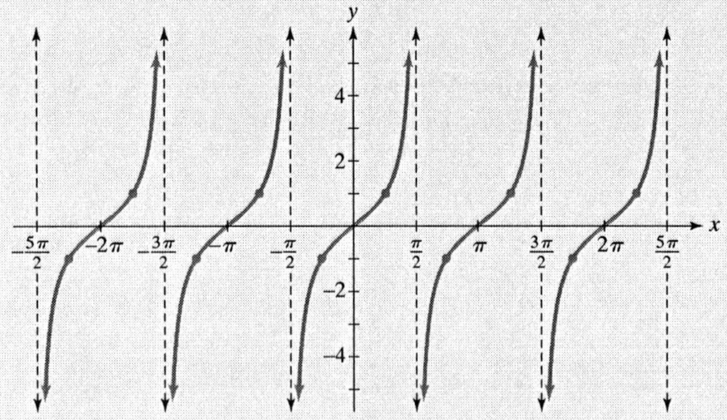

Characteristics

- **Period:** π
- **Domain:** All real numbers except odd multiples of $\frac{\pi}{2}$
- **Range:** All real numbers
- **Vertical asymptotes** at odd multiples of $\frac{\pi}{2}$
- **An x-intercept** occurs midway between each pair of consecutive asymptotes.
- **Odd function** with origin symmetry
- Points on the graph $\frac{1}{4}$ and $\frac{3}{4}$ of the way between consecutive asymptotes have y-coordinates of -1 and 1.

2 Graph variations of $y = \tan x$.

Graphing Variations of $y = \tan x$

We use the characteristics of the tangent curve to graph tangent functions of the form $y = A\tan(Bx - C)$.

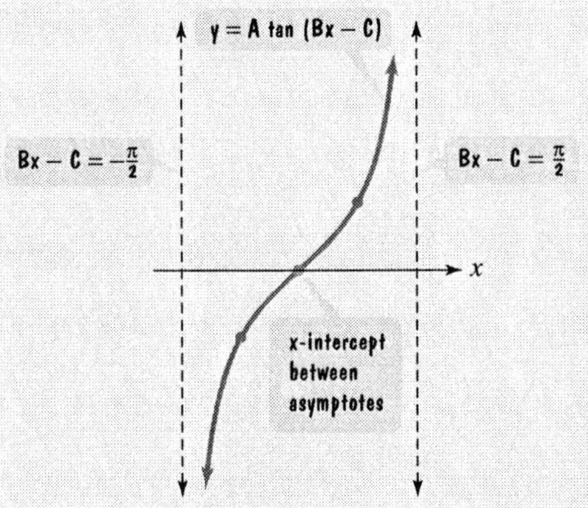

Graphing $y = A\tan(Bx - C)$

1. Find two consecutive asymptotes by setting the variable expression in the tangent equal to $-\frac{\pi}{2}$ and $\frac{\pi}{2}$ and solving
$$Bx - C = -\frac{\pi}{2} \text{ and } Bx - C = \frac{\pi}{2}.$$
2. Identify an x-intercept, midway between the consecutive asymptotes.
3. Find the points on the graph $\frac{1}{4}$ and $\frac{3}{4}$ of the way between the consecutive asymptotes. These points have y-coordinates of $-A$ and A.
4. Use steps 1–3 to graph one full period of the function. Add additional cycles to the left or right as needed.

EXAMPLE 1 Graphing a Tangent Function

Graph $y = 2\tan\frac{x}{2}$ for $-\pi < x < 3\pi$.

Solution Refer to Figure 5.58 as you read each step.

Step 1 Find two consecutive asymptotes. We must solve the equations

$$\frac{x}{2} = -\frac{\pi}{2} \quad \text{and} \quad \frac{x}{2} = \frac{\pi}{2}.$$ Set the variable expression in the tangent equal to $-\frac{\pi}{2}$ and $\frac{\pi}{2}$.

$$x = -\pi \qquad x = \pi$$ Multiply both sides of each equation by 2.

Thus, two consecutive asymptotes occur at $x = -\pi$ and $x = \pi$.

Step 2 Identify an x-intercept, midway between the consecutive asymptotes. Midway between $x = -\pi$ and $x = \pi$ is $x = 0$. An x-intercept is 0 and the graph passes through $(0, 0)$.

Step 3 Find points on the graph $\frac{1}{4}$ and $\frac{3}{4}$ of the way between the consecutive asymptotes. These points have y-coordinates of $-A$ and A. Because A, the coefficient of the tangent in $y = 2\tan\frac{x}{2}$ is 2, these points have y-coordinates of -2 and 2. The graph passes through $\left(-\frac{\pi}{2}, -2\right)$ and $\left(\frac{\pi}{2}, 2\right)$.

Step 4 Use steps 1–3 to graph one full period of the function. We use the two consecutive asymptotes, $x = -\pi$ and $x = \pi$, an x-intercept of 0, and points midway between the x-intercept and asymptotes with y-coordinates of -2 and 2. We graph one period of $y = 2\tan\frac{\pi}{2}$ from $-\pi$ to π. In order to graph for $-\pi < x < 3\pi$, we continue the pattern and extend the graph another full period to the right. The graph is shown in Figure 5.58.

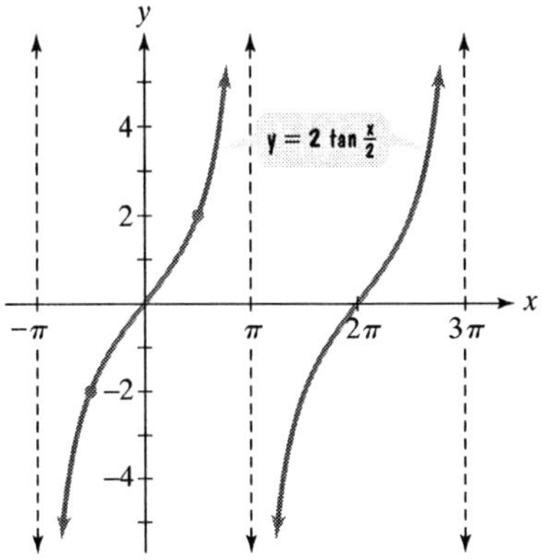

Figure 5.58 The graph is shown for two full periods.

Check Point 1 Graph $y = 3\tan 2x$ for $-\frac{\pi}{4} < x < \frac{3\pi}{4}$.

EXAMPLE 2 Graphing a Tangent Function

Graph two full periods of $y = \tan\left(x + \frac{\pi}{4}\right)$.

Solution The graph of $y = \tan\left(x + \frac{\pi}{4}\right)$ is the graph of $y = \tan x$ shifted horizontally to the left $\frac{\pi}{4}$ units. Refer to Figure 5.59 as you read each step.

Step 1 Find two consecutive asymptotes. We must solve the equations

$$x + \frac{\pi}{4} = -\frac{\pi}{2} \quad \text{and} \quad x + \frac{\pi}{4} = \frac{\pi}{2}.$$ Set the variable expression in the tangent equal to $-\frac{\pi}{2}$ and $\frac{\pi}{2}$.

$$x = -\frac{\pi}{4} - \frac{\pi}{2} \qquad x = -\frac{\pi}{4} + \frac{\pi}{2}$$ Subtract $\frac{\pi}{4}$ from both sides in each equation.

$$x = -\frac{3\pi}{4} \qquad x = \frac{\pi}{4}$$ Simplify.

Thus, two consecutive asymptotes occur at $x = -\frac{3\pi}{4}$ and $x = \frac{\pi}{4}$.

Step 2 Identify an x-intercept, midway between the consecutive asymptotes.

$$x\text{-intercept} = \frac{-\frac{3\pi}{4} + \frac{\pi}{4}}{2} = \frac{-\frac{2\pi}{4}}{2} = -\frac{2\pi}{8} = -\frac{\pi}{4}$$

An x-intercept is $-\frac{\pi}{4}$ and the graph passes through $\left(-\frac{\pi}{4}, 0\right)$.

Step 3 Find points on the graph $\frac{1}{4}$ and $\frac{3}{4}$ of the way between the consecutive asymptotes. These points have y-coordinates of $-A$ and A. Because A, the coefficient of the tangent in $y = \tan\left(x + \frac{\pi}{4}\right)$ is 1,these points have y-coordinates of -1 and 1. They are shown as blue dots in Figure 5.59.

Step 4 Use steps 1–3 to graph one full period of the function. We use the two consecutive asymptotes, $x = -\frac{3\pi}{4}$ and $x = \frac{\pi}{4}$, to graph one full period of $y = \tan\left(x + \frac{\pi}{4}\right)$ from $-\frac{3\pi}{4}$ to $\frac{\pi}{4}$. We graph two full periods by continuing the pattern and extending the graph another full period to the right. The graph is shown in Figure 5.59.

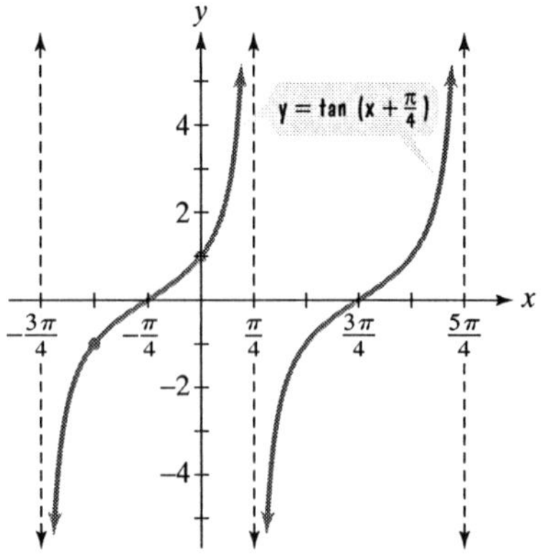

Figure 5.59 The graph is shown for two full periods.

Check Point 2 Graph two full periods of $y = \tan\left(x - \frac{\pi}{2}\right)$.

3 Understand the graph of $y = \cot x$.

The Graph of $y = \cot x$

Like the tangent function, the cotangent function, $y = \cot x$, has a period of π. The graph and its main characteristics are shown in the following box:

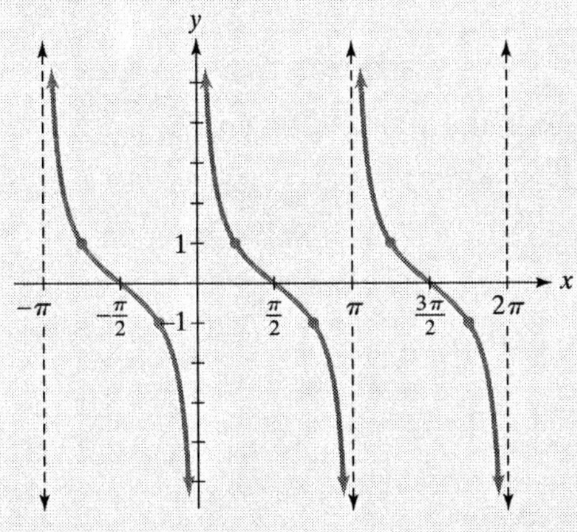

The Cotangent Curve: The Graph of $y = \cot x$ and Its Characteristics

Characteristics

- **Period:** π
- **Domain:** All real numbers except integral multiples of π
- **Range:** All real numbers
- **Vertical asymptotes** at integral multiples of π
- **An x-intercept** occurs midway between each pair of consecutive asymptotes.
- **Odd function** with origin symmetry
- Points on the graph $\frac{1}{4}$ and $\frac{3}{4}$ of the way between consecutive asymptotes have y-coordinates of 1 and -1.

4 Graph variations of $y = \cot x$.

Graphing Variations of $y = \cot x$

We use the characteristics of the cotangent curve to graph cotangent functions of the form $y = A\cot(Bx - C)$.

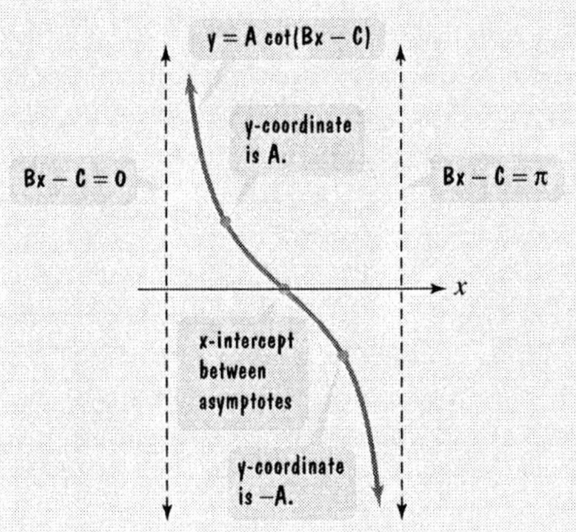

Graphing $y = A\cot(Bx - C)$

1. Find two consecutive asymptotes by setting the variable expression in the cotangent equal to 0 and π and solving
$$Bx - C = 0 \text{ and } Bx - C = \pi.$$
2. Identify an x-intercept, midway between the consecutive asymptotes.
3. Find the points on the graph $\frac{1}{4}$ and $\frac{3}{4}$ of the way between the consecutive asymptotes. These points have y-coordinates of A and $-A$.
4. Use steps 1–3 to graph one full period of the function. Add additional cycles to the left or right as needed.

EXAMPLE 3 Graphing a Cotangent Function

Graph $y = 3\cot 2x$.

Solution Refer to Figure 5.60, shown on the next page, as you read each step.

Step 1 Find two consecutive asymptotes. We must solve the equations

$$2x = 0 \quad \text{and} \quad 2x = \pi. \qquad \text{Set the variable expression in the cotangent equal to 0 and } \pi.$$

$$x = 0 \qquad x = \frac{\pi}{2} \qquad \text{Divide both sides of each equation by 2.}$$

Two consecutive asymptotes occur at $x = 0$ and $x = \frac{\pi}{2}$.

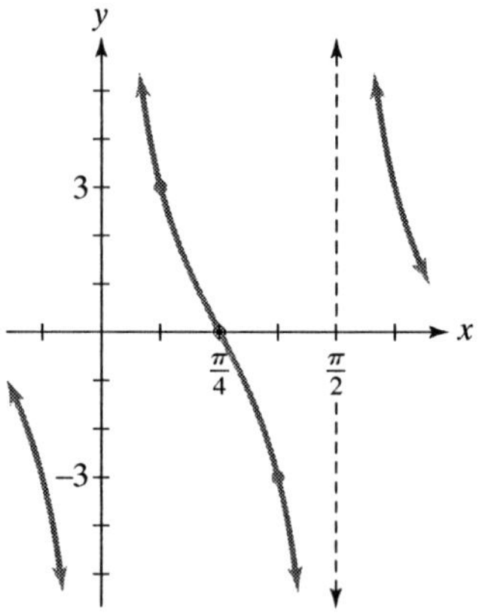

Figure 5.60 The graph of $y = 3\cot 2x$

Step 2 Identify an *x*-intercept, midway between the consecutive asymptotes. Midway between $x = 0$ and $x = \frac{\pi}{2}$ is $x = \frac{\pi}{4}$. An x-intercept is $\frac{\pi}{4}$ and the graph passes through $\left(\frac{\pi}{4}, 0\right)$.

Step 3 Find points on the graph $\frac{1}{4}$ and $\frac{3}{4}$ of the way between consecutive asymptotes. These points have *y*-coordinates of A and $-A$. Because A, the coefficient of the cotangent in $y = 3\cot 2x$ is 3, these points have y-coordinates of 3 and −3.

Step 4 Use steps 1–3 to graph one full period of the function. We use the two consecutive asymptotes, $x = 0$ and $x = \frac{\pi}{2}$, to graph one full period of $y = 3\cot 2x$. This curve is repeated to the left and right, as shown in Figure 5.60.

Check Point 3 Graph $y = \frac{1}{2}\cot\frac{\pi}{2}x$.

5 Understand the graphs of $y = \csc x$ and $y = \sec x$.

The Graphs of $y = \csc x$ and $y = \sec x$

We obtain the graphs of the cosecant and secant curves by using the reciprocal identities

$$\csc x = \frac{1}{\sin x} \quad \text{and} \quad \sec x = \frac{1}{\cos x}.$$

The identity on the left tells us that the value of the cosecant function $y = \csc x$ at a given value of x equals the reciprocal of the corresponding value of the sine function, provided that the value of the sine function is not 0. If the value of $\sin x$ is 0, then at each of these values of x, the cosecant function is not defined. A vertical asymptote is associated with each of these values on the graph of $y = \csc x$.

We obtain the graph of $y = \csc x$ by taking reciprocals of the y-values in the graph of $y = \sin x$. Vertical asymptotes of $y = \csc x$ occur at the x-intercepts of $y = \sin x$. Likewise, we obtain the graph of $y = \sec x$ by taking the reciprocal of $y = \cos x$. Vertical asymptotes of $y = \sec x$ occur at the x-intercepts of $y = \cos x$. The graphs of $y = \csc x$ and $y = \sec x$ and their key characteristics are shown in the following boxes. We have used dashed red lines to first graph $y = \sin x$ and $y = \cos x$, drawing vertical asymptotes through the x-intercepts.

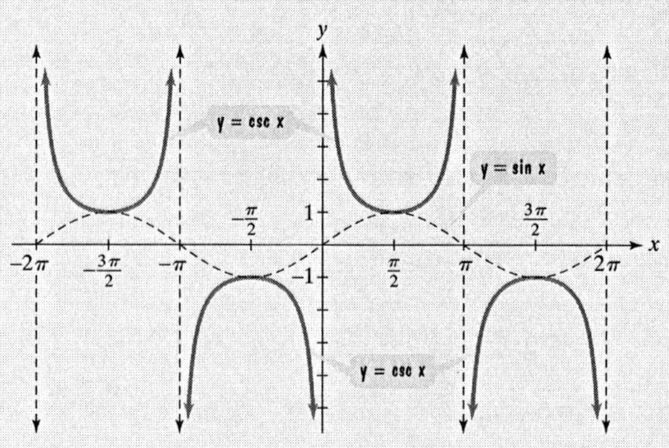

The Cosecant Curve: The Graph of $y = \csc x$ and Its Characteristics

Characteristics

- **Period:** 2π
- **Domain:** All real numbers except integral multiples of π
- **Range:** All real numbers y such that $y \le -1$ or $y \ge 1$
- **Vertical asymptotes** at integral multiples of π
- **Odd function,** $\csc(-x) = -\csc x$, with origin symmetry

The Secant Curve: The Graph of $y = \sec x$ and Its Characteristics

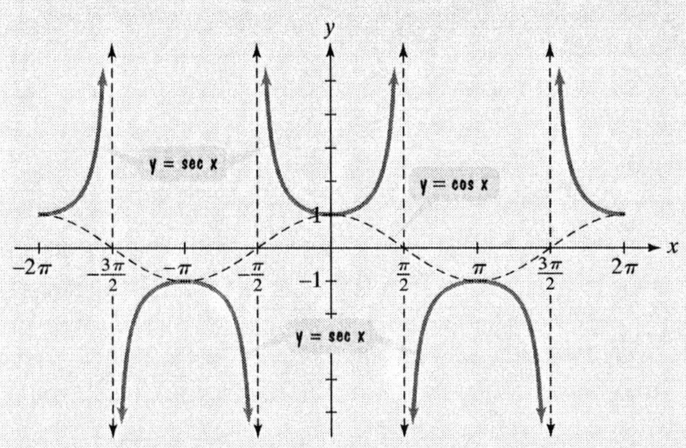

Characteristics

- **Period:** 2π
- **Domain:** All real numbers except odd multiples of $\frac{\pi}{2}$
- **Range:** All real numbers y such that $y \leq -1$ or $y \geq 1$
- **Vertical asymptotes** at odd multiples of $\frac{\pi}{2}$
- **Even function,** $\sec(-x) = \sec x$, with y-axis symmetry

6 Graph variations of $y = \csc x$ and $y = \sec x$.

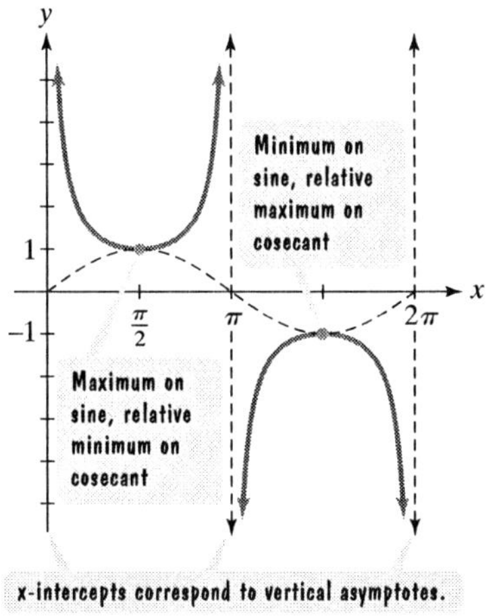

Figure 5.61

Graphing Variations of $y = \csc x$ and $y = \sec x$

We use graphs of reciprocal functions to obtain graphs of cosecant and secant functions. To graph a cosecant or secant curve, begin by graphing the reciprocal function. For example, to graph $y = 2\csc 2x$, we use the graph of $y = 2\sin 2x$. Likewise, to graph $y = -3\sec\frac{x}{2}$, we use the graph of $y = -3\cos\frac{x}{2}$.

Figure 5.61 illustrates how we use a sine curve to obtain a cosecant curve. Notice that

- x-intercepts on the red sine curve correspond to vertical asymptotes of the blue cosecant curve.
- A maximum point on the red sine curve corresponds to a minimum point on a continuous portion of the blue cosecant curve.
- A minimum point on the red sine curve corresponds to a maximum point on a continuous portion of the blue cosecant curve.

EXAMPLE 4 Using a Sine Curve to Obtain a Cosecant Curve

Use the graph of $y = 2\sin 2x$ in Figure 5.62 to obtain the graph of $y = 2\csc 2x$.

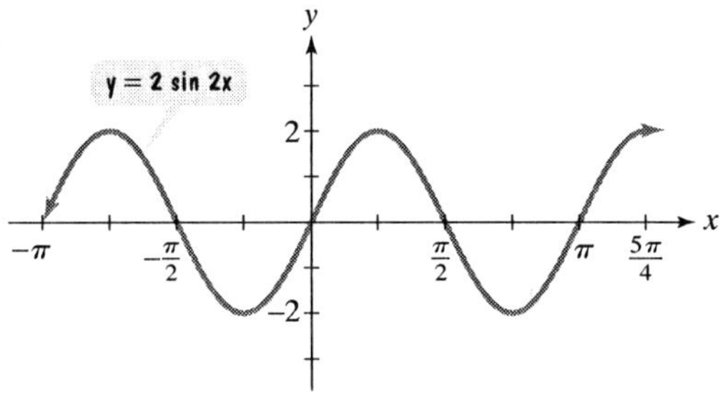

Figure 5.62

Solution On the next page, we begin our work by showing the given graph, the graph of $y = 2\sin 2x$, using dashed red lines.

The x-intercepts of $y = 2\sin 2x$ correspond to the vertical asymptotes of $y = 2\csc 2x$. Thus, we draw vertical asymptotes through the x-intercepts, shown in Figure 5.63. Using the asymptotes as guides, we sketch the graph of $y = 2\csc 2x$ in Figure 5.63.

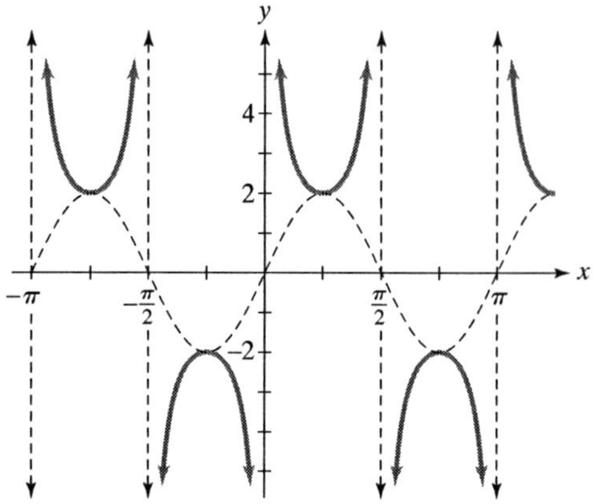

Figure 5.63 Using a sine curve to graph $y = 2\csc 2x$

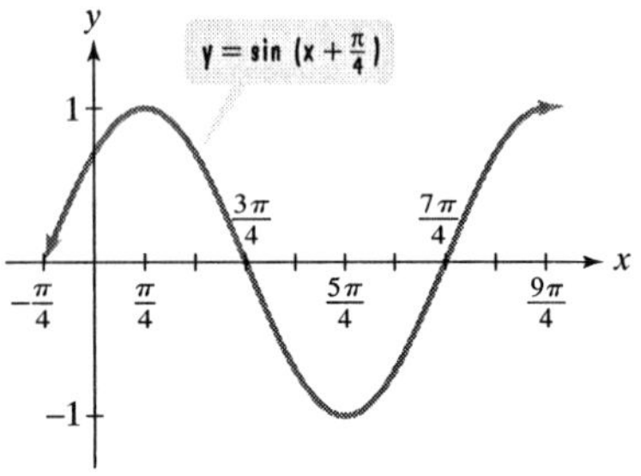

Check Point 4 Use the graph of $y = \sin\left(x + \frac{\pi}{4}\right)$, shown on the left, to obtain the graph of $y = \csc\left(x + \frac{\pi}{4}\right)$.

We use a cosine curve to obtain a secant curve in exactly the same way we used a sine curve to obtain a cosecant curve. Thus,

- x-intercepts on the cosine curve correspond to vertical asymptotes on the secant curve.
- A maximum point on the cosine curve corresponds to a minimum point on a continuous portion of the secant curve.
- A minimum point on the cosine curve corresponds to a maximum point on a continuous portion of the secant curve.

EXAMPLE 5 Graphing a Secant Function

Graph $y = -3\sec\frac{x}{2}$ for $-\pi < x < 5\pi$.

Solution We begin by graphing the reciprocal cosine function, $y = -3\cos\frac{x}{2}$. This equation is of the form $y = A\cos Bx$ with $A = -3$ and $B = \frac{1}{2}$.

amplitude: $|A| = |-3| = 3$ — The maximum y is 3 and the minimum is −3.

period: $\frac{2\pi}{B} = \frac{2\pi}{\frac{1}{2}} = 4\pi$ — Each cycle is completed in 4π radians.

We use quarter-periods, $\frac{4\pi}{4}$, or π, to find the x-values for the five key points. Starting with $x = 0$, the x-values are $0, \pi, 2\pi, 3\pi$, and 4π. Evaluating the function at each of these values of x, the key points are

$$(0, -3), (\pi, 0), (2\pi, 3), (3\pi, 0), \text{ and } (4\pi, -3).$$

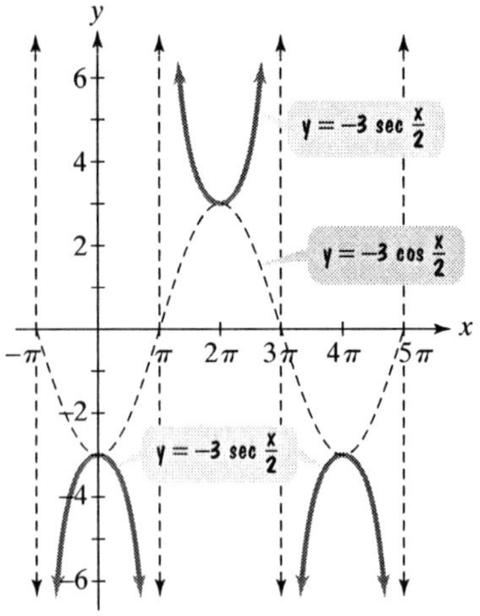

Figure 5.64 Using a cosine curve to graph $y = -3\sec\frac{x}{2}$

We use these key points to graph $y = -3\cos\frac{x}{2}$ from 0 to 4π. In order to graph for $-\pi \le x \le 5\pi$, extend the graph π units to the left and π units to the right. The graph is shown using a dashed red line in Figure 5.64. Now use this dashed red graph to obtain the graph of the reciprocal function. Draw vertical asymptotes through the x-intercepts. Using these asymptotes as guides, the graph of $y = -3\sec\frac{x}{2}$ is shown in blue in Figure 5.64.

Check Point 5 Graph $y = 2\sec 2x$ for $-\frac{3\pi}{4} < x < \frac{3\pi}{4}$.

The Six Curves of Trigonometry

Table 5.4 summarizes the graphs of the six trigonometric functions. Below each of the graphs is a description of the domain, range, and period of the function.

Table 5.4 Graphs of the Six Trigonometric Functions

<table>
<tr>
<td>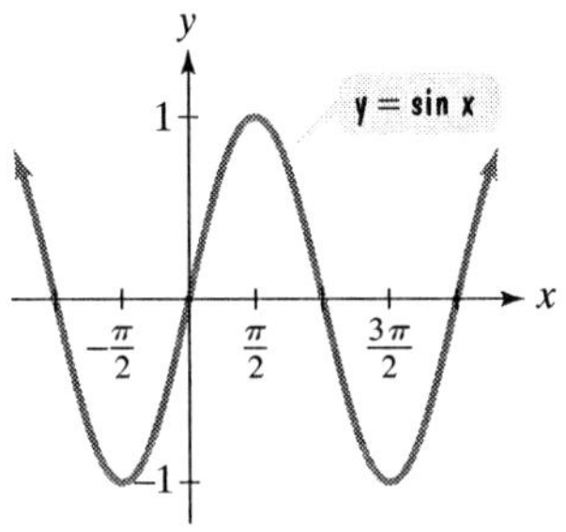

Domain: all real numbers
Range: $[-1, 1]$
Period: 2π</td>
<td>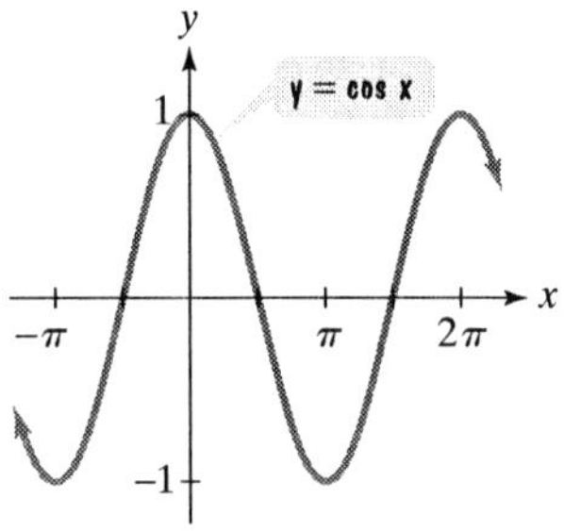

Domain: all real numbers
Range: $[-1, 1]$
Period: 2π</td>
<td>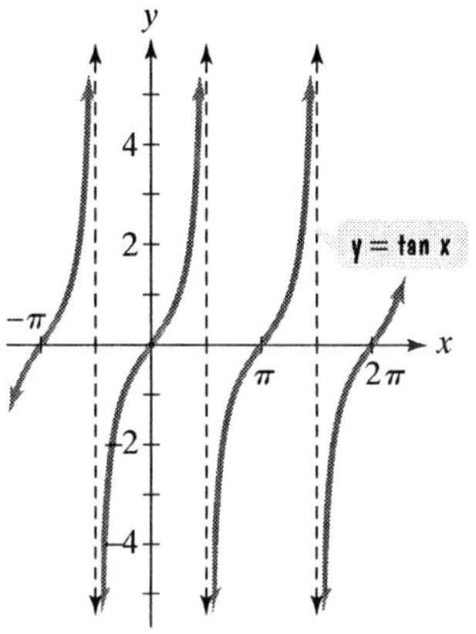

Domain: all real numbers except odd multiples of $\frac{\pi}{2}$
Range: all real numbers
Period: π</td>
</tr>
<tr>
<td>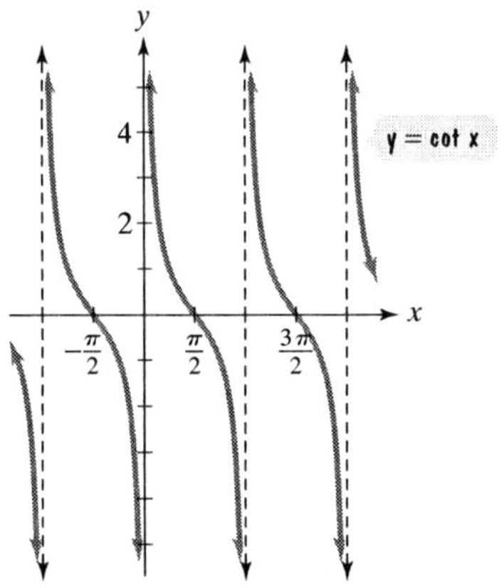

Domain: all real numbers except integral multiples of π
Range: all real numbers
Period: π</td>
<td>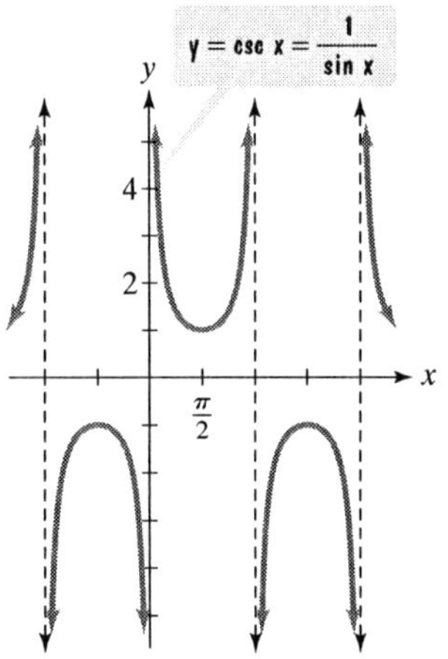

Domain: all real numbers except integral multiples of π
Range: $(-\infty, -1]$ or $[1, \infty)$
Period: 2π</td>
<td>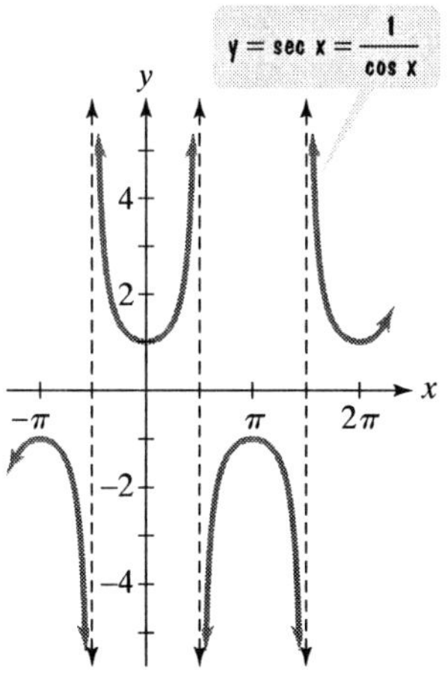

Domain: all real numbers except odd multiples of $\frac{\pi}{2}$
Range: $(-\infty, -1]$ or $[1, \infty)$
Period: 2π</td>
</tr>
</table>

EXERCISE SET 5.6

Practice Exercises

In Exercises 1–4, the graph of a tangent function is given. Select the equation for each graph from the following options:

$$y = \tan\left(x + \frac{\pi}{2}\right), \quad y = \tan(x + \pi), \quad y = -\tan x, \quad y = -\tan\left(x - \frac{\pi}{2}\right).$$

1.

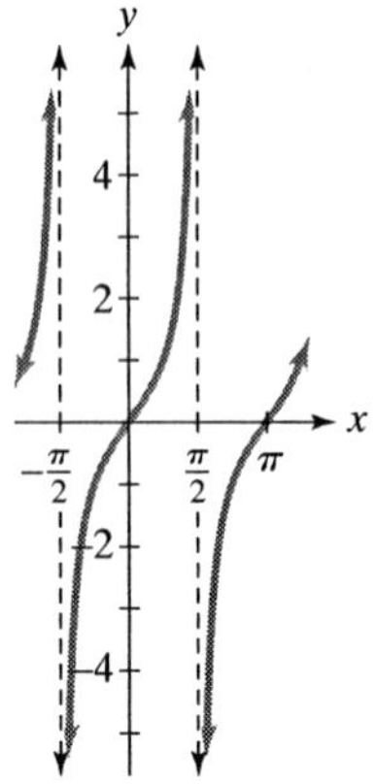

2.

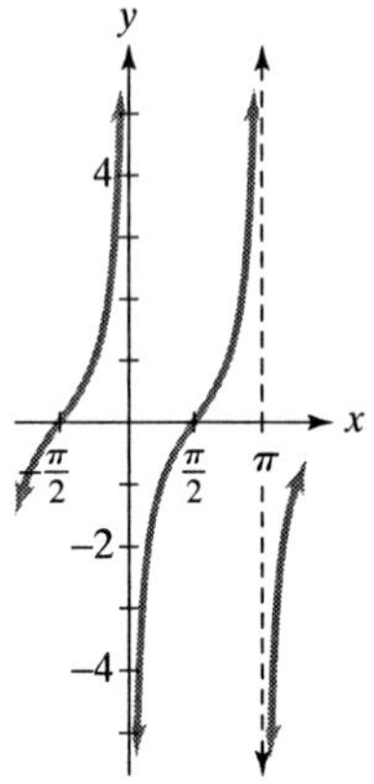

3.

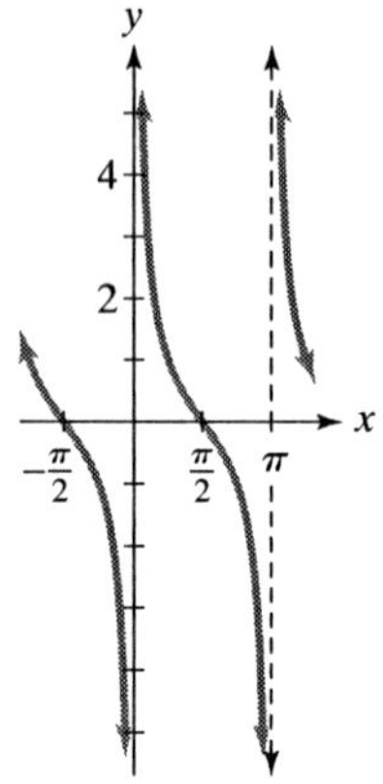

4.

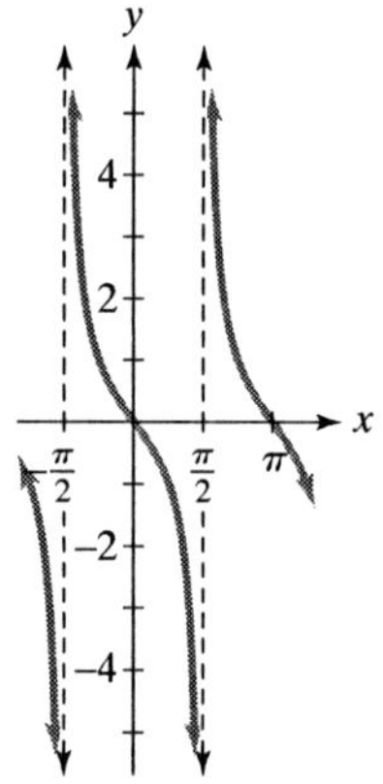

In Exercises 5–12, graph two periods of the given tangent function.

5. $y = 3\tan\frac{x}{4}$

6. $y = 2\tan\frac{x}{4}$

7. $y = \frac{1}{2}\tan 2x$

8. $y = 3\tan 2x$

9. $y = -2\tan\frac{1}{2}x$

10. $y = -3\tan\frac{1}{2}x$

11. $y = \tan(x - \pi)$

12. $y = \tan\left(x + \frac{\pi}{2}\right)$

In Exercises 13–16, the graph of a cotangent function is given. Select the equation for each graph from the following options:

$$y = \cot\left(x + \frac{\pi}{2}\right), \quad y = \cot(x + \pi), \quad y = -\cot x, \quad y = -\cot\left(x - \frac{\pi}{2}\right).$$

13.

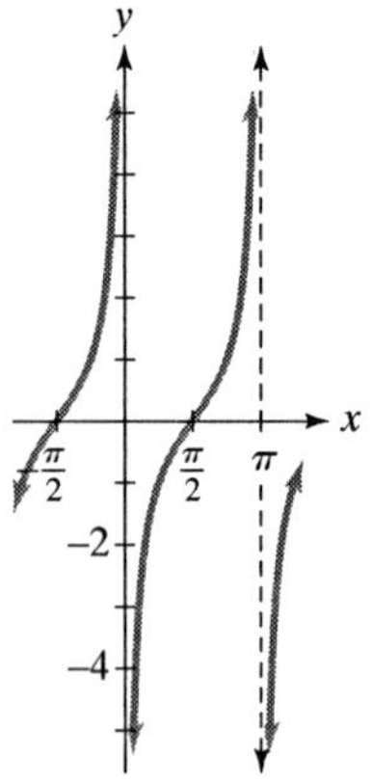

14.

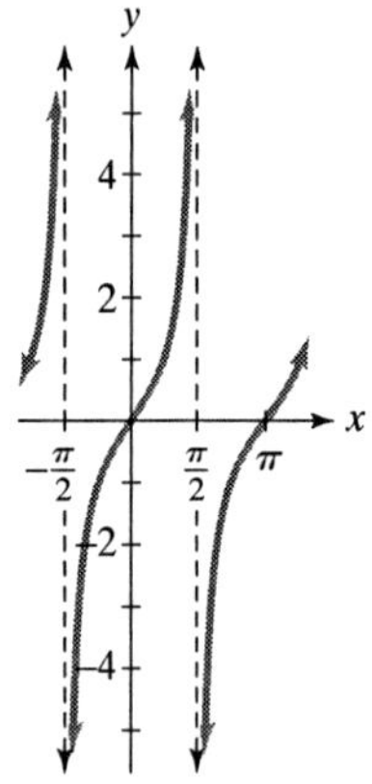

15.

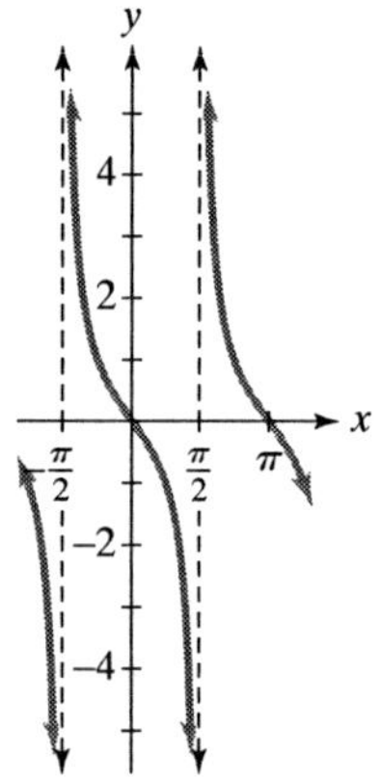

16.

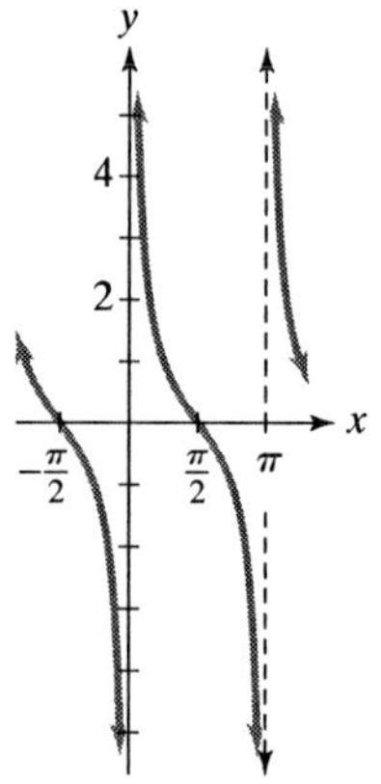

In Exercises 17–24, graph two periods of the given cotangent function.

17. $y = 2\cot x$

18. $y = \frac{1}{2}\cot x$

19. $y = \frac{1}{2}\cot 2x$

20. $y = 2\cot 2x$

21. $y = -3 \cot \frac{\pi}{2} x$

22. $y = -2 \cot \frac{\pi}{4} x$

23. $y = 3 \cot\left(x + \frac{\pi}{2}\right)$

24. $y = 3 \cot\left(x + \frac{\pi}{4}\right)$

In Exercises 25–28, use each graph to obtain the graph of the reciprocal function. Give the equation of the function for the graph that you obtain.

25.

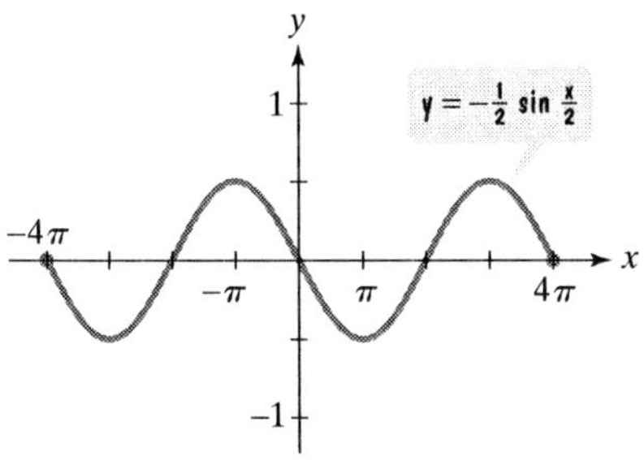

26.

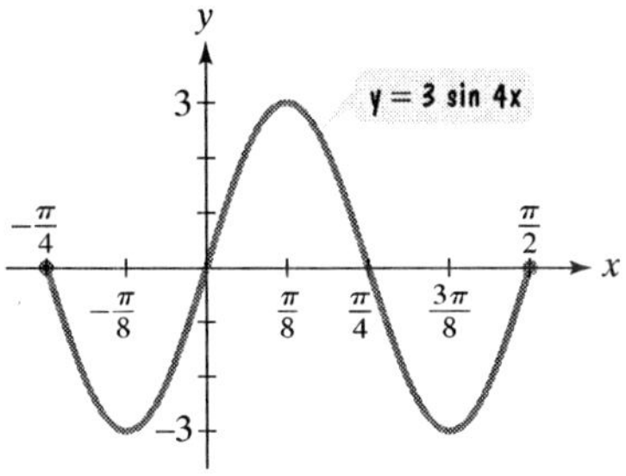

27.

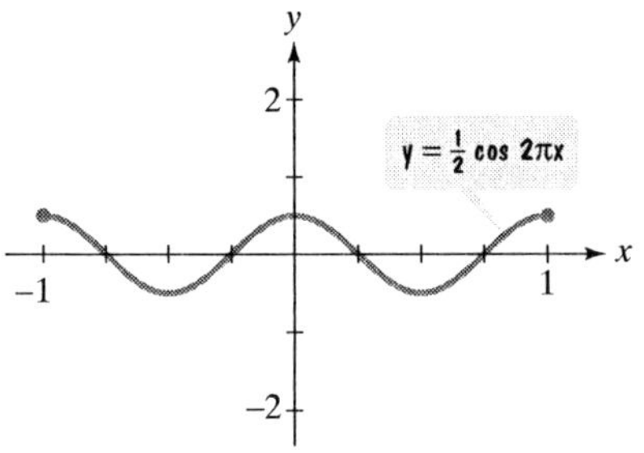

28.

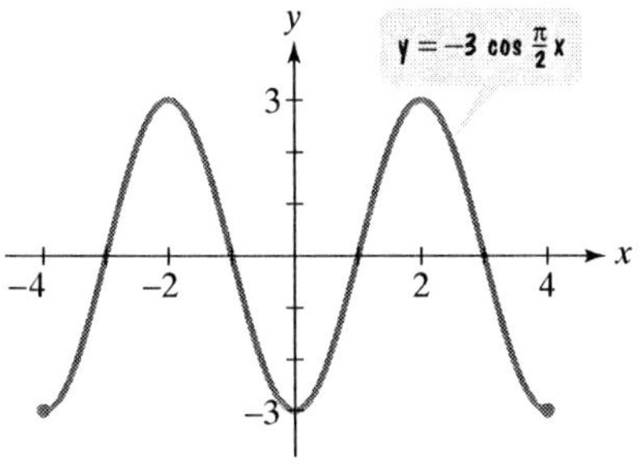

In Exercises 29–44, graph two periods of the given cosecant or secant function.

29. $y = 3 \csc x$

30. $y = 2 \csc x$

31. $y = \frac{1}{2} \csc \frac{x}{2}$

32. $y = \frac{3}{2} \csc \frac{x}{4}$

33. $y = 2 \sec x$

34. $y = 3 \sec x$

35. $y = \sec \frac{x}{3}$

36. $y = \sec \frac{x}{2}$

37. $y = -2 \csc \pi x$

38. $y = -\frac{1}{2} \csc \pi x$

39. $y = -\frac{1}{2} \sec \pi x$

40. $y = -\frac{3}{2} \sec \pi x$

41. $y = \csc(x - \pi)$

42. $y = \csc\left(x - \frac{\pi}{2}\right)$

43. $y = 2 \sec(x + \pi)$

44. $y = 2 \sec\left(x + \frac{\pi}{2}\right)$

Application Exercises

45. An ambulance with a rotating beacon of light is parked 12 feet from a building. The function

$$d = 12 \tan 2\pi t$$

describes the distance, d, in feet, of the rotating beacon from point C after t seconds.

a. Graph the function on the interval $[0, 2]$.

b. For what values of t in $[0, 2]$ is the function undefined? What does this mean in terms of the rotating beacon in the figure shown?

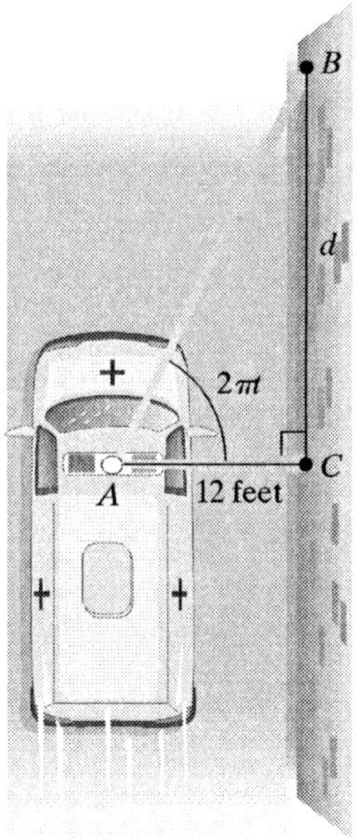

46. The angle of elevation from the top of a house to a jet flying 2 miles above the house is x radians. If d represents the horizontal distance, in miles, of the jet from the house, express d in terms of a trigonometric function of x. Then graph the function for $0 < x < \pi$.

47. Your best friend is marching with a band and has asked you to film her. The figure below shows that you have set yourself up 10 feet from the street where your friend will be passing from left to right. If d represents your distance, in feet, from your friend and x is the radian measure of the angle shown, express d in terms of a trigonometric function of x. Then graph the function for $-\frac{\pi}{2} < x < \frac{\pi}{2}$. Negative angles indicate that your marching buddy is on your left.

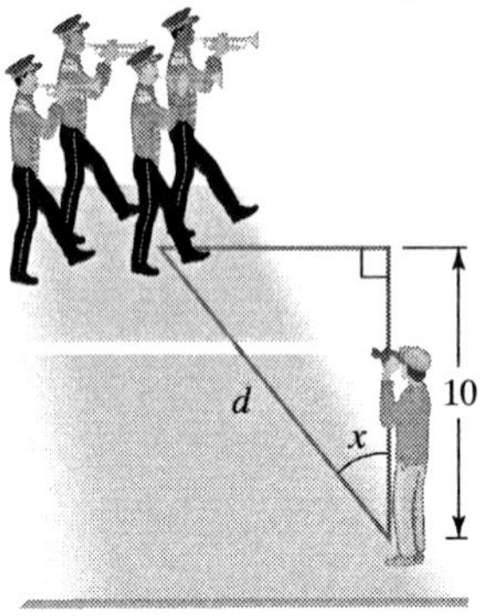

In Exercises 48–50, sketch a reasonable graph that models the given situation.

48. The number of hours of daylight per day in your hometown over a two-year period

49. The motion of a diving board vibrating 10 inches in each direction per second just after someone has dived off

50. The distance of a rotating beacon of light from a point on a wall (See the figure for Exercise 45.)

Writing in Mathematics

51. Without drawing a graph, describe the behavior of the basic tangent curve.

52. If you are given the equation of a tangent function, how do you find consecutive asymptotes?

53. If you are given the equation of a tangent function, how do you identify an x-intercept?

54. Without drawing a graph, describe the behavior of the basic cotangent curve.

55. If you are given the equation of a cotangent function, how do you find consecutive asymptotes?

56. Explain how to determine the range of $y = \csc x$ from the graph. What is the range?

57. Explain how to use a sine curve to obtain a cosecant curve. Why can the same procedure be used to obtain a secant curve from a cosine curve?

58. Scientists record brain activity by attaching electrodes to the scalp and then connecting these electrodes to a machine. The record of brain activity recorded with this machine is shown in the three graphs at the top of the next column. Which trigonometric functions would be most appropriate for describing the oscillations in brain activity? Describe similarities and differences among these functions when modeling brain activity when awake, during dreaming sleep, and during nondreaming sleep.

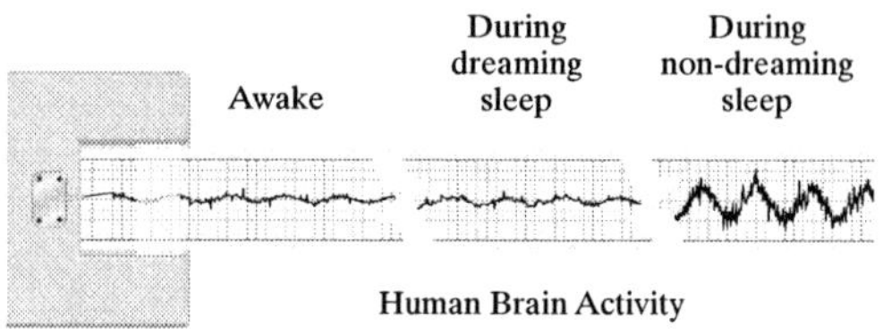

Human Brain Activity

Technology Exercises

In working Exercises 59–62, describe what happens at the asymptotes on the graphing utility. Compare the graphs in the connected and dot modes.

59. Use a graphing utility to verify any two of the tangent curves that you drew by hand in Exercises 5–12.

60. Use a graphing utility to verify any two of the cotangent curves that you drew by hand in Exercises 17–24.

61. Use a graphing utility to verify any two of the cosecant curves that you drew by hand in Exercises 29–44.

62. Use a graphing utility to verify any two of the secant curves that you drew by hand in Exercises 29–44.

In Exercises 63–68, use a graphing utility to graph each function. Use a range setting so that the graph is shown for at least two periods.

63. $y = \tan \frac{x}{4}$

64. $y = \tan 4x$

65. $y = \cot 2x$

66. $y = \cot \frac{x}{2}$

67. $y = \frac{1}{2} \tan \pi x$

68. $y = \frac{1}{2} \tan(\pi x + 1)$

In Exercises 69–72, use a graphing utility to graph each pair of functions in the same viewing rectangle. Use a range setting so that the graphs are shown for at least two periods.

69. $y = 0.8 \sin \frac{x}{2}$ and $y = 0.8 \csc \frac{x}{2}$

70. $y = -2.5 \sin \frac{\pi}{3} x$ and $y = -2.5 \csc \frac{\pi}{3} x$

71. $y = 4 \cos\left(2x - \frac{\pi}{6}\right)$ and $y = 4 \sec\left(2x - \frac{\pi}{6}\right)$

72. $y = -3.5 \cos\left(\pi x - \frac{\pi}{6}\right)$ and $y = -3.5 \sec\left(\pi x - \frac{\pi}{6}\right)$

73. Carbon dioxide particles in our atmosphere trap heat and raise the planet's temperature. The resultant gradually increasing temperature is called the greenhouse effect. Carbon dioxide accounts for about half of global warming. The function

$$y = 2.5 \sin 2\pi x + 0.0216x^2 + 0.654x + 316$$

models carbon dioxide concentration, y, in parts per million, where $x = 0$ represents January 1960; $x = \frac{1}{12}$, February 1960; $x = \frac{2}{12}$, March 1960; $\ldots$, $x = 1$, January 1961; $x = \frac{13}{12}$, February 1961; and so on. Use a graphing utility to graph the function in a [30, 45, 5] by [310, 420, 5] viewing rectangle. Describe what the graph reveals about carbon dioxide concentration from 1990 through 2005.

74. Graph $y = \sin \frac{1}{x}$ in a $[-0.2, 0.2, 0.01]$ by $[-1.2, 1.2, 0.01]$ viewing rectangle. What is happening as x approaches 0 from the left or the right? Explain this behavior.

Critical Thinking Exercises

In Exercises 75–76, write an equation for each blue graph.

75.

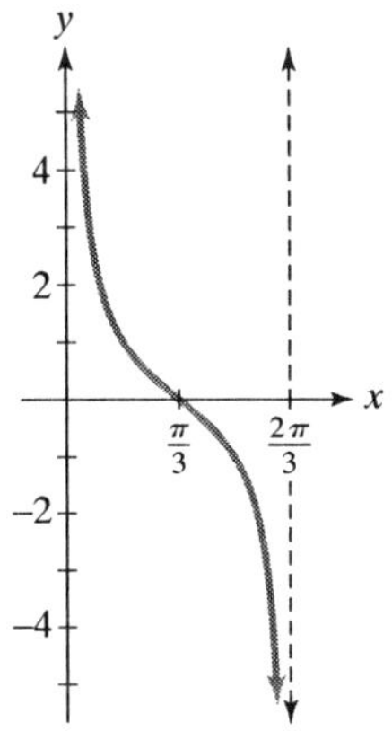

76.

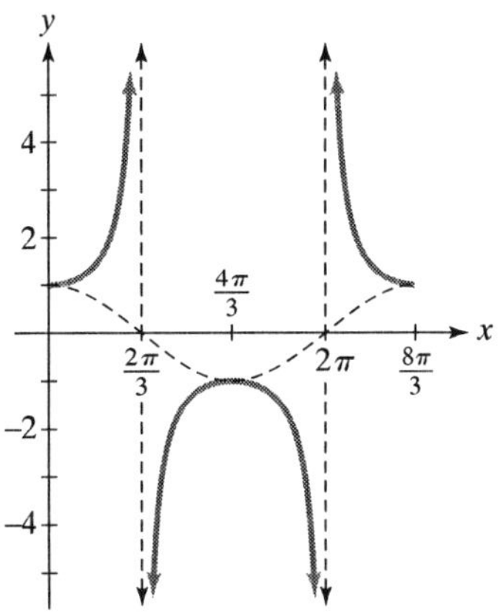

77. For $x > 0$, what effect does 2^{-x} in $y = 2^{-x} \sin x$ have on the graph of $y = \sin x$? What kind of behavior can be modeled by a function such as $y = 2^{-x} \sin x$?

SECTION 5.7 Inverse Trigonometric Functions

Objectives

1. Understand and use the inverse sine function.
2. Understand and use the inverse cosine function.
3. Understand and use the inverse tangent function.
4. Use a calculator to evaluate inverse trigonometric functions.
5. Find exact values of composite functions with inverse trigonometric functions.

You watched *The Matrix* on video and were impressed by the elaborate computer-generated effects. The movie is being shown again at a local theater, where you can experience its stunning visual force on a large screen. Where in the theater should you sit to maximize the film's visual impact? In this section's exercise set, you will see how an inverse trigonometric function can enhance your movie-going experiences.

Study Tip

Here are some helpful things to remember from our discussion of inverse functions in Section 2.7.

- If no horizontal line intersects the graph of a function more than once, the function is one-to-one and has an inverse function.
- If the point (a, b) is on the graph of f, then the point (b, a) is on the graph of the inverse function, denoted f^{-1}. The graph of f^{-1} is a reflection of the graph of f about the line $y = x$.

1 Understand and use the inverse sine function.

The Inverse Sine Function

Figure 5.65 shows the graph of $y = \sin x$. Can you see that every horizontal line that can be drawn between -1 and 1 intersects the graph infinitely many times? Thus, the sine function is not one-to-one and has no inverse function.

In Figure 5.66, we have taken a portion of the sine curve, restricting the domain of the sine function to $-\frac{\pi}{2} \le x \le \frac{\pi}{2}$. With this restricted domain, every horizontal line that can be drawn between -1 and 1 intersects the graph exactly once. Thus, the restricted function passes the horizontal line test and is one-to-one.

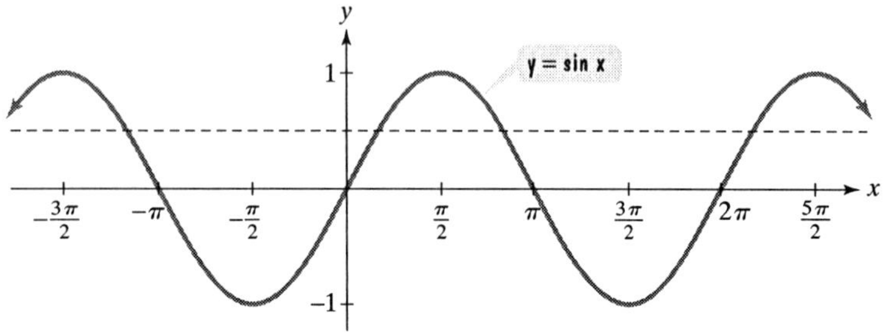

Figure 5.65 The horizontal line test shows that the sine function is not one-to-one and has no inverse function.

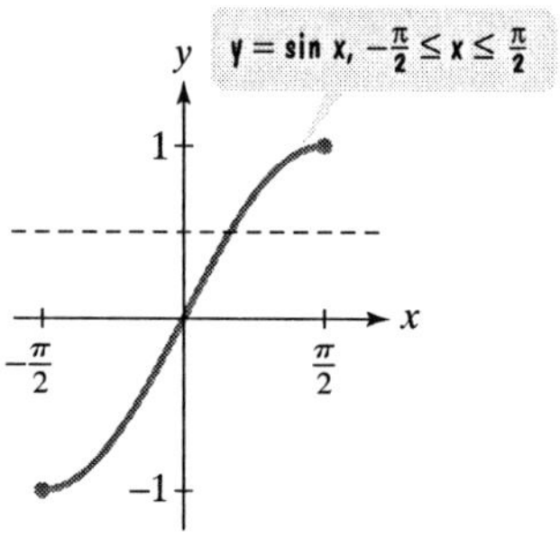

Figure 5.66 The restricted sine function passes the horizontal line test. It is one-to-one and has an inverse function.

On the restricted domain $-\frac{\pi}{2} \le x \le \frac{\pi}{2}$, $y = \sin x$ has an inverse function. The inverse of the restricted sine function is called the **inverse sine function.** Two notations are commonly used to denote the inverse sine function:

$$y = \sin^{-1} x \quad \text{or} \quad y = \arcsin x.$$

In this book, we will use $y = \sin^{-1} x$. This notation has the same symbol as the inverse function notation $f^{-1}(x)$.

The Inverse Sine Function

The **inverse sine function,** denoted by $\sin^{-1}$, is the inverse of the restricted sine function $y = \sin x$, $-\frac{\pi}{2} \le x \le \frac{\pi}{2}$. Thus,

$$y = \sin^{-1} x \quad \text{means} \quad \sin y = x$$

where $-\frac{\pi}{2} \le y \le \frac{\pi}{2}$ and $-1 \le x \le 1$. We read $y = \sin^{-1} x$ as "y equals the inverse sine at x."

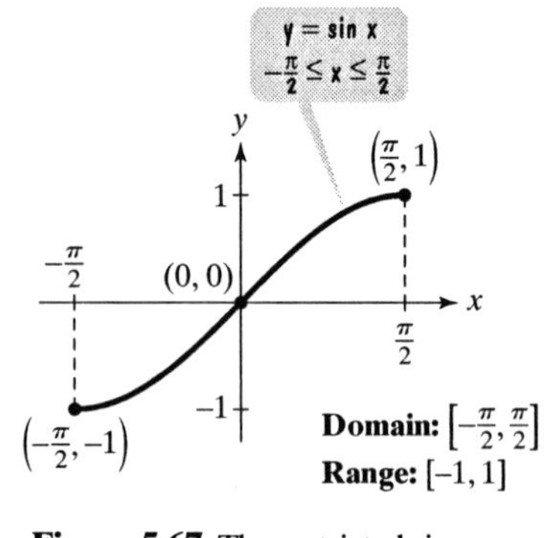

Figure 5.67 The restricted sine function

One way to graph $y = \sin^{-1} x$ is to take points on the graph of the restricted sine function and reverse the order of the coordinates. For example, Figure 5.67 shows that $\left(-\frac{\pi}{2}, -1\right)$, $(0, 0)$, and $\left(\frac{\pi}{2}, 1\right)$ are on the graph of the restricted sine function. Reversing the order of the coordinates gives $\left(-1, -\frac{\pi}{2}\right)$, $(0, 0)$, and $\left(1, \frac{\pi}{2}\right)$. We now use these three points to sketch the inverse sine function. The graph of $y = \sin^{-1} x$ is shown in Figure 5.68.

Another way to obtain the graph of $y = \sin^{-1} x$ is to reflect the graph of the restricted sine function about the line $y = x$, shown in Figure 5.69. The red graph is the restricted sine function and the blue graph is the graph of $y = \sin^{-1} x$.

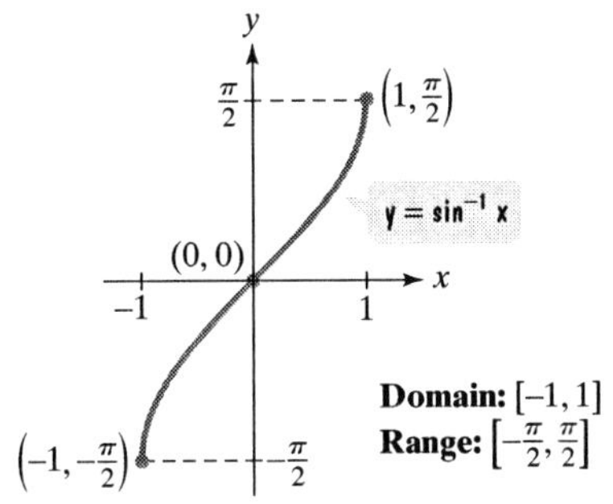

Figure 5.68 The graph of the inverse sine function

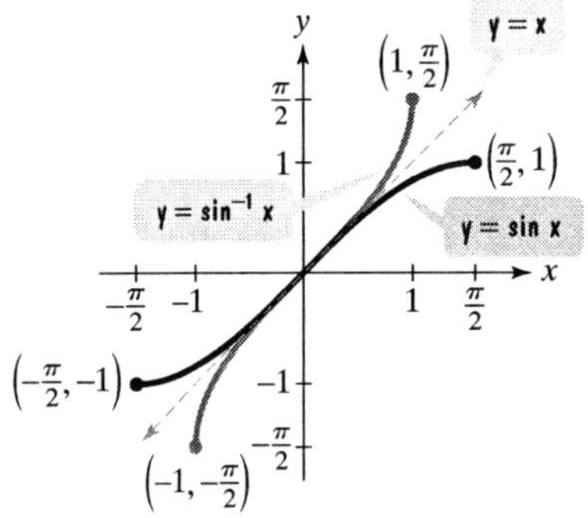

Figure 5.69 Using a reflection to obtain the graph of the inverse sine function

Exact values of $\sin^{-1} x$ can be found by thinking of $\sin^{-1} x$ as **the angle in the interval $\left[-\frac{\pi}{2}, \frac{\pi}{2}\right]$ whose sine is x.** For example, we can use the two points on the blue graph of the inverse sine function in Figure 5.69 and write

$$\sin^{-1}(-1) = -\frac{\pi}{2} \quad \text{and} \quad \sin^{-1} 1 = \frac{\pi}{2}.$$

The angle whose sine is −1 is $-\frac{\pi}{2}$.

The angle whose sine is 1 is $\frac{\pi}{2}$.

Because we are thinking of $\sin^{-1} x$ in terms of an angle, we will represent such an angle by θ.

Finding Exact Values of $\sin^{-1} x$.

1. Let $\theta = \sin^{-1} x$.
2. Rewrite $\theta = \sin^{-1} x$ as $\sin \theta = x$.
3. Use the exact values in Table 5.5,shown on the next page, to find the value of θ in $\left[-\frac{\pi}{2}, \frac{\pi}{2}\right]$ that satisfies $\sin \theta = x$.

Table 5.5 Exact Values for $\sin\theta, -\frac{\pi}{2} \le \theta \le \frac{\pi}{2}$

θ	$\sin\theta$
$-\frac{\pi}{2}$	-1
$-\frac{\pi}{3}$	$-\frac{\sqrt{3}}{2}$
$-\frac{\pi}{4}$	$-\frac{\sqrt{2}}{2}$
$-\frac{\pi}{6}$	$-\frac{1}{2}$
0	0
$\frac{\pi}{6}$	$\frac{1}{2}$
$\frac{\pi}{4}$	$\frac{\sqrt{2}}{2}$
$\frac{\pi}{3}$	$\frac{\sqrt{3}}{2}$
$\frac{\pi}{2}$	1

EXAMPLE 1 Finding the Exact Value of an Inverse Sine Function

Find the exact value of $\sin^{-1}\frac{\sqrt{2}}{2}$.

Solution

Step 1 Let $\theta = \sin^{-1}x$. Thus,

$$\theta = \sin^{-1}\frac{\sqrt{2}}{2}.$$

We must find the angle θ, $-\frac{\pi}{2} \le \theta \le \frac{\pi}{2}$, whose sine equals $\frac{\sqrt{2}}{2}$.

Step 2 Rewrite $\theta = \sin^{-1}x$ as $\sin\theta = x$. Using the definition of the inverse sine function, we rewrite $\theta = \sin^{-1}\frac{\sqrt{2}}{2}$ as

$$\sin\theta = \frac{\sqrt{2}}{2}.$$

Step 3 Use the exact values in Table 5.5 to find the value of θ in $\left[-\frac{\pi}{2}, \frac{\pi}{2}\right]$ that satisfies $\sin\theta = x$. Table 5.5 shows that the only angle in the interval $\left[-\frac{\pi}{2}, \frac{\pi}{2}\right]$ that satisfies $\sin\theta = \frac{\sqrt{2}}{2}$ is $\frac{\pi}{4}$. Thus, $\theta = \frac{\pi}{4}$. Because θ, in step 1, represents $\sin^{-1}\frac{\sqrt{2}}{2}$, we conclude that

$$\sin^{-1}\frac{\sqrt{2}}{2} = \frac{\pi}{4}.$$ The angle in $\left[-\frac{\pi}{2}, \frac{\pi}{2}\right]$ whose sine is $\frac{\sqrt{2}}{2}$ is $\frac{\pi}{4}$.

Check Point 1 Find the exact value of $\sin^{-1}\frac{\sqrt{3}}{2}$.

EXAMPLE 2 Finding the Exact Value of an Inverse Sine Function

Find the exact value of $\sin^{-1}\left(-\frac{1}{2}\right)$.

Solution

Step 1 Let $\theta = \sin^{-1}x$. Thus,

$$\theta = \sin^{-1}\left(-\frac{1}{2}\right).$$

We must find the angle θ, $-\frac{\pi}{2} \le \theta \le \frac{\pi}{2}$, whose sine equals $-\frac{1}{2}$.

Step 2 **Rewrite $\theta = \sin^{-1}x$ as $\sin\theta = x$.** We rewrite $\theta = \sin^{-1}\left(-\frac{1}{2}\right)$ and obtain

$$\sin\theta = -\frac{1}{2}.$$

Step 3 **Use the exact values in Table 5.5 to find the value of θ in $\left[-\frac{\pi}{2}, \frac{\pi}{2}\right]$ that satisfies $\sin\theta = x$.** Table 5.5, on page 524, shows that the only angle in the interval $\left[-\frac{\pi}{2}, \frac{\pi}{2}\right]$ that satisfies $\sin\theta = -\frac{1}{2}$ is $-\frac{\pi}{6}$. Thus,

$$\sin^{-1}\left(-\frac{1}{2}\right) = -\frac{\pi}{6}$$

Check Point 2 Find the exact value of $\sin^{-1}\left(-\frac{\sqrt{2}}{2}\right)$.

Some inverse sine expressions cannot be evaluated. Because the domain of the inverse sine function is $[-1, 1]$, it is only possible to evaluate $\sin^{-1}x$ for values of x in this domain. Thus, $\sin^{-1}3$ cannot be evaluated. There is no angle whose sine is 3.

2 Understand and use the inverse cosine function.

The Inverse Cosine Function

Figure 5.70 shows how we restrict the domain of the cosine function so that it becomes one-to-one and has an inverse function. Restrict the domain to the interval $[0, \pi]$, shown by the dark blue graph. Over this interval, the restricted cosine function passes the horizontal line test and has an inverse function.

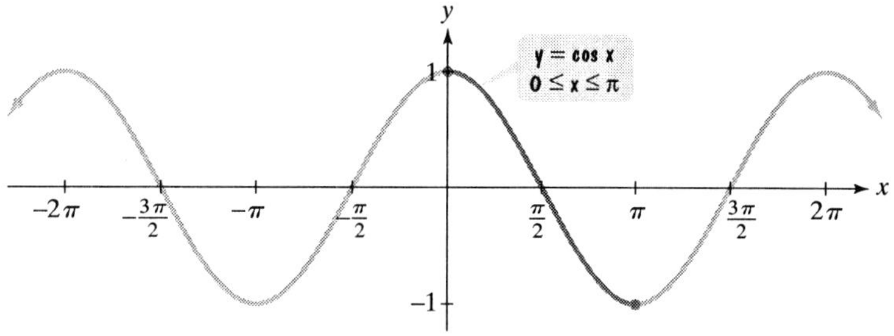

Figure 5.70 $y = \cos x$ is one-to-one on the interval $[0, \pi]$.

The Inverse Cosine Function

The **inverse cosine function,** denoted by $\cos^{-1}$, is the inverse of the restricted cosine function $y = \cos x$, $0 \le x \le \pi$. Thus,

$$y = \cos^{-1}x \quad \text{means} \quad \cos y = x$$

where $0 \le y \le \pi$ and $-1 \le x \le 1$.

One way to graph $y = \cos^{-1} x$ is to take points on the graph of the restricted cosine function and reverse the order of the coordinates. For example, Figure 5.71 shows that $(0, 1)$, $\left(\frac{\pi}{2}, 0\right)$, and $(\pi, -1)$ are on the graph of the restricted cosine function. Reversing the order of the coordinates gives $(1, 0)$, $\left(0, \frac{\pi}{2}\right)$, and $(-1, \pi)$. We now use these three points to sketch the inverse cosine function. The graph of $y = \cos^{-1} x$ is shown in Figure 5.72. You can also obtain this graph by reflecting the graph of the restricted cosine function about the line $y = x$.

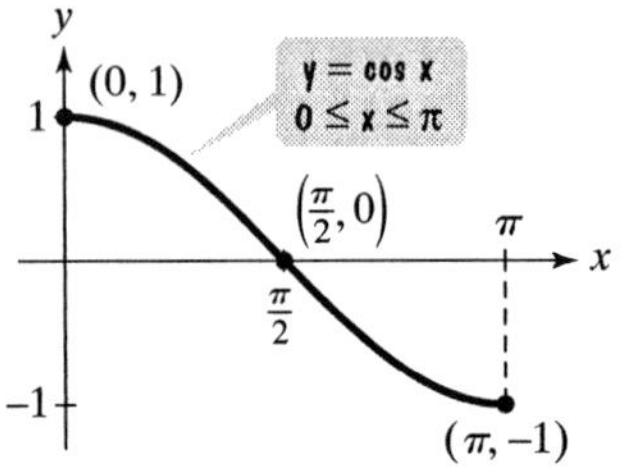

Domain: $[0, \pi]$
Range: $[-1, 1]$

Figure 5.71 The restricted cosine function

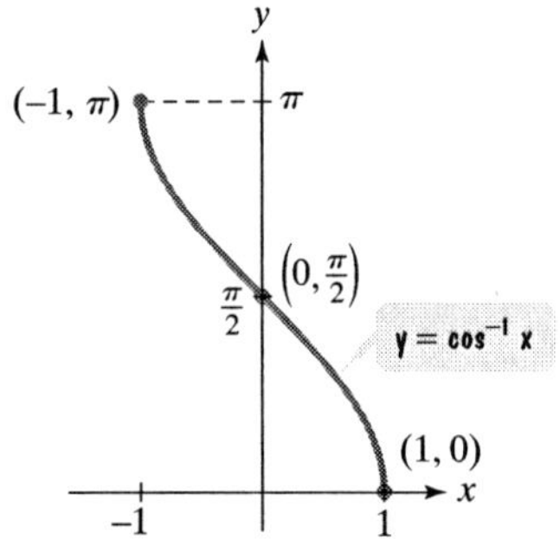

Domain: $[-1, 1]$
Range: $[0, \pi]$

Figure 5.72 The graph of the inverse cosine function

Table 5.6 Exact Values for $\cos\theta$, $0 \le \theta \le \pi$

θ	$\cos\theta$
0	1
$\frac{\pi}{6}$	$\frac{\sqrt{3}}{2}$
$\frac{\pi}{4}$	$\frac{\sqrt{2}}{2}$
$\frac{\pi}{3}$	$\frac{1}{2}$
$\frac{\pi}{2}$	0
$\frac{2\pi}{3}$	$-\frac{1}{2}$
$\frac{3\pi}{4}$	$-\frac{\sqrt{2}}{2}$
$\frac{5\pi}{6}$	$-\frac{\sqrt{3}}{2}$
π	-1

Exact values of $\cos^{-1} x$ can be found by thinking of $\cos^{-1} x$ as **the angle in the interval $[0, \pi]$ whose cosine is x.** This time we will use Table 5.6, which shows exact values for $\cos\theta$ for θ in the interval $[0, \pi]$.

EXAMPLE 3 Finding the Exact Value of an Inverse Cosine Function

Find the exact value of $\cos^{-1}\left(-\frac{\sqrt{3}}{2}\right)$.

Solution

Step 1 Let $\theta = \cos^{-1} x$. Thus,

$$\theta = \cos^{-1}\left(-\frac{\sqrt{3}}{2}\right).$$

We must find the angle θ, $0 \le \theta \le \pi$, whose cosine equals $-\frac{\sqrt{3}}{2}$.

Step 2 Rewrite $\theta = \cos^{-1} x$ as $\cos\theta = x$. We obtain

$$\cos\theta = -\frac{\sqrt{3}}{2}.$$

Step 3 Use the exact values in Table 5.6 to find the value of θ in $[0, \pi]$ that satisfies $\cos\theta = x$. The table shows that the only angle in the interval $[0, \pi]$ that satisfies $\cos\theta = -\frac{\sqrt{3}}{2}$ is $\frac{5\pi}{6}$. Thus, $\theta = \frac{5\pi}{6}$ and

$$\cos^{-1}\left(-\frac{\sqrt{3}}{2}\right) = \frac{5\pi}{6}.$$ The angle in $[0, \pi]$ whose cosine is $-\frac{\sqrt{3}}{2}$ is $\frac{5\pi}{6}$.

Check Point 3 Find the exact value of $\cos^{-1}\left(-\frac{1}{2}\right)$.

3 Understand and use the inverse tangent function.

The Inverse Tangent Function

Figure 5.73 shows how we restrict the domain of the tangent function so that it becomes one-to-one and has an inverse function. Restrict the domain to the interval $\left(-\frac{\pi}{2}, \frac{\pi}{2}\right)$, shown by the solid blue graph. Over this interval, the restricted tangent function passes the horizontal line test and has an inverse function.

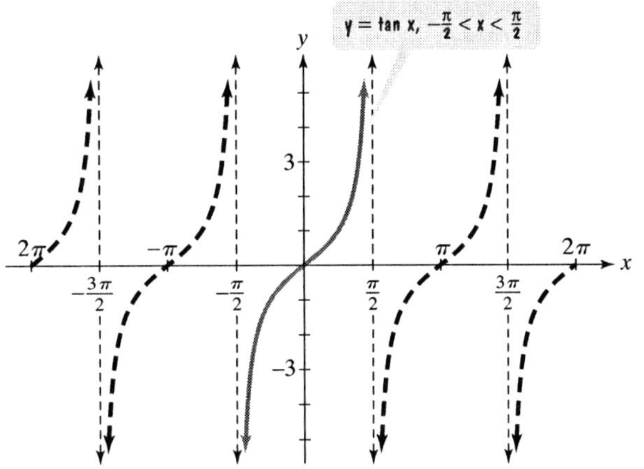

Figure 5.73 $y = \tan x$ is one-to-one on the interval $\left(-\frac{\pi}{2}, \frac{\pi}{2}\right)$.

The Inverse Tangent Function

The **inverse tangent function,** denoted by $\tan^{-1}$, is the inverse of the restricted tangent function $y = \tan x$, $-\frac{\pi}{2} < x < \frac{\pi}{2}$. Thus,

$$y = \tan^{-1} x \quad \text{means} \quad \tan y = x$$

where $-\frac{\pi}{2} < y < \frac{\pi}{2}$ and $-\infty < x < \infty$.

We graph $y = \tan^{-1}x$ by taking points on the graph of the restricted function and reversing the order of the coordinates. Figure 5.74 shows that $\left(-\frac{\pi}{4}, -1\right)$, $(0, 0)$, and $\left(\frac{\pi}{4}, 1\right)$ are on the graph of the restricted tangent function. Reversing the order gives $\left(-1, -\frac{\pi}{4}\right)$, $(0, 0)$, and $\left(1, \frac{\pi}{4}\right)$. We now use these three points to graph the inverse tangent function. The graph of $y = \tan^{-1}x$ is shown in Figure 5.75. Notice that the vertical asymptotes become horizontal asymptotes for the graph of the inverse function.

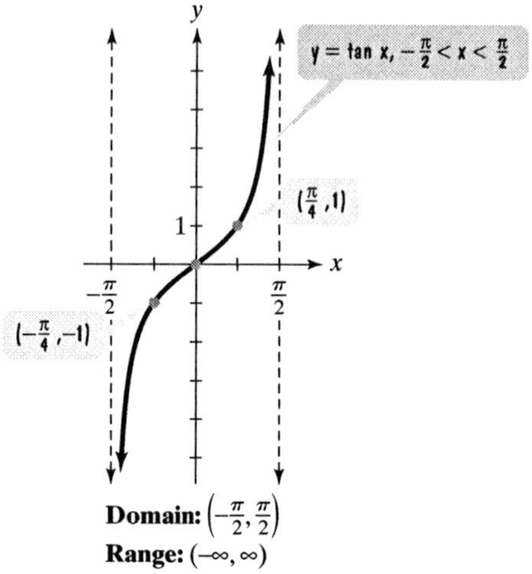

Figure 5.74 The restricted tangent function

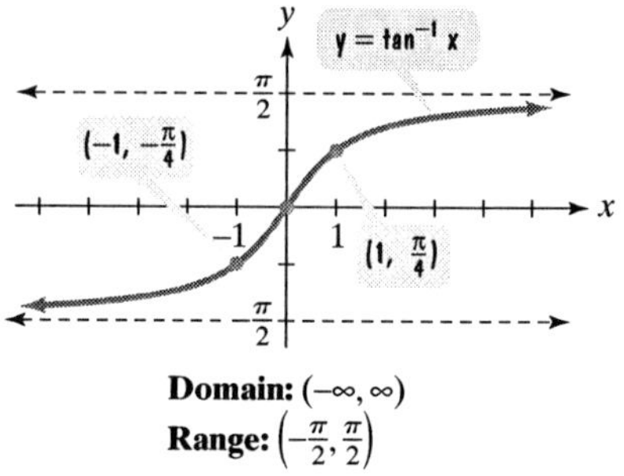

Figure 5.75 The graph of the inverse tangent function

Exact values of $\tan^{-1}x$ can be found by thinking of $\tan^{-1}x$ as **the angle in the interval** $\left(-\frac{\pi}{2}, \frac{\pi}{2}\right)$ **whose tangent is** x. We use Table 5.7, which shows exact values for $\tan\theta$ for θ in the interval $\left(-\frac{\pi}{2}, \frac{\pi}{2}\right)$.

Table 5.7 Exact Values for $\tan\theta, -\frac{\pi}{2} < \theta < \frac{\pi}{2}$

θ	$\tan\theta$
$-\frac{\pi}{2}$	Undefined
$-\frac{\pi}{3}$	$-\sqrt{3}$
$-\frac{\pi}{4}$	-1
$-\frac{\pi}{6}$	$-\frac{\sqrt{3}}{3}$
0	0
$\frac{\pi}{6}$	$\frac{\sqrt{3}}{3}$
$\frac{\pi}{4}$	1
$\frac{\pi}{3}$	$\sqrt{3}$
$\frac{\pi}{2}$	Undefined

EXAMPLE 4 Finding the Exact Value of an Inverse Tangent Function

Find the exact value of $\tan^{-1}\sqrt{3}$.

Solution

Step 1 Let $\theta = \tan^{-1}x$. Thus,

$$\theta = \tan^{-1}\sqrt{3}.$$

We must find the angle θ, $-\frac{\pi}{2} < \theta < \frac{\pi}{2}$, whose tangent equals $\sqrt{3}$.

Step 2 Rewrite $\theta = \tan^{-1}x$ as $\tan\theta = x$. We obtain $\tan\theta = \sqrt{3}$.

Step 3 Use the exact values in Table 5.7 to find the value of θ in $\left(-\frac{\pi}{2}, \frac{\pi}{2}\right)$ that satisfies $\tan\theta = x$. The table shows that the only angle in the interval $\left(-\frac{\pi}{2}, \frac{\pi}{2}\right)$ that satisfies $\tan\theta = \sqrt{3}$ is $\frac{\pi}{3}$. Thus, $\theta = \frac{\pi}{3}$ and

$$\tan^{-1}\sqrt{3} = \frac{\pi}{3}. \quad \text{The angle in } \left(-\frac{\pi}{2}, \frac{\pi}{2}\right) \text{ whose tangent is } \sqrt{3} \text{ is } \frac{\pi}{3}.$$

Check Point 4 Find the exact value of $\tan^{-1}(-1)$.

Table 5.8 summarizes the graphs of the three basic inverse trigonometric functions. Below each of the graphs is a description of the function's domain and range.

Table 5.8 Graphs of the Three Basic Inverse Trigonometric Functions

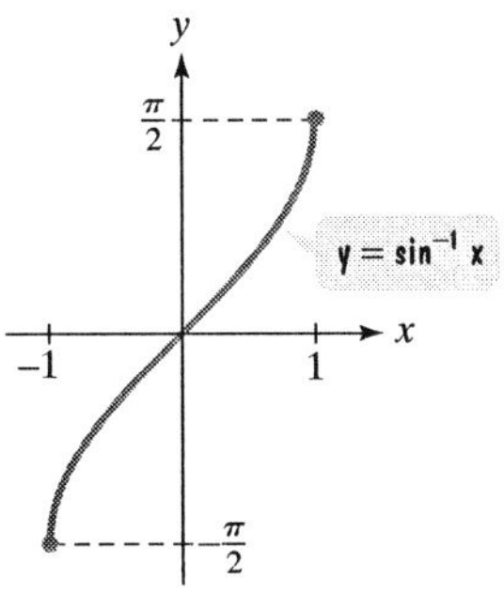

Domain: $[-1, 1]$
Range: $\left[-\frac{\pi}{2}, \frac{\pi}{2}\right]$

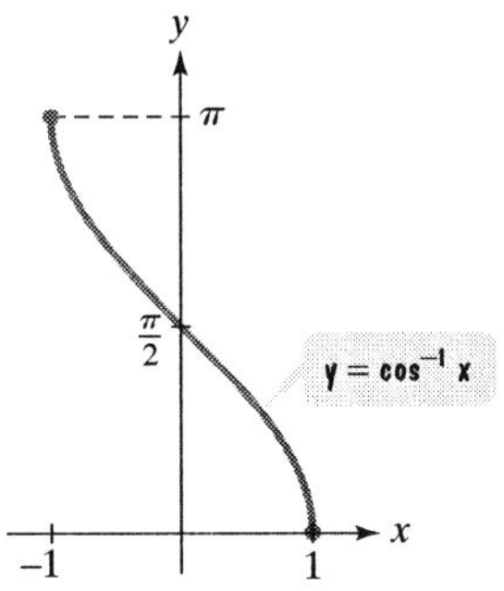

Domain: $[-1, 1]$
Range: $[0, \pi]$

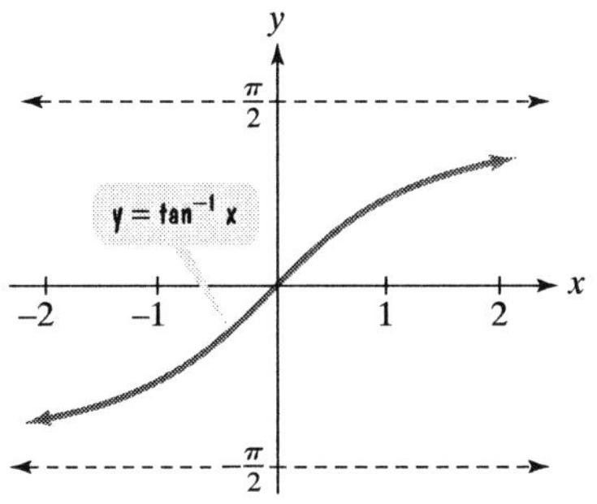

Domain: $(-\infty, \infty)$
Range: $\left(-\frac{\pi}{2}, \frac{\pi}{2}\right)$

4 Use a calculator to evaluate inverse trigonometric functions.

Using a Calculator to Evaluate Inverse Trigonometric Functions

Calculators give approximate values of inverse trigonometric functions. Use the secondary keys marked [SIN⁻¹], [COS⁻¹], and [TAN⁻¹]. These keys are not buttons that you actually press. They are the secondary functions for the buttons labeled [SIN], [COS], and [TAN], respectively. Consult your manual for the location of this feature.

EXAMPLE 5 Calculators and Inverse Trigonometric Functions

Use a calculator to find the value to four decimal places of:

a. $\sin^{-1}\frac{1}{4}$ **b.** $\tan^{-1}(-9.65)$.

Solution

Scientific Calculator Solution

Function	**Mode**	**Keystrokes**	**Display, rounded to four places**
a. $\sin^{-1}\frac{1}{4}$	Radian	1 [÷] 4 [=] [SIN⁻¹]	0.2527
b. $\tan^{-1}(-9.65)$	Radian	9.65 [+/−] [TAN⁻¹]	−1.4675

Graphing Calculator Solution

Function	**Mode**	**Keystrokes**	**Display, rounded to four places**
a. $\sin^{-1}\frac{1}{4}$	Radian	[SIN⁻¹] [(] 1 [÷] [4] [)] [ENTER]	0.2527
b. $\tan^{-1}(-9.65)$	Radian	[TAN⁻¹] [(−)] 9.65 [ENTER]	−1.4675

Check Point 5 Use a calculator to find the value to four decimal places of:

a. $\cos^{-1}\frac{1}{3}$ **b.** $\tan^{-1}(-35.85)$.

What happens if you attempt to evaluate an inverse trigomometric function at a value that is not in its domain? In real number mode, most calculators will display an error message. For example, an error message can result if you attempt to approximate $\cos^{-1}3$. There is no angle whose cosine is 3. The domain of the inverse cosine function is $[-1, 1]$, and 3 does not belong to this domain.

5 Find exact values of composite functions with inverse trigonometric functions.

Composition of Functions Involving Inverse Trigonometric Functions

In our discussion of functions and their inverses in Section 2.7, we saw that

$$f(f^{-1}(x)) = x \quad \text{and} \quad f^{-1}(f(x)) = x.$$

x must be in the domain of f^{-1}.

x must be in the domain of f.

We apply these properties to the sine, cosine, tangent, and their inverse functions to obtain the following properties:

Inverse Properties

The Sine Function and Its Inverse

$\sin(\sin^{-1}x) = x$ for every x in the interval $[-1, 1]$

$\sin^{-1}(\sin x) = x$ for every x in the interval $\left[-\frac{\pi}{2}, \frac{\pi}{2}\right]$

The Cosine Function and Its Inverse

$\cos(\cos^{-1}x) = x$ for every x in the interval $[-1, 1]$

$\cos^{-1}(\cos x) = x$ for every x in the interval $[0, \pi]$

The Tangent Function and Its Inverse

$\tan(\tan^{-1}x) = x$ for every real number x

$\tan^{-1}(\tan x) = x$ for every x in the interval $\left(-\frac{\pi}{2}, \frac{\pi}{2}\right)$

The restrictions on x in the inverse properties are a bit tricky. For example,

$$\sin^{-1}\left(\sin\frac{\pi}{4}\right) = \frac{\pi}{4}$$

$\sin^{-1}(\sin x) = x$ for x in $\left[-\frac{\pi}{2}, \frac{\pi}{2}\right]$. Observe that $\frac{\pi}{4}$ is in this interval.

Can we use $\sin^{-1}(\sin x) = x$ to find the exact value of $\sin^{-1}\left(\sin\frac{5\pi}{4}\right)$? Is $\frac{5\pi}{4}$ in the interval $\left[-\frac{\pi}{2}, \frac{\pi}{2}\right]$? No. Thus, to evaluate $\sin^{-1}\left(\sin\frac{5\pi}{4}\right)$, we must first find $\sin\frac{5\pi}{4}$.

$\frac{5\pi}{4}$ is in quadrant III, where the sine is negative.

$$\sin\frac{5\pi}{4} = -\sin\frac{\pi}{4} = -\frac{\sqrt{2}}{2}$$

The reference angle for $\frac{5\pi}{4}$ is $\frac{\pi}{4}$.

We evaluate $\sin^{-1}\left(\sin\frac{5\pi}{4}\right)$ as follows:

$$\sin^{-1}\left(\sin\frac{5\pi}{4}\right) = \sin^{-1}\left(-\frac{\sqrt{2}}{2}\right) = -\frac{\pi}{4}$$ If necessary, see Table 5.5 on page 524.

To determine how to evaluate the composition of functions involving inverse trigonometric functions, first examine the value of x. You can use the inverse properties in the box on the previous page only if x is in the specified interval.

EXAMPLE 6 Evaluating Compositions of Functions and Their Inverses

Find the exact value, if possible, of:

a. $\cos(\cos^{-1} 0.6)$ **b.** $\sin^{-1}\left(\sin\frac{3\pi}{2}\right)$ **c.** $\cos(\cos^{-1} 2\pi)$.

Solution

a. The inverse property $\cos(\cos^{-1} x) = x$ applies for every x in $[-1, 1]$. To evaluate $\cos(\cos^{-1} 0.6)$, observe that $x = 0.6$. This value of x lies in $[-1, 1]$, which is the domain of the inverse cosine function. This means that we can use the inverse property $\cos(\cos^{-1} x) = x$. Thus,

$$\cos(\cos^{-1} 0.6) = 0.6.$$

b. The inverse property $\sin^{-1}(\sin x) = x$ applies for every x in $\left[-\frac{\pi}{2}, \frac{\pi}{2}\right]$. To evaluate $\sin^{-1}\left(\sin\frac{3\pi}{2}\right)$, observe that $x = \frac{3\pi}{2}$. This value of x does not lie in $\left[-\frac{\pi}{2}, \frac{\pi}{2}\right]$. To evaluate this expression, we first find $\sin\frac{3\pi}{2}$.

$$\sin^{-1}\left(\sin\frac{3\pi}{2}\right) = \sin^{-1}(-1) = -\frac{\pi}{2}$$ The angle in $\left[-\frac{\pi}{2}, \frac{\pi}{2}\right]$ whose sine is -1 is $-\frac{\pi}{2}$.

c. The inverse property $\cos(\cos^{-1} x) = x$ applies for every x in $[-1, 1]$. To attempt to evaluate $\cos(\cos^{-1} 2\pi)$, observe that $x = 2\pi$. This value of x does not lie in $[-1, 1]$, which is the domain of the inverse cosine function. Thus, the expression $\cos(\cos^{-1} 2\pi)$ is not defined because $\cos^{-1} 2\pi$ is not defined.

Check Point 6 Find the exact value, if possible, of:

a. $\cos(\cos^{-1} 0.7)$ **b.** $\sin^{-1}(\sin \pi)$ **c.** $\cos(\cos^{-1} \pi)$.

We can use points on terminal sides of angles in standard position to find exact values of expressions involving the composition of a function and a different inverse function. Here are two examples:

$$\cos\left(\tan^{-1}\frac{5}{12}\right) \qquad \cot\left[\sin^{-1}\left(-\frac{1}{3}\right)\right].$$

Inner part involves the angle in $\left(-\frac{\pi}{2}, \frac{\pi}{2}\right)$ whose tangent is $\frac{5}{12}$.

Inner part involves the angle in $\left[-\frac{\pi}{2}, \frac{\pi}{2}\right]$ whose sine is $-\frac{1}{3}$.

The inner part of each expression involves an angle. To evaluate such expressions, we represent such angles by θ. Then we use a sketch that illustrates our representation. Examples 7 and 8 show how to carry out such evaluations.

EXAMPLE 7 Evaluating a Composite Trigonometric Expression

Find the exact value of $\cos\left(\tan^{-1}\frac{5}{12}\right)$.

Solution We let θ represent the angle in $\left(-\frac{\pi}{2}, \frac{\pi}{2}\right)$ whose tangent is $\frac{5}{12}$. Thus,

$$\theta = \tan^{-1}\frac{5}{12}.$$

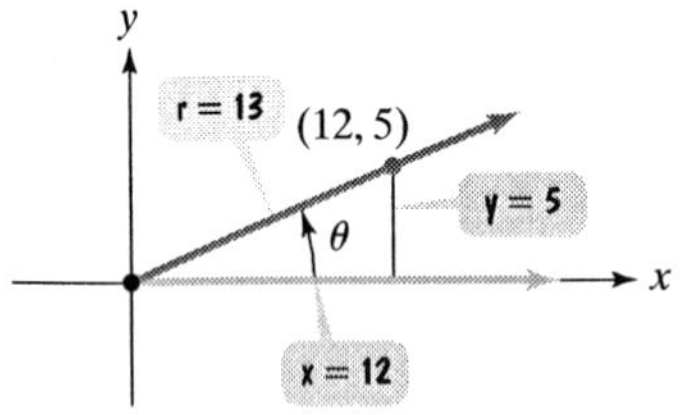

Figure 5.76 Representing $\tan\theta = \frac{5}{12}$

Using the definition of the inverse tangent function, we can rewrite this as

$$\tan\theta = \frac{5}{12}.$$

Because $\tan\theta$ is positive, θ must be an angle in $\left(0, \frac{\pi}{2}\right)$. Thus, θ is a first-quadrant angle. Figure 5.76 shows a right triangle in quadrant I with

$$\tan\theta = \frac{5}{12}.$$

Side opposite θ, or y

Side adjacent to θ, or x

The hypotenuse of the triangle, r, or the distance from the origin to (12, 5), is found using $r = \sqrt{x^2 + y^2}$.

$$r = \sqrt{x^2 + y^2} = \sqrt{12^2 + 5^2} = \sqrt{144 + 25} = \sqrt{169} = 13$$

We use the values for x and r to find the exact value of $\cos\left(\tan^{-1}\frac{5}{12}\right)$.

$$\cos\left(\tan^{-1}\frac{5}{12}\right) = \cos\theta = \frac{\text{side adjacent to } \theta, \text{ or } x}{\text{hypotenuse, or } r} = \frac{12}{13}$$

Check Point 7 Find the exact value of $\sin\left(\tan^{-1}\frac{3}{4}\right)$.

EXAMPLE 8 Evaluating a Composite Trigonometric Expression

Find the exact value of $\cot\left[\sin^{-1}\left(-\frac{1}{3}\right)\right]$.

Solution We let θ represent the angle in $\left[-\frac{\pi}{2}, \frac{\pi}{2}\right]$ whose sine is $-\frac{1}{3}$. Thus,

$$\theta = \sin^{-1}\left(-\frac{1}{3}\right) \quad \text{and} \quad \sin\theta = -\frac{1}{3}.$$

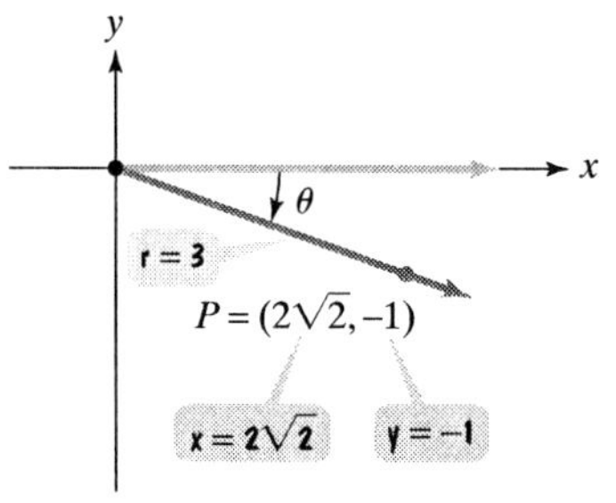

Figure 5.77 Representing $\sin\theta = -\frac{1}{3}$

Because $\sin\theta$ is negative in $\sin\theta = -\frac{1}{3}$, θ must be an angle in $\left[-\frac{\pi}{2}, 0\right)$. Thus, θ is a negative angle that lies in quadrant IV. Figure 5.77 shows angle θ in quadrant IV with

In quadrant IV, y is negative.

$$\sin\theta = -\frac{1}{3} = \frac{y}{r} = \frac{-1}{3}.$$

Thus, $y = -1$ and $r = 3$. The value of x can be found using $r = \sqrt{x^2 + y^2}$ or $x^2 + y^2 = r^2$.

$$x^2 + (-1)^2 = 3^2$$
$$x^2 + 1 = 9$$
$$x^2 = 8$$
$$x = \sqrt{8} = \sqrt{4 \cdot 2} = 2\sqrt{2}$$ Remember that x is positive in quadrant IV.

We use values for x and y to find the exact value of $\cot\left[\sin^{-1}\left(-\frac{1}{3}\right)\right]$.

$$\cot\left[\sin^{-1}\left(-\frac{1}{3}\right)\right] = \cot\theta = \frac{x}{y} = \frac{2\sqrt{2}}{-1} = -2\sqrt{2}$$

Check Point 8 Find the exact value of $\cos\left[\sin^{-1}\left(-\frac{1}{2}\right)\right]$.

Some composite functions with inverse trigonometric functions can be simplified to algebraic expressions. To simplify such an expression, we represent the inverse trigonometric function in the expression by θ. Then we use a right triangle.

EXAMPLE 9 Simplifying an Expression Involving $\sin^{-1}x$

If $0 < x \le 1$, write $\cos(\sin^{-1}x)$ as an algebraic expression in x.

Solution We let θ represent the angle in $\left[-\frac{\pi}{2}, \frac{\pi}{2}\right]$ whose sine is x. Thus,

$$\theta = \sin^{-1}x, \quad \text{and} \quad \sin\theta = x.$$

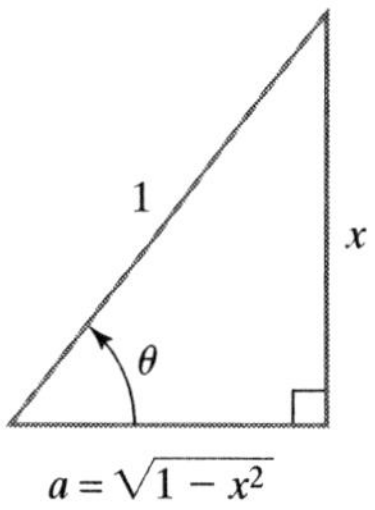

Figure 5.78 Representing $\sin\theta = x$

Because $0 < x \le 1$, $\sin\theta$ is positive. Thus, θ is a first-quadrant angle and can be represented as an acute angle of a right triangle. Figure 5.78 shows a right triangle with

$$\sin\theta = x = \frac{x}{1}.$$ Side opposite θ; Hypotenuse

The third side, a in Figure 5.78, can be found using the Pythagorean Theorem.

$$a^2 + x^2 = 1^2$$ Apply the Pythagorean Theorem to the right triangle in Figure 5.78.

$$a^2 = 1 - x^2$$ Subtract x^2 from both sides.

$$a = \sqrt{1 - x^2}$$ Solve for a.

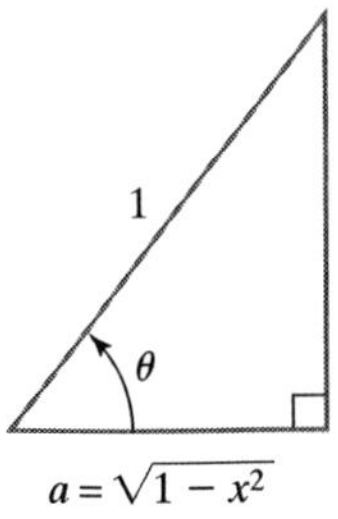

Figure 5.78, repeated

We use the right triangle in Figure 5.78 to write $\cos(\sin^{-1} x)$ as an algebraic expression.

$$\cos(\sin^{-1} x) = \cos\theta = \frac{\text{side adjacent to } \theta}{\text{hypotenuse}} = \frac{\sqrt{1-x^2}}{1} = \sqrt{1-x^2}$$

Check Point 9 If $x > 0$, write $\sec(\tan^{-1} x)$ as an algebraic expression in x.

The inverse secant function, $y = \sec^{-1} x$, is used in calculus. However, inverse cotangent and inverse cosecant functions are rarely used. Two of these remaining inverse trigonometric functions are briefly developed in the exercise set that follows.

EXERCISE SET 5.7

Practice Exercises

In Exercises 1–18, find the exact value of each expression.

1. $\sin^{-1}\frac{1}{2}$ **2.** $\sin^{-1} 0$ **3.** $\sin^{-1}\frac{\sqrt{2}}{2}$

4. $\sin^{-1}\frac{\sqrt{3}}{2}$ **5.** $\sin^{-1}\left(-\frac{1}{2}\right)$ **6.** $\sin^{-1}\left(-\frac{\sqrt{3}}{2}\right)$

7. $\cos^{-1}\frac{\sqrt{3}}{2}$ **8.** $\cos^{-1}\frac{\sqrt{2}}{2}$ **9.** $\cos^{-1}\left(-\frac{\sqrt{2}}{2}\right)$

10. $\cos^{-1}\left(-\frac{\sqrt{3}}{2}\right)$ **11.** $\cos^{-1} 0$ **12.** $\cos^{-1} 1$

13. $\tan^{-1}\frac{\sqrt{3}}{3}$ **14.** $\tan^{-1} 1$ **15.** $\tan^{-1} 0$

16. $\tan^{-1}(-1)$ **17.** $\tan^{-1}(-\sqrt{3})$ **18.** $\tan^{-1}\left(-\frac{\sqrt{3}}{3}\right)$

In Exercises 19–30, use a calculator to find the value of each expression rounded to two decimal places.

19. $\sin^{-1} 0.3$ **20.** $\sin^{-1} 0.47$

21. $\sin^{-1}(-0.32)$ **22.** $\sin^{-1}(-0.625)$

23. $\cos^{-1}\frac{3}{8}$ **24.** $\cos^{-1}\frac{4}{9}$

25. $\cos^{-1}\frac{\sqrt{5}}{7}$ **26.** $\cos^{-1}\frac{\sqrt{7}}{10}$

27. $\tan^{-1}(-20)$ **28.** $\tan^{-1}(-30)$

29. $\tan^{-1}(-\sqrt{473})$ **30.** $\tan^{-1}(-\sqrt{5061})$

In Exercises 31–46, find the exact value of each expression, if possible. Do not use a calculator.

31. $\sin(\sin^{-1} 0.9)$ **32.** $\cos(\cos^{-1} 0.57)$

33. $\sin^{-1}\left(\sin\frac{\pi}{3}\right)$ **34.** $\cos^{-1}\left(\cos\frac{2\pi}{3}\right)$

35. $\sin^{-1}\left(\sin\frac{5\pi}{6}\right)$ **36.** $\cos^{-1}\left(\cos\frac{4\pi}{3}\right)$

37. $\tan(\tan^{-1} 125)$ **38.** $\tan(\tan^{-1} 380)$

39. $\tan^{-1}\left[\tan\left(-\frac{\pi}{6}\right)\right]$ **40.** $\tan^{-1}\left[\tan\left(-\frac{\pi}{3}\right)\right]$

41. $\tan^{-1}\left(\tan\frac{2\pi}{3}\right)$ **42.** $\tan^{-1}\left(\tan\frac{3\pi}{4}\right)$

43. $\sin^{-1}(\sin\pi)$ **44.** $\cos^{-1}(\cos 2\pi)$

45. $\sin(\sin^{-1}\pi)$ **46.** $\cos(\cos^{-1} 3\pi)$

In Exercises 47–60, use a sketch to find the exact value of each expression.

47. $\cos(\sin^{-1}\frac{4}{5})$ **48.** $\sin(\tan^{-1}\frac{7}{24})$

49. $\tan(\cos^{-1}\frac{5}{13})$ **50.** $\cot(\sin^{-1}\frac{5}{13})$

51. $\tan[\sin^{-1}(-\frac{3}{5})]$ **52.** $\cos[\sin^{-1}(-\frac{4}{5})]$

53. $\sin\left(\cos^{-1}\frac{\sqrt{2}}{2}\right)$ **54.** $\cos(\sin^{-1}\frac{1}{2})$

55. $\sec[\sin^{-1}(-\frac{1}{4})]$ **56.** $\sec[\sin^{-1}(-\frac{1}{2})]$

57. $\tan[\cos^{-1}(-\frac{1}{3})]$ **58.** $\tan[\cos^{-1}(-\frac{1}{4})]$

59. $\csc\left[\cos^{-1}\left(-\frac{\sqrt{3}}{2}\right)\right]$ **60.** $\sec\left[\sin^{-1}\left(-\frac{\sqrt{2}}{2}\right)\right]$

In Exercises 61–70, use a right triangle to write each expression as an algebraic expression. Assume that x is positive and in the domain of the given inverse trigonometric function.

61. $\tan(\cos^{-1} x)$ **62.** $\sin(\tan^{-1} x)$

63. $\cos(\sin^{-1} 2x)$ **64.** $\sin(\cos^{-1} 2x)$

65. $\cos\left(\sin^{-1}\frac{1}{x}\right)$

66. $\sec\left(\cos^{-1}\frac{1}{x}\right)$

67. $\cot\left(\tan^{-1}\frac{x}{\sqrt{3}}\right)$

68. $\cot\left(\tan^{-1}\frac{x}{\sqrt{2}}\right)$

69. $\sec\left(\sin^{-1}\frac{x}{\sqrt{x^2+4}}\right)$

70. $\cot\left(\sin^{-1}\frac{\sqrt{x^2-9}}{x}\right)$

71. a. Graph the restricte-d secant function, $y = \sec x$, by restricting x to the intervals $\left[0, \frac{\pi}{2}\right)$ and $\left(\frac{\pi}{2}, \pi\right]$.

b. Use the horizontal line test to explain why the restricted secant function has an inverse function.

c. Use the graph of the restricted secant function to graph $y = \sec^{-1} x$.

72. a. Graph the restricted cotangent function, $y = \cot x$, by restricting x to the interval $(0, \pi)$.

b. Use the horizontal line test to explain why the restricted cotangent function has an inverse function.

c. Use the graph of the restricted cotangent function to graph $y = \cot^{-1} x$.

Application Exercises

73. Your neighborhood movie theater has a 25-foot-high screen located 8 feet above your eye level. If you sit too close to the screen, your viewing angle is too small, resulting in a distorted picture. By contrast, if you sit too far back, the image is quite small, diminishing the movie's visual impact. If you sit x feet back from the screen, your viewing angle, θ, is given by

$$\theta = \tan^{-1}\frac{33}{x} - \tan^{-1}\frac{8}{x}.$$

Find the viewing angle, in radians, at distances of 5 feet, 10 feet, 15 feet, 20 feet, and 25 feet.

74. The function $\theta = \tan^{-1}\frac{33}{x} - \tan^{-1}\frac{8}{x}$, described in Exercise 73, is graphed below in a $[0, 50, 10]$ by $[0, 1, 0.1]$ viewing rectangle. Use the graph to describe what happens to your viewing angle as you move farther back from the screen. How far back from the screen, to the nearest foot, should you sit to maximize your viewing angle? Verify this observation by finding the viewing angle one foot closer to the screen and one foot farther from the screen for this ideal viewing distance.

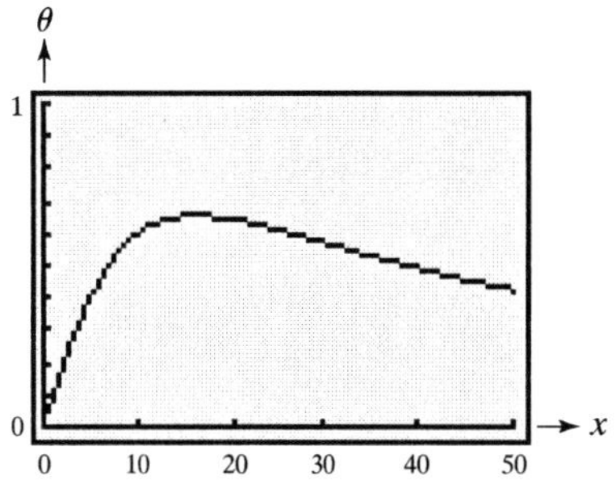

The formula

$$\theta = 2\tan^{-1}\frac{21.634}{x}$$

gives the viewing angle, θ, in radians, for a camera whose lens is x millimeters wide. Use this formula to solve Exercises 75–76.

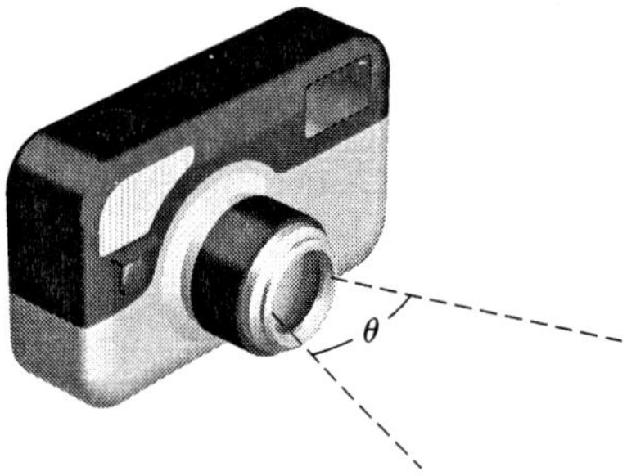

75. Find the viewing angle, in radians and in degrees (to the nearest tenth of a degree), of a 28-millimeter lens.

76. Find the viewing angle, in radians and degrees (to the nearest tenth of a degree), of a 300-millimeter telephoto lens.

For years, mathematicians were challenged by the following problem: What is the area of a region under a curve between two values of x? The problem was solved in the seventeenth century with the development of integral calculus. Using calculus, the area of the region under $y = \frac{1}{x^2+1}$, above the x-axis, and between $x = a$ and $x = b$ is $\tan^{-1} b - \tan^{-1} a$. Use this result, shown in the figure, on the next page, to find *the area of the region under* $y = \dfrac{1}{x^2 + 1}$, *above the x-axis, and between the values of a and b given in Exercises 77–78.*

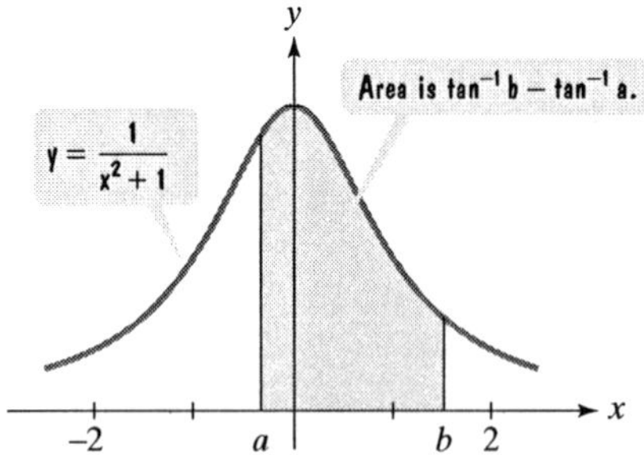

77. $a = 0$ and $b = 2$

78. $a = -2$ and $b = 1$

Writing in Mathematics

79. Explain why, without restrictions, no trigonometric function has an inverse function.

80. Describe the restriction on the sine function so that it has an inverse function.

81. How can the graph of $y = \sin^{-1} x$ be obtained from the graph of the restricted sine function?

82. Without drawing a graph, describe the behavior of the graph of $y = \sin^{-1} x$. Mention the function's domain and range in your description.

83. Describe the restriction on the cosine function so that it has an inverse function.

84. Without drawing a graph, describe the behavior of the graph of $y = \cos^{-1} x$. Mention the function's domain and range in your description.

85. Describe the restriction on the tangent function so that it has an inverse function.

86. Without drawing a graph, describe the behavior of the graph of $y = \tan^{-1} x$. Mention the function's domain and range in your description.

87. If $\sin^{-1}\left(\sin \dfrac{\pi}{3}\right) = \dfrac{\pi}{3}$, is $\sin^{-1}\left(\sin \dfrac{5\pi}{6}\right) = \dfrac{5\pi}{6}$? Explain your answer.

88. Explain how a right triangle can be used to find the exact value of $\sec\left(\sin^{-1}\frac{4}{5}\right)$.

89. Find the height of the screen and the number of feet that it is located above eye level in your favorite movie theater. Modify the formula given in Exercise 73 so that it applies to your theater. Then describe where in the theater you should sit so that a movie creates the greatest visual impact.

Technology Exercises

In Exercises 90–93, graph each pair of functions in the same viewing rectangle. Use your knowledge of the domain and range for the inverse trigonometric functions to select an appropriate viewing rectangle. What does the graph of the second equation in each exercise do to the graph of the first equation?

90. $y = \sin^{-1} x$ and $y = \sin^{-1} x + 2$

91. $y = \cos^{-1} x$ and $y = \cos^{-1}(x - 1)$

92. $y = \tan^{-1} x$ and $y = -2\tan^{-1} x$

93. $y = \sin^{-1} x$ and $y = \sin^{-1}(x + 2) + 1$

94 Graph $y = \tan^{-1} x$ and its two horizontal asymptotes in a $[-3, 3, 1]$ by $\left[-\pi, \pi, \dfrac{\pi}{2}\right]$ viewing rectangle. Then change the range setting to $[-50, 50, 5]$ by $\left[-\pi, \pi, \dfrac{\pi}{2}\right]$. What do you observe?

95. Graph $y = \sin^{-1} x + \cos^{-1} x$ in a $[-2, 2, 1]$ by $[0, 3, 1]$ viewing rectangle. What appears to be true about the sum of the inverse sine and inverse cosine for values between -1 and 1 inclusive?

Critical Thinking Exercises

96. Solve $y = 2\sin^{-1}(x - 5)$ for x in terms of y.

97. Solve for x: $2\sin^{-1} x = \dfrac{\pi}{4}$.

98. Prove that if $x > 0$, $\tan^{-1} x + \tan^{-1}\dfrac{1}{x} = \dfrac{\pi}{2}$.

99. Derive the formula for θ, your viewing angle at the movie theater, in Exercise 73. *Hint*: Use the figure shown and represent the acute angle on the left in the smaller right triangle by α. Find expressions for $\tan\alpha$ and $\tan(\alpha + \theta)$.

SECTION 5.8 Applications of Trigonometric Functions

Objectives

1. Solve a right triangle.
2. Solve problems involving bearings.
3. Model simple harmonic motion.

In the late 1960s, popular musicians were searching for new sounds. Film composers were looking for ways to create unique sounds as well. From these efforts, synthesizers that electronically reproduce musical sounds were born. From providing the backbone of today's most popular music to providing the strange sounds for the most experimental music, synthesizers are at the forefront of music technology.

If we did not understand the periodic nature of sinusoidal functions, the synthesizers used in almost all forms of music would not exist. In this section, we look at applications of trigonometric functions in right triangles and in modeling periodic phenomena such as sound.

1 Solve a right triangle.

Solving Right Triangles

Solving a right triangle means finding the missing lengths of its sides and the measurements of its angles. We will label right triangles so that side a is opposite angle A, side b is opposite angle B, and side c, the hypotenuse, is opposite right angle C. Figure 5.79 illustrates this labeling.

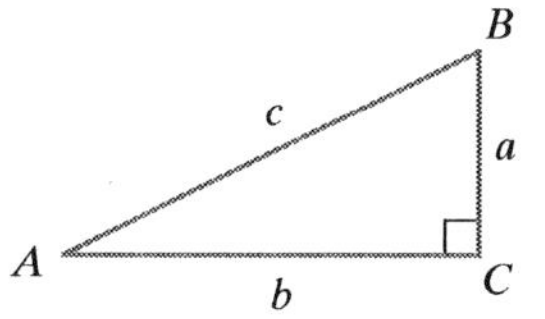

Figure 5.79 Labeling right triangles

When solving a right triangle, we will use the sine, cosine, and tangent functions, rather than their reciprocals. Example 1 shows how to solve a right triangle when we know the length of a side and the measure of an acute angle.

EXAMPLE 1 Solving a Right Triangle

Solve the right triangle shown in Figure 5.80.

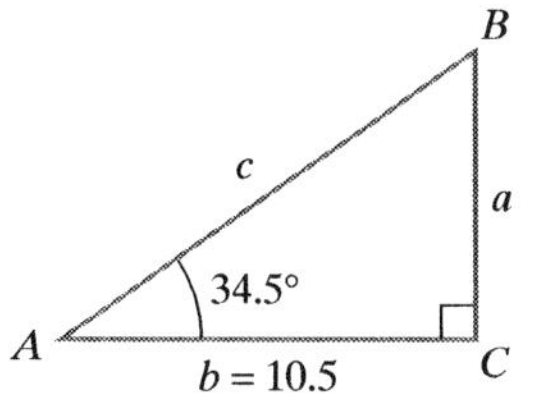

Figure 5.80 Find B, a, and c.

Solution We begin by finding the measure of angle B. We do not need a trigonometric function to do so. Because $C = 90°$ and the sum of a triangle's angles is 180°, we see that $A + B = 90°$. Thus,

$$B = 90° - A = 90° - 34.5° = 55.5°.$$

Now we need to find a. Because we have a known angle, an unknown opposite side, and a known adjacent side, we use the tangent function.

$$\tan 34.5° = \frac{a}{10.5}$$

Side opposite the 34.5° angle

Side adjacent to the 34.5° angle

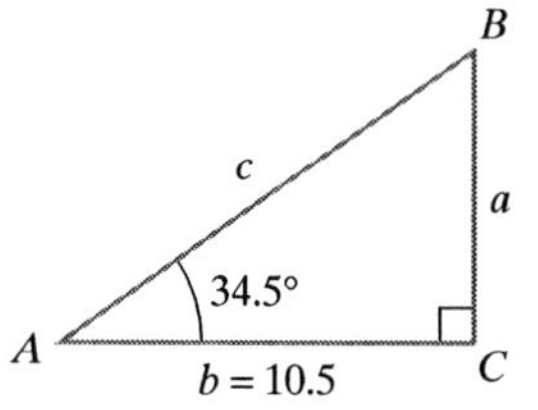

Figure 5.80, repeated

Now we multiply both sides of this equation by 10.5 and solve for a.

$$a = 10.5 \tan 34.5° \approx 7.22$$

Finally, we need to find c. Because we have a known angle, a known adjacent side, and an unknown hypotenuse, we use the cosine function.

$$\cos 34.5° = \frac{10.5}{c}$$

Side adjacent to the 34.5° angle

hypotenuse

Now we multiply both sides of this equation by c and then solve for c.

$$c \cos 34.5° = 10.5 \quad \text{Multiply both sides by } c.$$

$$c = \frac{10.5}{\cos 34.5°} \approx 12.74 \quad \text{Divide both sides by } \cos 34.5° \text{ and solve for } c.$$

In summary, $B = 55.5°$, $a \approx 7.22$, and $c \approx 12.74$.

Discovery

There is often more than one correct way to solve a right triangle. In Example 1, find a using angle $B = 55.5°$. Find c using the Pythagorean Theorem.

Check Point 1 In Figure 5.79, let $A = 62.7°$ and $a = 8.4$. Solve the right triangle, rounding lengths to two decimal places.

Trigonometry was first developed to measure heights and distances that were inconvenient or impossible to measure. In solving application problems, begin by making a sketch involving a right triangle that illustrates the problem's conditions. Then put your knowledge of solving right triangles to work and find the required distance or height.

EXAMPLE 2 Finding the Side of a Triangle

From a point on level ground 125 feet from the base of a tower, the angle of elevation is 57.2°. Approximate the height of the tower to the nearest foot.

Solution A sketch is shown in Figure 5.81, where a represents the height of the tower. In the right triangle, we have a known angle, an unknown opposite side, and a known adjacent side. Therefore, we use the tangent function.

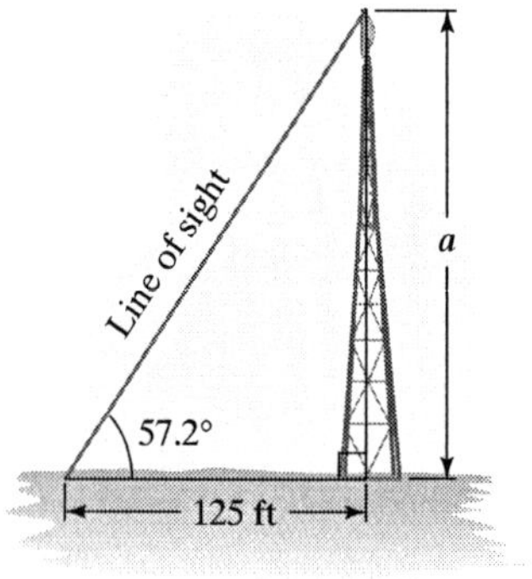

Figure 5.81 Determining height without using direct measurement

$$\tan 57.2° = \frac{a}{125}$$

Side opposite the 57.2° angle

Side adjacent to the 57.2° angle

Now we multiply both sides of this equation by 125 and solve for a.

$$a = 125 \tan 57.2° \approx 194$$

The tower is approximately 194 feet high.

Check Point 2 From a point on level ground 80 feet from the base of the Eiffel Tower, the angle of elevation is 85.4°. Approximate the height of the Eiffel Tower to the nearest foot.

Example 3 illustrates how to find the measure of an acute angle of a right triangle if the lengths of two sides are known.

EXAMPLE 3 Finding the Angle of a Triangle

A kite flies at a height of 30 feet when 65 feet of string is out. If the string is in a straight line, find the angle that it makes with the ground. Round to the nearest tenth of a degree.

Solution A sketch is shown in Figure 5.82, where A represents the angle the string makes with the ground. In the right triangle, we have an unknown angle, a known opposite side, and a known hypotenuse. Therefore, we use the sine function.

$$\sin A = \frac{30}{65}$$ Side opposite A / hypotenuse

$$A = \sin^{-1}\frac{30}{65} \approx 27.5°$$

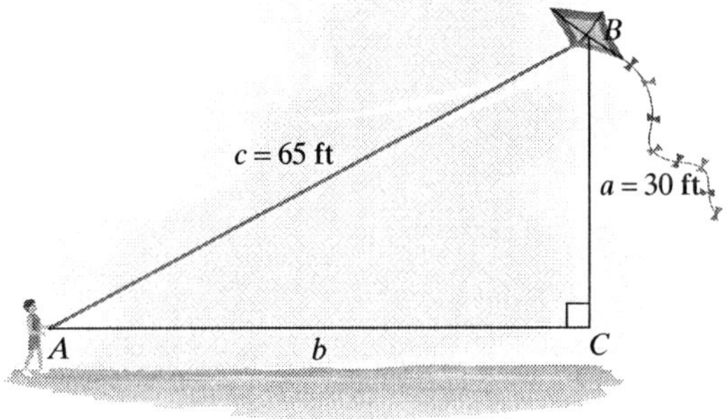

Figure 5.82 Flying a kite

The string makes an angle of approximately 27.5° with the ground.

Check Point 3 A guy wire is 13.8 yards long and is attached from the ground to a pole 6.7 yards above the ground. Find the angle, to the nearest tenth of a degree, that the wire makes with the ground.

EXAMPLE 4 Using Two Right Triangles to Solve a Problem

You are taking your first hot-air balloon ride. Your friend is standing on level ground, 100 feet away from your point of launch, making a video of the terrified look on your rapidly ascending face. How rapidly? At one instant, the angle of elevation from the video camera to your face is 31.7°. One minute later, the angle of elevation is 76.2°. How far did you travel during that minute?

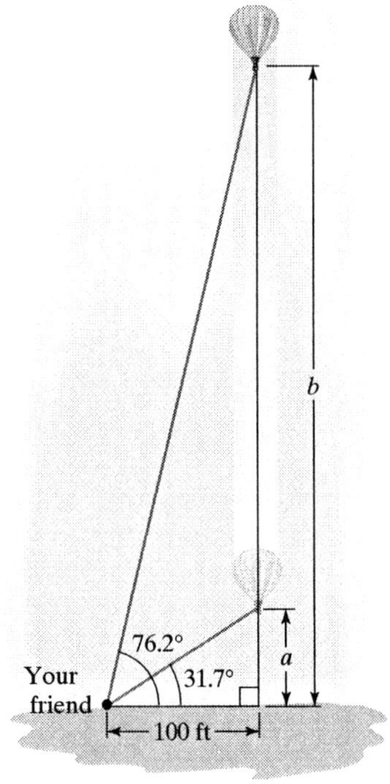

Figure 5.83 Ascending in a hot-air balloon

Solution A sketch that illustrates the problem is shown in Figure 5.83. We need to determine $b - a$, the distance traveled during the one-minute period. We find a using the small right triangle. Because we have a known angle, an unknown opposite side, and a known adjacent side, we use the tangent function.

$$\tan 31.7° = \frac{a}{100}$$ Side opposite the 31.7° angle / Side adjacent to the 31.7° angle

$$a = 100 \tan 31.7° \approx 61.8$$

We find b using the tangent function in the large right triangle. Turn back the page and verify this observation in Figure 5.83.

$$\tan 76.2° = \frac{b}{100}$$

Side opposite the 76.2° angle

Side adjacent to the 76.2° angle

$$b = 100 \tan 76.2° \approx 407.1$$

The balloon traveled 407.1 − 61.8, or approximately 345.3 feet, during the minute.

Check Point 4 You are standing on level ground 800 feet from Mt. Rushmore, looking at the sculpture of Abraham Lincoln's face. The angle of elevation to the bottom of the sculpture is 32° and the angle of elevation to the top is 35°. Find the height of the sculpture of Lincoln's face to the nearest tenth of a foot.

2 Solve problems involving bearings.

Trigonometry and Bearings

In navigation and surveying problems, the term *bearing* is used to specify the location of one point relative to another. The **bearing** from point O to point P is the acute angle between ray OP and a north-south line. Figure 5.84 illustrates some examples of bearings. The north-south line and the east-west line intersect at right angles.

Study Tip

The bearing from O to P can also be described using the phrase "the bearing of P from O."

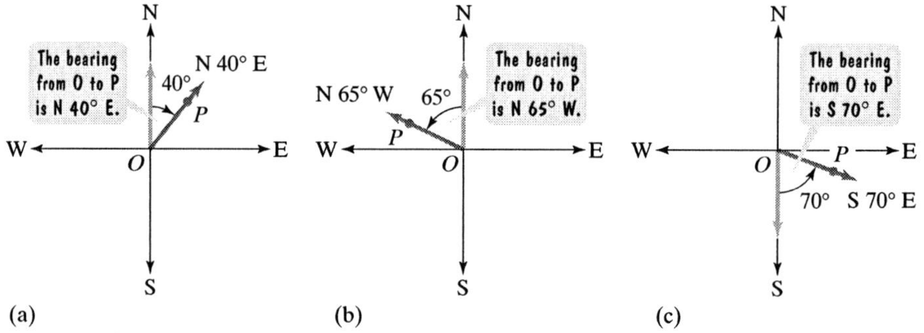

Figure 5.84 An illustration of three bearings

Each bearing has three parts: a letter (N or S), the measure of an acute angle, and a letter (E or W). Here's how we write a bearing:

- If the acute angle is measured from the *north side* of the north-south line, then we write N first. [See Figure 5.84(a).] If the acute angle is measured from the *south side* of the north-south line, then we write S first. [See Figure 5.84(c).]
- Second, we write the measure of the acute angle.
- If the acute angle is measured on the *east side* of the north-south line, then we write E last. [See Figure 5.84(a)]. If the acute angle is measured on the *west side* of the north-south line, then we write W last. [See Figure 5.84(b).]

EXAMPLE 5 Understanding Bearings

Use Figure 5.85 on the next page to find:

a. the bearing from O to B.

b. the bearing from O to A.

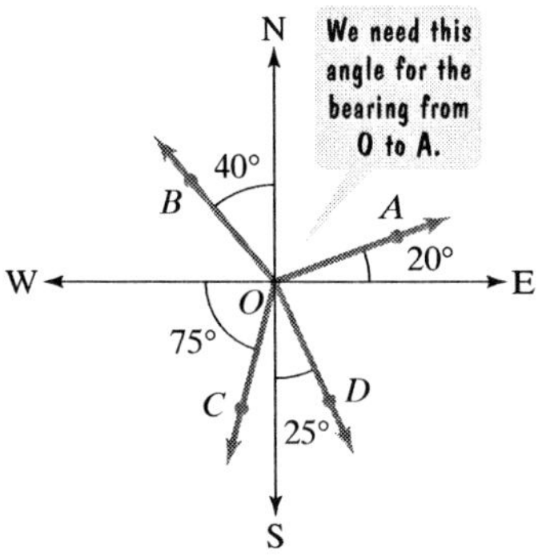

Figure 5.85 Finding bearings

Solution

a. To find the bearing from O to B, we need the acute angle between the ray OB and the north-south line through O. The measurement of this angle is given to be 40°. Figure 5.85 shows that the angle is measured from the north side of the north-south line and lies west of the north-south line. Thus, the bearing from O to B is N 40° W.

b. To find the bearing from O to A, we need the acute angle between the ray OA and the north-south line through O. This angle is specified by the voice balloon in Figure 5.85. Because of the given 20° angle, this angle measures $90° - 20°$, or 70°. This angle is measured from the north side of the north-south line. This angle is also east of the north-south line. Thus, the bearing from O to A is N 70° E.

Check Point 5 Use Figure 5.85 to find:

a. the bearing from O to D.

b. the bearing from O to C.

EXAMPLE 6 Finding the Bearing of a Boat

A boat leaves the entrance to a harbor and travels 25 miles on a bearing of N 42° E. Figure 5.86 shows that the captain then turns the boat 90°clockwise and travels 18 miles on a bearing of S 48° E. At that time:

a. How far is the boat from the harbor entrance?

b. What is the bearing of the boat from the harbor entrance?

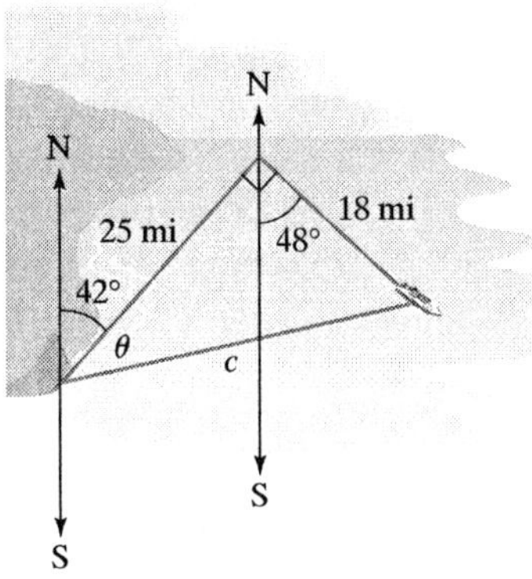

Figure 5.86 Finding a boat's bearing from the harbor entrance

Solution

a. The boat's distance from the harbor entrance is represented by c in Figure 5.86. Because we know the length of two sides of the right triangle, we find c using the Pythagorean Theorem. We have

$$c^2 = a^2 + b^2 = 25^2 + 18^2 = 949$$

$$c = \sqrt{949} \approx 30.8.$$

The boat is approximately 30.8 miles from the harbor entrance.

b. The bearing of the boat from the harbor entrance means the bearing from the entrance to the boat. Look at the north-south line passing through the harbor entrance on the left in Figure 5.86. The acute angle from this line to the ray on which the boat lies is $42° + \theta$. Because we are measuring the angle from the north side of the line and the boat is east of the harbor, its bearing from the harbor entrance is N$(42° + \theta)$E. To find θ, we use the right triangle shown in Figure 5.86 and the tangent function.

$$\tan\theta = \frac{\text{side opposite } \theta}{\text{side adjacent to } \theta} = \frac{18}{25}$$

$$\theta = \tan^{-1}\frac{18}{25}$$

We can use a calculator in degree mode to find the value of θ: $\theta \approx 35.8°$. Thus, $42° + \theta = 42° + 35.8° = 77.8°$. The bearing of the boat from the harbor entrance is N 77.8° E.

Study Tip

When making a diagram showing bearings, draw a north-south line through each point at which a change in course occurs. The north side of the line lies above each point. The south side of the line lies below each point.

Check Point 6 You leave the entrance to a system of hiking trails and hike 2.3 miles on a bearing of S 31° W. Then the trail turns 90° clockwise and you hike 3.5 miles on a bearing of N 59° W. At that time:

a. How far are you from the entrance to the trail system?

b. What is your bearing from the entrance to the trail system?

3 Model simple harmonic motion.

Simple Harmonic Motion

Because of their periodic nature, trigonometric functions are used to model phenomena that occur again and again. This includes vibratory or oscillatory motion, such as the motion of a vibrating guitar string, the swinging of a pendulum, or the bobbing of an object attached to a spring. Trigonometric functions are also used to describe radio waves from your favorite FM station, television waves from your not-to-be-missed weekly sitcom, and sound waves from your most-prized CDs.

To see how trigonometric functions are used to model vibratory motion, consider this: A ball is attached to a spring hung from the ceiling. You pull the ball down 4 inches and then release it. If we neglect the effects of friction and air resistance, the ball will continue bobbing up and down on the end of the spring. These up-and-down oscillations are called **simple harmonic motion.**

To better understand this motion, we use a d-axis, where d represents distance. This axis is shown in Figure 5.87. On this axis, the position of the ball before you pull it down is $d = 0$. This rest position is called the **equilibrium position.** Now you pull the ball down 4 inches to $d = -4$ and release it. Figure 5.88 shows a sequence of "photographs" taken at one-second time intervals illustrating the distance of the ball from its rest position, d.

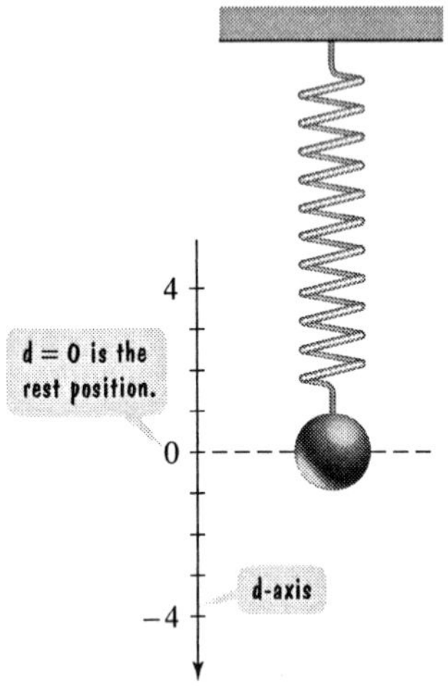

Figure 5.87 Using a d-axis to describe a ball's distance from its rest position

The curve in Figure 5.88 shows how the ball's distance from its rest position changes over time. The curve is sinusoidal and the motion can be described using a cosine or a sine function.

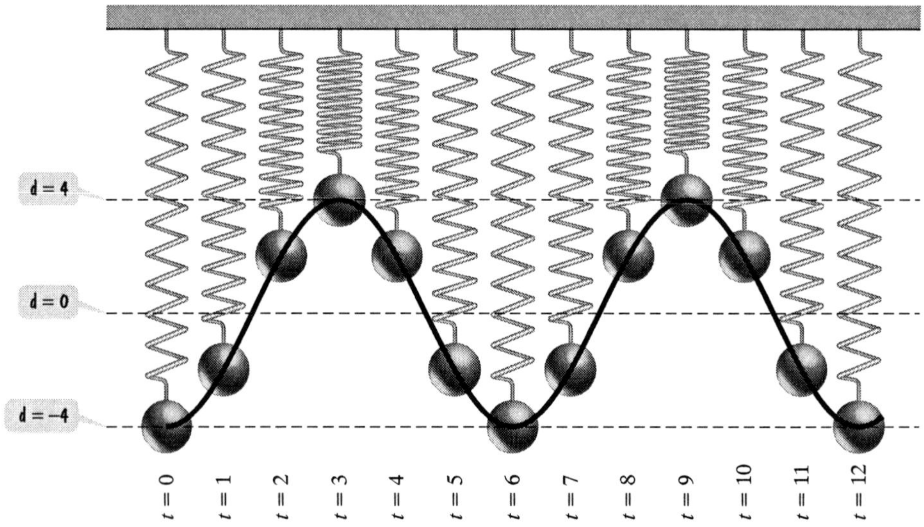

Figure 5.88 A sequence of "photographs" showing the bobbing ball's distance from the rest position, taken at one-second intervals

Simple Harmonic Motion

An object that moves on a coordinate axis is in **simple harmonic motion** if its distance from the origin, d, at time t is given by either

$$d = a \cos \omega t \quad \text{or} \quad d = a \sin \omega t.$$

The motion has **amplitude** $|a|$, the maximum displacement of the object from its rest position. The **period** of the motion is $\frac{2\pi}{\omega}$, where $\omega > 0$. The period gives the time it takes for the motion to go through one complete cycle.

In describing simple harmonic motion, the equation with the cosine function is used if the object is at its greatest distance from rest position, the origin, at $t = 0$. By contrast, the equation with the sine function is used if the object is at its rest position, the origin, at $t = 0$.

Diminishing Motion with Increasing Time

Due to friction and other resistive forces, the motion of an oscillating object decreases over time. The function

$$d = 3e^{-0.1t} \cos 2t$$

models this type of motion. The graph of the function is shown in a $t = [0, 10, 1]$ by $d = [-3, 3, 1]$ viewing rectangle. Notice how the amplitude is decreasing with time as the moving object loses energy.

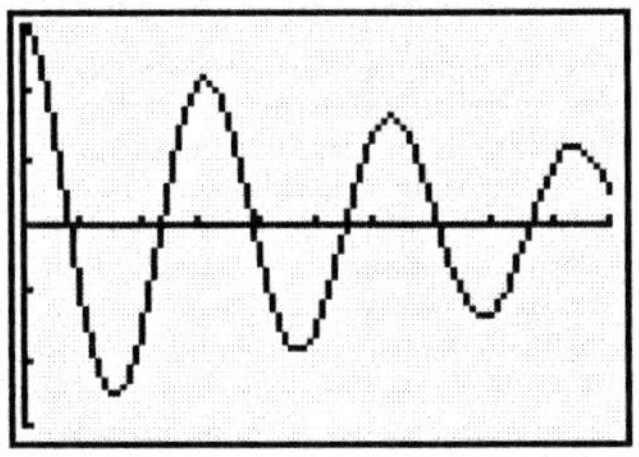

EXAMPLE 7 Finding an Equation for an Object in Simple Harmonic Motion

A ball on a spring is pulled 4 inches below its rest position and then released. The period of the motion is 6 seconds. Write the equation for the ball's simple harmonic motion.

Solution We need to write an equation that describes d, the distance of the ball from its rest position, after t seconds. (The motion is illustrated by the "photo" sequence in Figure 5.88 on page 542.) When the object is released ($t = 0$), the ball's distance from its rest position is 4 inches down. Because it is *down* 4 inches, d is negative: When $t = 0$, $d = -4$. Notice the greatest distance from rest position occurs at $t = 0$. Thus, we will use the equation with the cosine function,

$$d = a \cos \omega t,$$

to model the ball's simple harmonic motion.

Now we determine values for a and ω. Recall that $|a|$ is the maximum displacement. Because the ball initially moves down, $a = -4$.

The value of ω in $d = a \cos \omega t$ can be found using the formula for the period.

$$\text{period} = \frac{2\pi}{\omega} = 6 \qquad \text{We are given that the period of the motion is 6 seconds.}$$

$$2\pi = 6\omega \qquad \text{Multiply both sides by } \omega.$$

$$\omega = \frac{2\pi}{6} = \frac{\pi}{3} \qquad \text{Divide both sides by 6 and solve for } \omega.$$

We see that $a = -4$ and $\omega = \frac{\pi}{3}$. Substitute these values into $d = a \cos \omega t$. The equation for the ball's simple harmonic motion is

$$d = -4 \cos \frac{\pi}{3} t.$$

Check Point 7 A ball on a spring is pulled 6 inches below its rest position and then released. The period for the motion is 4 seconds. Write the equation for the ball's simple harmonic motion.

The period of the harmonic motion in Example 7 was 6 seconds. It takes 6 seconds for the moving object to complete one cycle. Thus, $\frac{1}{6}$ of a cycle is completed every second. We call $\frac{1}{6}$ the *frequency* of the moving object. **Frequency** describes the number of complete cycles per unit time and is the reciprocal of the period.

Frequency of an Object in Simple Harmonic Motion

An object in simple harmonic motion given by

$$d = a\cos\omega t \text{ or } d = a\sin\omega t$$

has **frequency** f given by

$$f = \frac{\omega}{2\pi}, \omega > 0.$$

Equivalently,

$$f = \frac{1}{\text{period}}.$$

Modeling Music

Sounds are caused by vibrating objects that result in variations in pressure in the surrounding air. Areas of high and low pressure moving through the air are modeled by the harmonic motion formulas. When these vibrations reach our eardrums, the eardrums' vibrations send signals to our brains which create the sensation of hearing.

French mathematician John Fourier (1768–1830) proved that all musical sounds—instrumental and vocal—could be modeled by sums involving sine functions. Modeling musical sounds with sinusoidal functions is used by synthesizers to electronically produce sounds unobtainable from ordinary musical instruments.

EXAMPLE 8 Analyzing Simple Harmonic Motion

Figure 5.89 shows a mass on a smooth table attached to a spring. The mass moves in simple harmonic motion described by

$$d = 10\cos\frac{\pi}{6}t$$

with t measured in seconds and d in centimeters. Find (**a.**) the maximum displacement, (**b.**) the frequency, and (**c.**) the time required for one cycle.

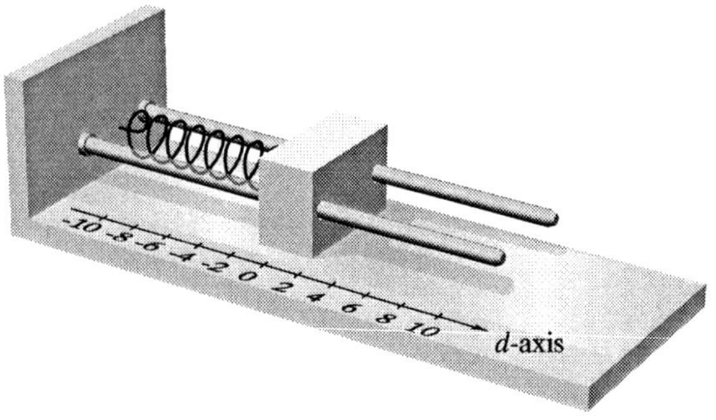

Figure 5.89 A mass attached to a spring, moving in simple harmonic motion

Solution We begin by identifying values for a and ω.

$$d = 10\cos\frac{\pi}{6}t$$

The form of this equation is $d = a\cos\omega t$ with $a = 10$ and $\omega = \frac{\pi}{6}$.

a. The maximum displacement from the rest position is the amplitude. Because $a = 10$, the maximum displacement is 10 centimeters.

b. The frequency, f, is

$$f = \frac{\omega}{2\pi} = \frac{\frac{\pi}{6}}{2\pi} = \frac{\pi}{6} \cdot \frac{1}{2\pi} = \frac{1}{12}.$$

The frequency is $\frac{1}{12}$ cycle (or oscillation) per second.

c. The time required for one cycle is the period.

$$\text{period} = \frac{2\pi}{\omega} = \frac{2\pi}{\frac{\pi}{6}} = 2\pi \cdot \frac{6}{\pi} = 12$$

The time required for one cycle is 12 seconds.

Check Point 8 An object moves in simple harmonic motion described by $d = 12 \cos \frac{\pi}{4} t$, where t is measured in seconds and d in centimeters. Find (**a.**) the maximum displacement, (**b.**) the frequency, and (**c.**) the time required for one cycle.

Resisting Damage of Simple Harmonic Motion

Simple harmonic motion from an earthquake caused this highway in Oakland, California, to collapse. By studying the harmonic motion of the soil under the highway, engineers learn to build structures that can resist damage.

EXERCISE SET 5.8

Practice Exercises

In Exercises 1–12, solve the right triangle shown in the figure. Round lengths to two decimal places and express angles to the nearest tenth of a degree.

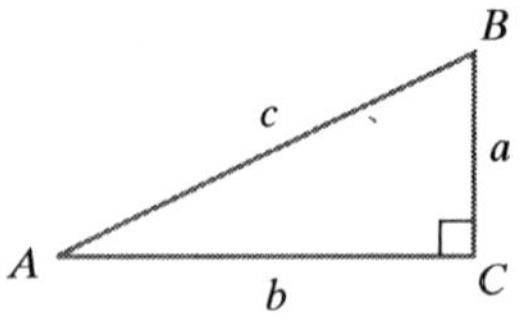

1. $A = 23.5°, b = 10$
2. $A = 41.5°, b = 20$
3. $A = 52.6°, c = 54$
4. $A = 54.8°, c = 80$
5. $B = 16.8°, b = 30.5$
6. $B = 23.8°, b = 40.5$
7. $a = 30.4, c = 50.2$
8. $a = 11.2, c = 65.8$
9. $a = 10.8, b = 24.7$
10. $a = 15.3, b = 17.6$
11. $b = 2, c = 7$
12. $b = 4, c = 9$

Use the figure shown to solve Exercises 13–16.

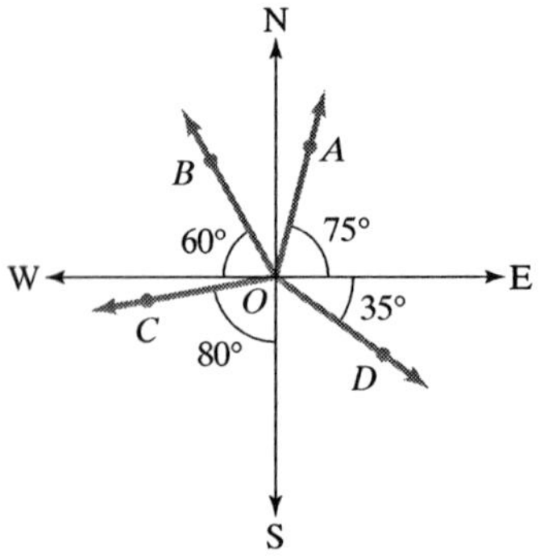

13. Find the bearing from O to A.
14. Find the bearing from O to B.
15. Find the bearing from O to C.
16. Find the bearing from O to D.

In Exercises 17–20, an object is attached to a coiled spring. The object is pulled down (negative direction from the rest position) and then released. Write an equation for the distance of the object from its rest position after t seconds.

	Distance from rest position at $t = 0$	Amplitude	Period
17.	6 centimeters	6 centimeters	4 seconds
18.	8 inches	8 inches	2 seconds
19.	0	3 inches	1.5 seconds
20.	0	5 centimeters	2.5 seconds

In Exercises 21–28, an object moves in simple harmonic motion described by the given equation, where t is measured in seconds and d in inches. In each exercise, find:

a. *the maximum displacement*

b. *the frequency*

c. *the time required for one cycle.*

21. $d = 5\cos\frac{\pi}{2}t$
22. $d = 10\cos 2\pi t$
23. $d = -6\cos 2\pi t$
24. $d = -8\cos\frac{\pi}{2}t$
25. $d = \frac{1}{2}\sin 2t$
26. $d = \frac{1}{3}\sin 2t$
27. $d = -5\sin\frac{2\pi}{3}t$
28. $d = -4\sin\frac{3\pi}{2}t$

Application Exercises

29. The tallest television transmitting tower in the world is in North Dakota. From a point on level ground 5280 feet (one mile) from the base of the tower, the angle of elevation is 21.3°. Approximate the height of the tower to the nearest foot.

30. From a point on level ground 30 yards from the base of a building, the angle of elevation is 38.7°. Approximate the height of the building to the nearest foot.

31. The Statue of Liberty is approximately 305 feet tall. If the angle of elevation of a ship to the top of the statue is 23.7°, how far, to the nearest foot, is the ship from the statue's base?

32. A 200-foot cliff drops vertically into the ocean. If the angle of elevation of a ship to the top of the cliff is 22.3°, how far off shore, to the nearest foot, is the ship?

33. A helicopter hovers 1000 feet above a small island. The figure shows that the angle of depression from the

helicopter to point P is 36°. How far off the coast, to the nearest foot, is the island?

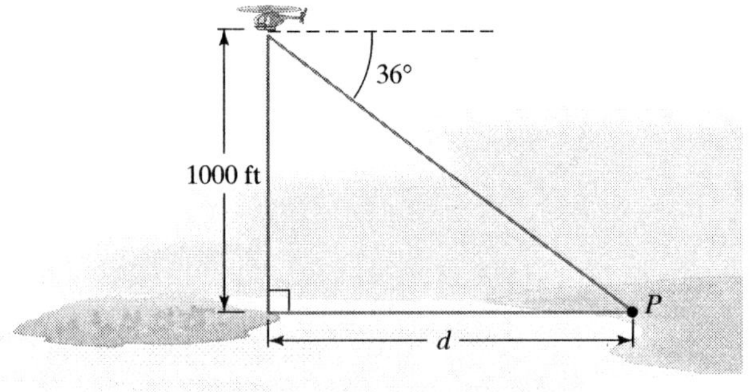

34. A police helicopter is flying at 800 feet. A stolen car is sighted at an angle of depression of 72°. Find the distance of the stolen car, to the nearest foot, from a point directly below the helicopter.

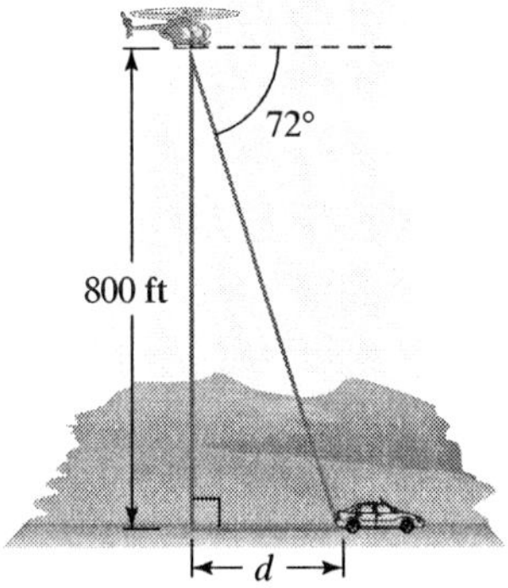

35. A wheelchair ramp is to be built beside the steps to the campus library. Find the angle of elevation of the 23-foot ramp, to the nearest tenth of a degree, if its final height is 6 feet.

36. A building that is 250 feet high casts a shadow 40 feet long. Find the angle of elevation, to the nearest tenth of a degree, of the sun at this time.

37. A hot-air balloon is rising vertically. The angle of elevation from a point on level ground 125 feet from the balloon to a point directly under the passenger compartment changes from 19.2° to 31.7°. How far, to the nearest tenth of a foot, does the balloon rise during this period?

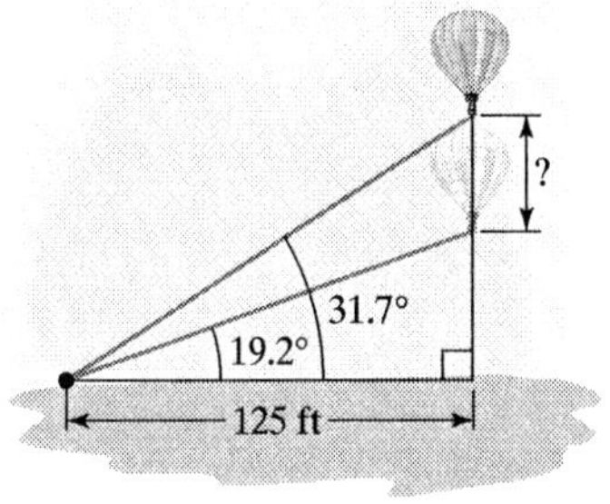

38. A flagpole is situated on top of a building. The angle of elevation from a point on level ground 330 feet from the building to the top of the flagpole is 63°. The angle of elevation from the same point to the bottom of the flagpole is 53°. Find the height of the flagpole to the nearest tenth of a foot.

39. A boat leaves the entrance to a harbor and travels 150 miles on a bearing of N 53° E. How many miles north and how many miles east from the harbor has the boat traveled?

40. A boat leaves the entrance to a harbor and travels 40 miles on a bearing of S 64° E. How many miles south and how many miles east from the harbor has the boat traveled?

41. A forest ranger sights a fire directly to the south. A second ranger, 7 miles east of the first ranger, also sights the fire. The bearing from the second ranger to the fire is S 28° W. How far, to the nearest tenth of a mile, is the first ranger from the fire?

42. A ship sights a lighthouse directly to the south. A second ship, 9 miles east of the first ship, also sights the lighthouse. The bearing from the second ship to the lighthouse is S 34° W. How far, to the nearest tenth of a mile, is the first ship from the lighthouse?

43. You leave your house and run 2 miles due west followed by 1.5 miles due north. At that time, what is your bearing from your house?

44. A ship is 9 miles east and 6 miles south of a harbor. What bearing should be taken to sail directly to the harbor?

45. A jet leaves a runway whose bearing is N 35° E from the control tower. After flying 5 miles, the jet turns 90° and flies on a bearing of S 55° E for 7 miles. At that time, what is the bearing of the jet from the control tower?

46. A ship leaves port with a bearing of S 40° W. After traveling 7 miles, the ship turns 90° and travels on a bearing of N 50° W for 11 miles. At that time, what is the bearing of the ship from port?

47. An object in simple harmonic motion has a frequency of $\frac{1}{2}$ oscillation per minute and an amplitude of 6 feet. Write an equation in the form $d = a \sin \omega t$ for the object's simple harmonic motion.

48. An object in simple harmonic motion has a frequency of $\frac{1}{4}$ oscillation per minute and an amplitude of 8 feet. Write an equation in the form $d = a \sin \omega t$ for the object's simple harmonic motion.

49. A piano tuner uses a tuning fork. If middle C has a frequency of 264 vibrations per second, write an equation in the form $d = \sin \omega t$ for the simple harmonic motion.

50. A radio station, 98.1 on the FM dial, has radio waves with a frequency of 98.1 million cycles per second. Write an equation in the form $d = \sin \omega t$ for the simple harmonic motion of the radio waves.

Writing in Mathematics

51. What does it mean to solve a right triangle?

52. Explain how to find one of the acute angles of a right triangle if two sides are known.

53. Describe a situation in which a right triangle and a trigonometric function are used to measure a height or distance that would otherwise be inconvenient or impossible to measure.

54. What is meant by the bearing from point O to point P? Give an example with your description.

55. What is simple harmonic motion? Give an example with your description.

56. Explain the period and the frequency of simple harmonic motion. How are they related?

57. Explain how the photograph of the damaged highway on page 545 illustrates simple harmonic motion.

Technology Exercises

The functions in Exercises 58–59 model motion in which the amplitude decreases with time due to friction or other resistive forces. Graph each function in the given viewing rectangle. How many complete oscillations occur on the time interval $0 \le x \le 10$?

58. $y = 4e^{-0.1} \cos 2x \quad [0, 10, 1]$ by $[-4, 4, 1]$

59. $y = -6e^{-0.09x} \cos 2\pi x \quad [0, 10, 1]$ by $[-6, 6, 1]$

Critical Thinking Exercises

60. The figure at the top of the next column shows a satellite circling 112 miles above Earth. When the satellite is directly above point B, angle A measures 76.6°. Find Earth's radius to the nearest mile.

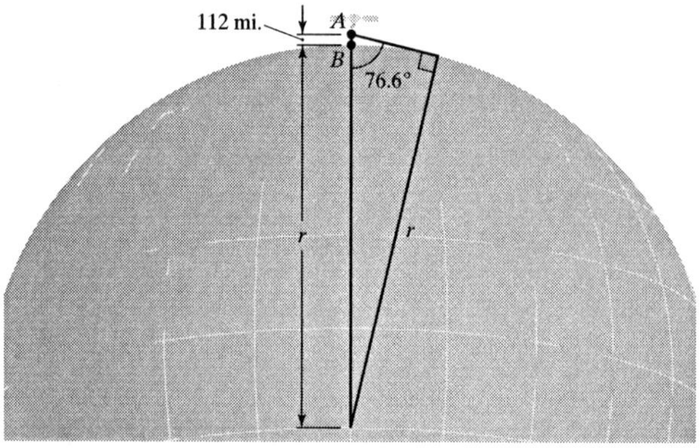

61. The figure shows that the angle of elevation to the top of the building changes from 20° to 40° as an observer advances 75 feet toward the building. Find the height of the building to the nearest foot.

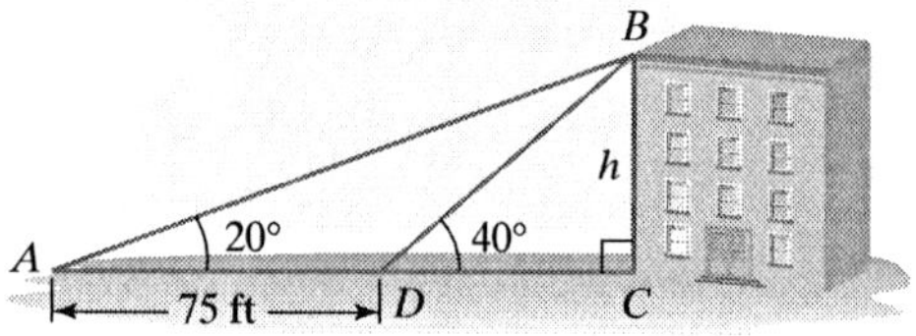

Group Exercise

62. Music and mathematics have been linked over the centuries. Group members should research and present a seminar to the class on music and mathematics. Be sure to include the role of trigonometric functions in the music-mathematics link.

Chapter Summary, Review, and Test

Summary

DEFINITIONS AND CONCEPTS	EXAMPLES
5.1 Angles and Their Measure	
a. An angle consists of two rays with a common endpoint, the vertex.	
b. An angle is in standard position if its vertex is at the origin and its initial side lies along the positive x-axis. Figure 5.3 on page 440 shows positive and negative angles in standard position.	Ex. 1, p. 441
c. A quadrantal angle is one with its terminal side on the x-axis or the y-axis.	
d. Angles can be measured in degrees. $1°$ is $\frac{1}{360}$ of a complete rotation.	
e. Acute angles measure less than 90°, right angles 90°, obtuse angles more than 90° but less than 180°, and straight angles 180°.	

DEFINITIONS AND CONCEPTS | **EXAMPLES**

f. Two angles with the same initial and terminal sides are called coterminal angles. Ex. 2, p. 442

g. Two angles are complements if their sum is 90° and supplements if their sum is 180°. Only positive angles are used. Ex. 3, p. 443

h. Angles can be measured in radians. One radian is the measure of the central angle when the intercepted arc and radius have the same length. In general, the radian measure of a central angle is the length of the intercepted arc divided by the circle's radius: $\theta = \frac{s}{r}$. Ex. 4, p. 445

i. To convert from degrees to radians, multiply degrees by $\frac{\pi \text{ radians}}{180°}$. To convert from radians to degrees, multiply radians by $\frac{180°}{\pi \text{ radians}}$. Ex. 5, p. 446; Ex. 6, p. 446

j. The arc length formula, $s = r\theta$, is described in the box on page 447. Ex. 7, p. 448

k. The definitions of linear speed, $v = \frac{s}{t}$, and angular speed, $\omega = \frac{\theta}{t}$, are given in the box on page 448.

l. Linear speed is expressed in terms of angular speed by $v = r\omega$, where v is the linear speed of a point a distance r from the center of rotation and ω is the angular speed in radians per unit of time. Ex. 8, p. 449

5.2 Right Triangle Trigonometry

a. The right triangle definitions of the six trigonometric functions are given in the box on page 454. Ex. 1, p. 455

b. Function values for 30°, 45°, and 60° can be obtained using these special triangles. Ex. 2, p. 456; Ex. 3, p. 457

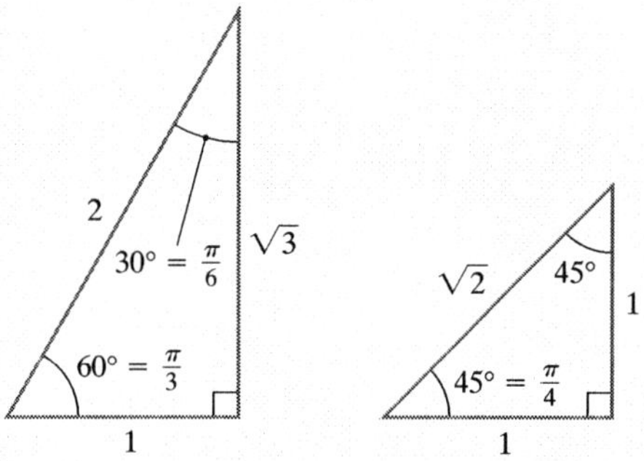

c. Fundamental Identities

1. Reciprocal Identities Ex. 4, p. 459; Ex. 5, p. 460

$$\sin\theta = \frac{1}{\csc\theta} \quad \cos\theta = \frac{1}{\sec\theta} \quad \tan\theta = \frac{1}{\cot\theta}$$

$$\csc\theta = \frac{1}{\sin\theta} \quad \sec\theta = \frac{1}{\cos\theta} \quad \cot\theta = \frac{1}{\tan\theta}$$

2. Quotient Identities

$$\tan\theta = \frac{\sin\theta}{\cos\theta} \quad \cot\theta = \frac{\cos\theta}{\sin\theta}$$

3. Pythagorean Identities

$$\sin^2\theta + \cos^2 = 1$$
$$1 + \tan^2\theta = \sec^2\theta$$
$$1 + \cot^2\theta = \csc^2\theta$$

d. The value of a trigonometric function of θ is equal to the cofunction of the complement of θ. Cofunction identities are listed in the box on page 461. Ex. 6, p. 461

DEFINITIONS AND CONCEPTS	EXAMPLES
5.3 Trigonometric Functions of Any Angle	
a. Definitions of the trigonometric functions of any angle are given in the box on page 470.	Ex. 1, p. 471; Ex. 2, p. 471
b. Signs of the trigonometric functions: All functions are positive in quadrant I. If θ lies in quadrant II, $\sin\theta$ and $\csc\theta$ are positive. If θ lies in quadrant III, $\tan\theta$ and $\cot\theta$ are positive. If θ lies in quadrant IV, $\cos\theta$ and $\sec\theta$ are positive.	Ex. 3, p. 474; Ex. 4, p. 474
c. If θ is a nonacute angle in standard position that lies in a quadrant, its reference angle is the positive acute angle θ' formed by the terminal side of θ and the x-axis. The reference angle for a given angle can be found by making a sketch that shows the angle in standard position. Figure 5.30 on page 475 shows reference angles for θ in quadrants II, III, and IV.	Ex. 5, p. 475
d. The values of the trigonometric functions of a given angle are the same as the values of the functions of the reference angle, except possibly for the sign. A procedure for using reference angles to evaluate trigonometric functions is given in the box on page 476.	Ex. 6, p. 476
5.4 Trigonometric Functions of Real Numbers; Periodic Functions	
a. Definitions of the trigonometric functions in terms of a unit circle are given in the box on page 481.	Ex. 1, p. 481
b. The cosine and secant functions are even: $\cos(-t) = \cos t, \quad \sec(-t) = \sec t.$ The other trigonometric functions are odd: $\sin(-t) = -\sin t, \quad \csc(-t) = -\csc t,$ $\tan(-t) = -\tan t, \quad \cot(-t) = -\cot t.$	Ex. 2, p. 483
c. If $f(t + p) = f(t)$, function f is periodic. The smallest p for which f is periodic is the period of f. The tangent and cotangent functions have period π. The other four trigonometric functions have period 2π.	Ex. 3, p. 484
5.5 and 5.6 Graphs of the Trigonometric Functions	
a. Graphs of the six trigonometric functions, with a description of the domain, range, and period of each function, are given in Table 5.4 on page 517.	
b. The graph of $y = A\sin(Bx - C)$ can be obtained using amplitude $= \lvert A\rvert$, period $= \frac{2\pi}{B}$ and phase shift $= \frac{C}{B}$. See the illustration in the box on page 494. **c.** The graph of $y = A\cos(Bx - C)$ can be obtained using amplitude $= \lvert A\rvert$, period $= \frac{2\pi}{B}$, and phase shift $= \frac{C}{B}$. See the illustration in the box on page 500.	Ex. 1, p. 489; Ex. 2, p. 491; Ex. 3, p. 492; Ex. 4, p. 495 Ex. 5, p. 498; Ex. 6, p. 500
d. The constant D in $y = A\sin(Bx - C) + D$ and $y = A\cos(Bx - C) + D$ causes vertical shifts in the graphs in the preceding items (b) and (c). If $D > 0$, the shift is D units upward and if $D < 0$, the shift is D units downward. Oscillation is about $y = D$.	Ex. 7, p. 502
e. The graph of $y = A\tan(Bx - C)$ is obtained using the procedure in the box on page 511. Consecutive asymptotes $\left(\text{solve } Bx - C = -\frac{\pi}{2} \text{ and } Bx - C = \frac{\pi}{2}\right)$ and an x-intercept midway between them play a key role in the graphing process.	Ex. 1, p. 511; Ex. 2, p. 512
f. The graph of $y = A\cot(Bx - C)$ is obtained using the procedure in the box on page 513. Consecutive asymptotes (solve $Bx - C = 0$ and $Bx - C = \pi$) and an x-intercept midway between them play a key role in the graphing process.	Ex. 3, p. 513
g. To graph a cosecant curve, begin by graphing the reciprocal sine curve. Draw vertical asymptotes through x-intercepts, using asymptotes as guides to sketch the graph. To graph a secant curve, first graph the reciprocal cosine curve and use the same procedure.	Ex. 4, p. 515; Ex. 5, p. 516

DEFINITIONS AND CONCEPTS	EXAMPLES
5.7 Inverse Trigonometric Functions	
a. On the restricted domain $-\frac{\pi}{2} \le x \le \frac{\pi}{2}$, $y = \sin x$ has an inverse function, defined in the box on page 522. Think of $\sin^{-1} x$ as the angle in $\left[-\frac{\pi}{2}, \frac{\pi}{2}\right]$ whose sine is x.	Ex. 1, p. 524; Ex. 2, p. 524
b. On the restricted domain $0 \le x \le \pi$, $y = \cos x$ has an inverse function, defined in the box on page 525. Think of $\cos^{-1} x$ as the angle in $[0, \pi]$ whose cosine is x.	Ex. 3, p. 526
c. On the restricted domain $-\frac{\pi}{2} < x < \frac{\pi}{2}$, $y = \tan x$ has an inverse function, defined in the box on page 527. Think of $\tan^{-1} x$ as the angle in $\left(-\frac{\pi}{2}, \frac{\pi}{2}\right)$ whose tangent is x.	Ex. 4, p. 528
d. Graphs of the three basic inverse trigonometric functions, with a description of the domain and range of each function, are given in Table 5.8 on page 529.	
e. Inverse properties are given in the box on page 530. Points on terminal sides of angles in standard position and right triangles are used to find exact values of the composition of a function and a different inverse function.	Ex. 6, p. 531; Ex. 7, p. 532; Ex. 8, p. 532; Ex. 9, p. 533
5.8 Applications of Trigonometric Functions	
a. Solving a right triangle means finding the missing lengths of its sides and the measurements of its angles. The Pythagorean Theorem, two acute angles whose sum is 90°, and appropriate trigonometric functions are used in this process.	Ex. 1, p. 537; Ex. 2, p. 538; Ex. 3, p. 539; Ex. 4, p. 539
b. The bearing from point O to point P is the acute angle between ray OP and a north-south line.	Ex. 5, p. 540; Ex. 6, p. 541
c. Simple harmonic motion, described in the box on page 543, is modeled by $d = a\cos\omega t$ or $d = a\sin\omega t$, with amplitude $= \lvert a \rvert$, period $= \frac{2\pi}{\omega}$ and frequency $= \frac{\omega}{2\pi} = \frac{1}{\text{period}}$.	Ex. 7, p. 543; Ex. 8, p. 544

Review Exercises

5.1

In Exercises 1–4, draw each angle in standard position.

1. 190° **2.** −135°

3. $\frac{5\pi}{6}$ **4.** $-\frac{2\pi}{3}$

In Exercises 5–6, find a positive angle less than 360° that is coterminal with the given angle.

5. 400° **6.** −85°

In Exercises 7–8, if possible, find the complement and the supplement of the given angle.

7. 73° **8.** $\frac{2\pi}{3}$

9. Find the radian measure of the central angle of a circle of radius 6 centimeters that intercepts an arc of length 27 centimeters.

In Exercises 10–12, convert each angle in degrees to radians. Express your answer as a multiple of π.

10. 15° **11.** 120° **12.** 315°

In Exercises 13–15, convert each angle in radians to degrees.

13. $\frac{5\pi}{3}$ **14.** $\frac{7\pi}{5}$ **15.** $-\frac{5\pi}{6}$

16. Find the length of the arc on a circle of radius 10 feet intercepted by a 135° central angle. Express arc length in terms of π. Then round your answer to two decimal places.

17. The angular speed of a propeller on a wind generator is 10.3 revolutions per minute. Express this angular speed in radians per minute.

18. The propeller of an airplane has a radius of 3 feet. The propeller is rotating at 2250 revolutions per minute. Find the linear speed, in feet per minute, of the tip of the propeller.

5.2

19. Use the triangle to find each of the six trigonometric functions of θ.

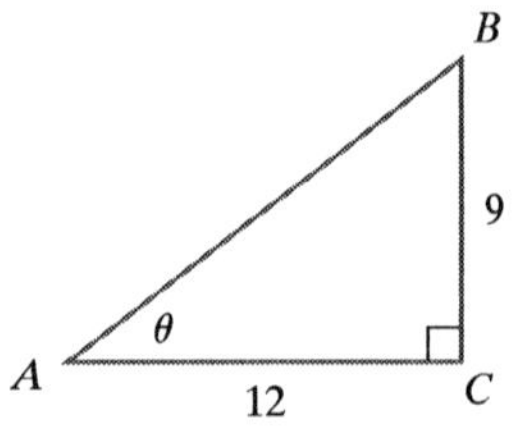

In Exercises 20–23, find the exact value of each expression. Do not use a calculator.

20. $\tan 60°$ **21.** $\cos \frac{\pi}{4}$

22. $\sec \frac{\pi}{6}$ **23.** $\sin^2 \frac{\pi}{5} + \cos^2 \frac{\pi}{5}$

24. If θ is an acute angle and $\sin \theta = \frac{2}{\sqrt{7}}$, use the identity $\sin^2 \theta + \cos^2 \theta = 1$ to find $\cos \theta$.

In Exercises 25–26, find a cofunction with the same value as the given expression.

25. $\sin 70°$ **26.** $\cos \frac{\pi}{2}$

In Exercises 27–29, find the measure of the side of the right triangle whose length is designated by a lowercase letter. Round answers to the nearest whole number.

27.

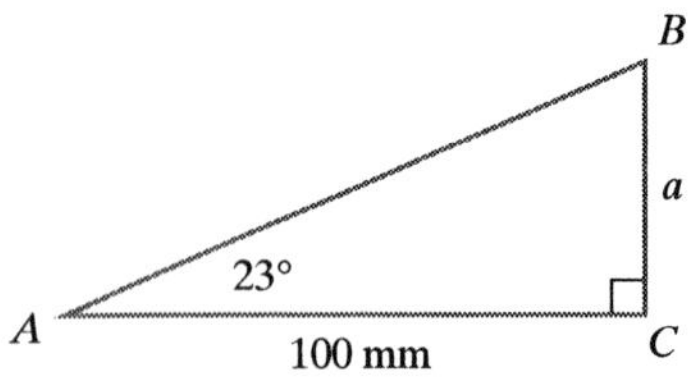

28.

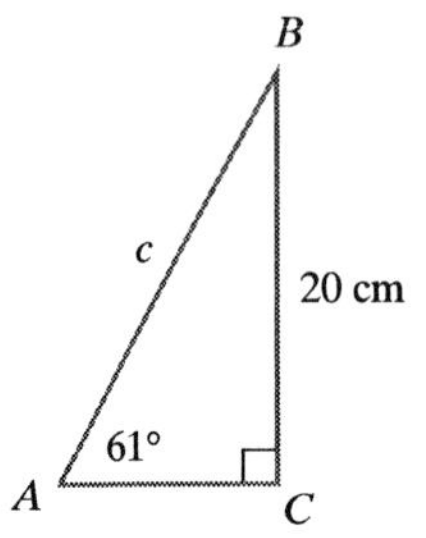

29.

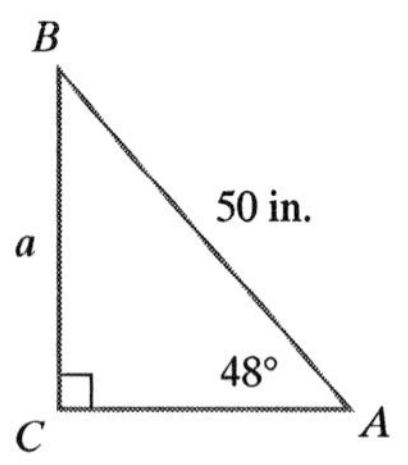

30. A hiker climbs for a half mile up a slope whose inclination is 17°. How many feet of altitude, to the nearest foot, does the hiker gain?

31. To find the distance across a lake, a surveyor took the measurements in the figure shown. What is the distance across the lake? Round to the nearest meter.

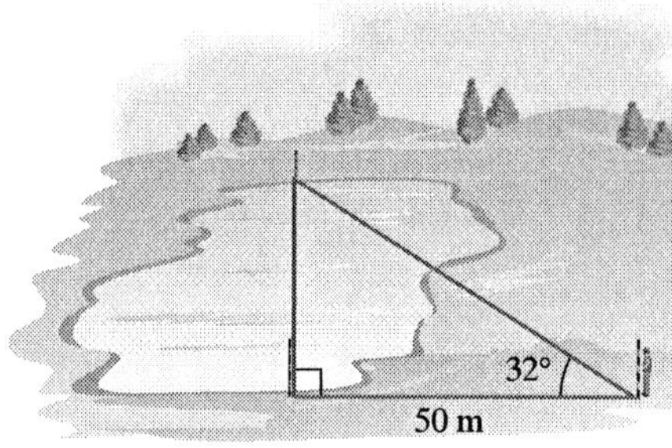

32. When a six-foot pole casts a four-foot shadow, what is the angle of elevation of the sun? Round to the nearest whole degree.

5.3 and 5.4

In Exercises 33–34, a point on the terminal side of angle θ is given. Find the exact value of each of the six trigonometric functions of θ, or state that the function is undefined.

33. $(-1, -5)$ **34.** $(0, -1)$

In Exercises 35–36, let θ be an angle in standard position. Name the quadrant in which θ lies.

35. $\tan \theta > 0$ and $\sec \theta > 0$

36. $\tan \theta > 0$ and $\cos \theta < 0$

In Exercises 37–38, find the exact value of each of the remaining trigonometric functions of θ.

37. $\cos \theta = \frac{2}{5}$, $\sin \theta < 0$ **38.** $\tan \theta = -\frac{1}{3}$, $\sin \theta > 0$

In Exercises 39–41, find the reference angle for each angle.

39. $265°$ **40.** $\frac{5\pi}{8}$ **41.** $-410°$

In Exercises 42–50, find the exact value of each expression. Do not use a calculator.

42. $\sin 240°$ **43.** $\tan 120°$ **44.** $\sec \frac{7\pi}{4}$

45. $\cos \frac{11\pi}{6}$ **46.** $\cot(-210°)$ **47.** $\csc\left(-\frac{2\pi}{3}\right)$

48. $\sin\left(-\frac{\pi}{3}\right)$ **49.** $\sin 495°$ **50.** $\tan \frac{13\pi}{4}$

5.5

In Exercises 51–56, determine the amplitude and period of each function. Then graph one period of the function.

51. $y = 3 \sin 4x$ **52.** $y = -2 \cos 2x$

53. $y = 2 \cos \frac{1}{2} x$ **54.** $y = \frac{1}{2} \sin \frac{\pi}{3} x$

55. $y = -\sin \pi x$ **56.** $y = 3 \cos \frac{x}{3}$

In Exercises 57–61, determine the amplitude, period, and phase shift of each function. Then graph one period of the function.

57. $y = 2\sin(x - \pi)$

58. $y = -3\cos(x + \pi)$

59. $y = \frac{3}{2}\cos\left(2x + \frac{\pi}{4}\right)$

60. $y = \frac{5}{2}\sin\left(2x + \frac{\pi}{2}\right)$

61. $y = -3\sin\left(\frac{\pi}{3}x - 3\pi\right)$

In Exercises 62–63, use a vertical shift to graph one period of the function.

62. $y = \sin 2x + 1$

63. $y = 2\cos\frac{1}{3}x - 2$

64. The equation

$$y = 98.6 + 0.3\sin\left(\frac{\pi}{12}x - \frac{11\pi}{12}\right)$$

models variation in body temperature, y, in °F, x hours after midnight.

- **a.** What is body temperature at midnight?
- **b.** What is the period of the body temperature cycle?
- **c.** When is body temperature highest? What is the body temperature at this time?
- **d.** When is body temperature lowest? What is the body temperature at this time?
- **e.** Graph one period of the body temperature function.

5.6

In Exercises 65–71, graph two full periods of the given tangent or cotangent function.

65. $y = 4\tan 2x$

66. $y = -2\tan\frac{\pi}{4}x$

67. $y = \tan(x + \pi)$

68. $y = -\tan\left(x - \frac{\pi}{4}\right)$

69. $y = 2\cot 3x$

70. $y = -\frac{1}{2}\cot\frac{\pi}{2}x$

71. $y = 2\cot\left(x + \frac{\pi}{2}\right)$

In Exercises 72–75, graph two full periods of the given cosecant or secant function.

72. $y = 3\sec 2\pi x$

73. $y = -2\csc \pi x$

74. $y = 3\sec(x + \pi)$

75. $y = \frac{5}{2}\csc(x - \pi)$

5.7

In Exercises 76–93, find the exact value of each expression. Do not use a calculator.

76. $\sin^{-1} 1$

77. $\cos^{-1} 1$

78. $\tan^{-1} 1$

79. $\sin^{-1}\left(-\frac{\sqrt{3}}{2}\right)$

80. $\cos^{-1}\left(-\frac{1}{2}\right)$

81. $\tan^{-1}\left(-\frac{\sqrt{3}}{3}\right)$

82. $\cos\left(\sin^{-1}\frac{\sqrt{2}}{2}\right)$

83. $\sin(\cos^{-1} 0)$

84. $\tan\left[\sin^{-1}\left(-\frac{1}{2}\right)\right]$

85. $\tan\left[\cos^{-1}\left(-\frac{\sqrt{3}}{2}\right)\right]$

86. $\csc\left(\tan^{-1}\frac{\sqrt{3}}{3}\right)$

87. $\cos\left(\tan^{-1}\frac{3}{4}\right)$

88. $\sin\left(\cos^{-1}\frac{3}{5}\right)$

89. $\tan\left[\sin^{-1}\left(-\frac{3}{5}\right)\right]$

90. $\tan\left[\cos^{-1}\left(-\frac{4}{5}\right)\right]$

91. $\sin^{-1}\left(\sin\frac{\pi}{3}\right)$

92. $\sin^{-1}\left(\sin\frac{2\pi}{3}\right)$

93. $\sin^{-1}\left(\cos\frac{2\pi}{3}\right)$

In Exercises 94–95, use a right triangle to write each expression as an algebraic expression. Assume that x is positive and in the domain of the given inverse trigonometric function.

94. $\cos\left(\tan^{-1}\frac{x}{2}\right)$

95. $\sec\left(\sin^{-1}\frac{1}{x}\right)$

5.8

In Exercises 96–99, solve the right triangle shown in the figure. Round lengths to two decimal places and express angles to the nearest tenth of a degree.

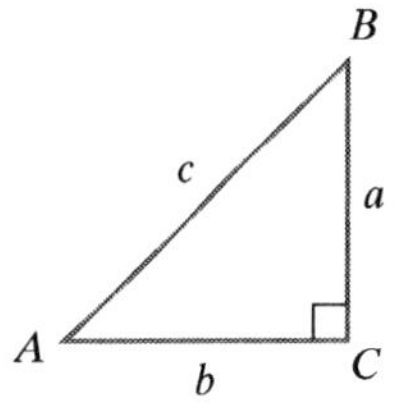

96. $A = 22.3°, c = 10$

97. $B = 37.4°, b = 6$

98. $a = 2, c = 7$

99. $a = 1.4, b = 3.6$

100. From a point on level ground 80 feet from the base of a building, the angle of elevation is 25.6°. Approximate the height of the building to the nearest foot.

101. Two buildings with flat roofs are 60 yards apart. The height of the shorter building is 40 yards. From its roof, the angle of elevation to the edge of the roof of the taller building is 40°. Find the height of the taller building to the nearest yard.

102. You want to measure the height of an antenna on the top of a 125-foot building. From a point in front of the building, you measure the angle of elevation to the top of the building to be 68° and the angle of elevation to the top of the antenna to be 71°. How tall is the antenna, to the nearest tenth of a foot?

In Exercises 103–104, use the figures shown to find the bearing from O to A.

103.

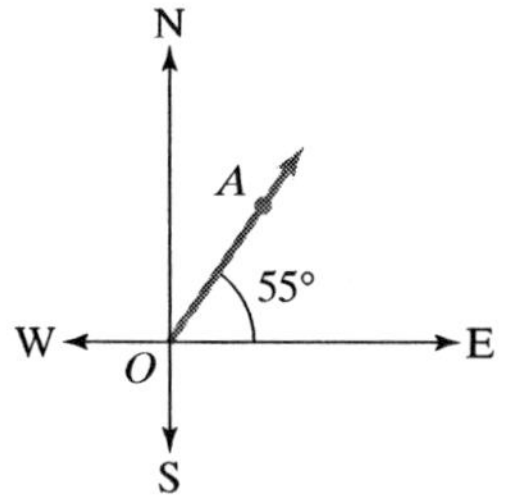

104.

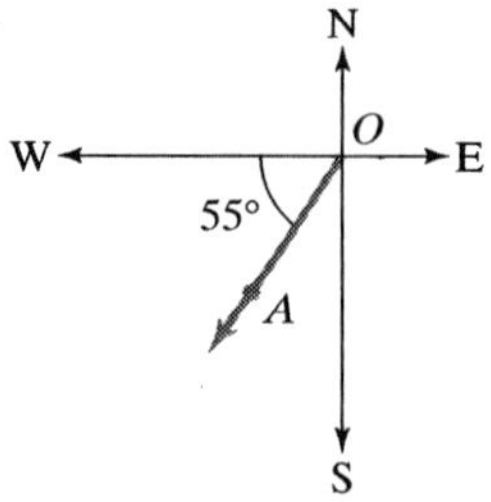

105. A ship is due west of a lighthouse. A second ship is 12 miles south of the first ship. The bearing from the second ship to the lighthouse is N 64° E. How far, to the nearest tenth of a mile, is the first ship from the lighthouse?

106. From city A to city B, a plane flies 850 miles at a bearing of N 58° E. From city B to city C, the plane flies 960 miles at a bearing of S 32° E.

a. Find, to the nearest tenth of a mile, the distance from city A to city C.

b. What is the bearing from city A to city C?

In Exercises 107–108, an object moves in simple harmonic motion described by the given equation, where t is measured in seconds and d in centimeters. In each exercise, find:

a. *the maximum displacement*

b. *the frequency*

c. *the time required for one cycle.*

107. $d = 20 \cos \frac{\pi}{4} t$ **108.** $d = \frac{1}{2} \sin 4t$

In Exercises 109–110, an object is attached to a coiled spring. The object is pulled down (negative direction from the rest position) and then released. Write an equation for the distance of the object from its rest position after t seconds.

	Distance from rest position at $t = 0$	Amplitude	Period
109.	30 inches	30 inches	2 seconds
110.	0	$\frac{1}{4}$ inch	5 seconds

Chapter 5 Test

1. Convert 135° to exact radian measure.

2. Find the supplement of the angle whose radian measure is $\frac{9\pi}{13}$. Express the answer in terms of π.

3. Find the length of the arc on a circle of radius 20 feet intercepted by a 75° central angle. Express arc length in terms of π. Then round your answer to two decimal places.

4. If $(-2, 5)$ is a point on the terminal side of angle θ, find the exact value of each of the six trigonometric functions of θ.

5. Determine the quadrant in which θ lies if $\cos\theta < 0$ and $\cot\theta > 0$.

6. If $\cos\theta = \frac{1}{3}$ and $\tan\theta < 0$, find the exact value of each of the remaining trigonometric functions of θ.

In Exercises 7–9, find the exact value of each expression. Do not use a calculator.

7. $\tan \frac{\pi}{6} \cos \frac{\pi}{3} - \cos \frac{\pi}{2}$ **8.** $\tan 300°$

9. $\sin \frac{7\pi}{4}$

In Exercises 10–13, graph one period of each function.

10. $y = 3 \sin 2x$ **11.** $y = -2 \cos\left(x - \frac{\pi}{2}\right)$

12. $y = 2 \tan \frac{x}{2}$ **13.** $y = -\frac{1}{2} \csc \pi x$

14. Find the exact value of $\tan\left[\cos^{-1}\left(-\frac{1}{2}\right)\right]$.

15. Solve the right triangle in the figure shown. Round lengths to one decimal place.

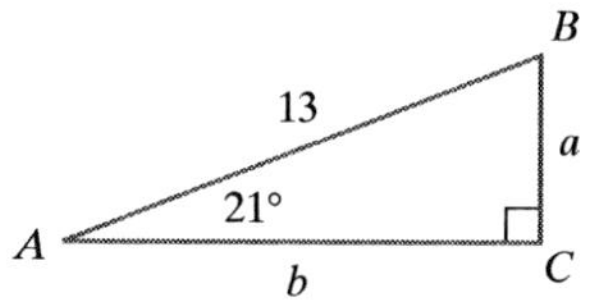

16. The angle of elevation of a building from a point on the ground 30 yards from its base is 37°. Find the height of the building to the nearest yard.

17. A 73-foot rope from the top of a circus tent pole is anchored to the flat ground 43 feet from the bottom of the pole. Find the angle, to the nearest tenth of a degree, that the rope makes with the pole.

18. Use the figure to find the bearing from O to P.

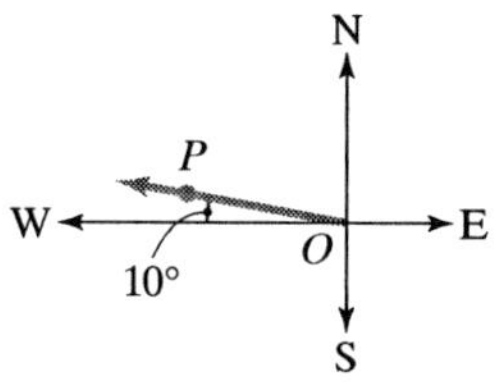

19. An object moves in simple harmonic motion described by $d = -6 \cos \pi t$, where t is measured in seconds and d in inches. Find (**a.**) the maximum displacement, (**b.**) the frequency, and (**c.**) the time required for one oscillation.

20. Why are trigonometric functions ideally suited to model phenomena that repeat in cycles?

Cumulative Review Exercises (Chapters 1–5)

Solve each equation or inequality in Exercises 1–6.

1. $x^2 = 18 + 3x$ **2.** $x^3 + 5x^2 - 4x - 20 = 0$

3. $\log_2 x + \log_2(x - 2) = 3$ **4.** $\sqrt{x - 3} + 5 = x$

5. $x^3 - 4x^2 + x + 6 = 0$ **6.** $|2x - 5| \leq 11$

7. If $f(x) = \sqrt{x - 6}$, find $f^{-1}(x)$.

8. Divide $20x^3 - 6x^2 - 9x + 10$ by $5x + 2$.

9. Write as a single logarithm and evaluate: $\log 25 + \log 40$.

10. Convert $\frac{14\pi}{9}$ radians to degrees.

11. Find the maximum number of positive and negative real roots of the equation $3x^4 - 2x^3 + 5x^2 + x - 9 = 0$.

In Exercises 12–16, graph each equation.

12. $f(x) = \frac{x}{x^2 - 1}$ **13.** $(x - 2)^2 + y^2 = 1$

14. $y = (x - 1)(x + 2)^2$

15. $y = \sin\left(2x + \frac{\pi}{2}\right)$, from 0 to 2π

16. $y = 2 \tan 3x$; graph two complete cycles.

17. You invest in a new play. The cost includes an overhead of \$30,000, plus production costs of \$2500 per performance. A sold-out performance brings you \$3125. How many sold-out performances must be played in order for you to break even?

18. Use the exponential growth model $A = A_0 e^{kt}$ to solve this exercise. Data from the Federal Communication Commission show that the use of toll-free 800 numbers has grown exponentially. In 1991 there were 10.2 billion such calls and by 1998, there were 86.7 billion.

a. Find the exponential function that models the data.

b. By which year will the number of toll-free 800 numbers reach 200 billion?

19. The rate of heat lost through insulation varies inversely as the thickness of the insulation. The rate of heat lost through a 3.5-inch thickness of insulation is 2200 Btu per hour. What is the rate of heat lost through a 5-inch thickness of the same insulation?

20. A tower is 200 feet tall. To the nearest degree, find the angle of elevation from a point 50 feet from the base of the tower to the top of the tower.

Analytic Trigonometry

Chapter 6

This chapter emphasizes the algebraic aspects of trigonometry. We derive important categories of identities involving trigonometric functions. These identities are used to simplify and analyze expressions that model phenomena as diverse as the distance achieved when throwing an object and musical sounds on a touch-tone phone. For example, we can find out critical information about an athlete's performance by using an identity to analyze an expression involving throwing distance. You will learn how to use trigonometric identities to better understand your periodic world.

You enjoy watching your friend participate in the shot put at college track and field events. After a few full turns in a circle, she throws ("puts") an 8-pound, 13-ounce shot from the shoulder. The range of her throwing distance continues to improve. Knowing that you are studying trigonometry, she asks if there is some way that a trigonometric expression might help achieve her best possible distance in the event.

SECTION 6.1 Verifying Trigonometric Identities

Objective

1. Use the fundamental trigonometric identities to verify identities.

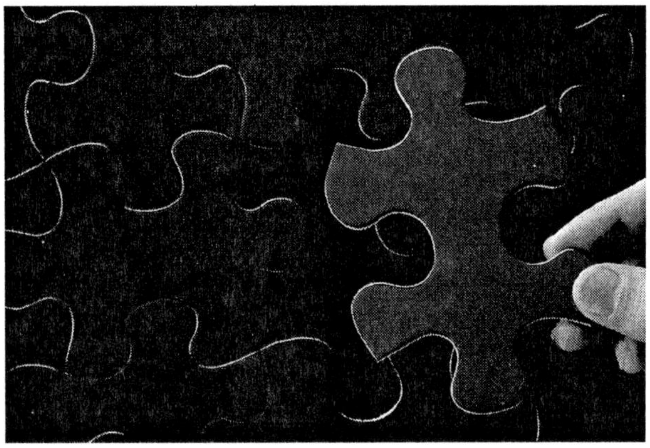

Do you enjoy solving puzzles? The process is a natural way to develop problem-solving skills that are important in every area of our lives. Engaging in problem solving for sheer pleasure releases chemicals in the brain that enhance our feeling of well-being. Perhaps this is why puzzles have fascinated people for over 12,000 years.

Thousands of relationships exist among the six trigonometric functions. Verifying these relationships is like solving a puzzle. Why? There are no rigid rules for the process. Thus, proving a trigonometric relationship requires you to be creative in your approach to problem solving. By learning to establish these relationships, you will become a better, more confident problem solver. Furthermore, you may enjoy the feeling of satisfaction that accompanies solving each "puzzle."

The Fundamental Identities

In Chapter 5, we used right triangles to establish relationships among the trigonometric functions. Although we limited domains to acute angles, the fundamental identities listed in the following box are true for all values of x for which the expressions are defined.

Fundamental Trigonometric Identities

Reciprocal Identities

$$\sin x = \frac{1}{\csc x} \quad \cos x = \frac{1}{\sec x} \quad \tan x = \frac{1}{\cot x}$$

$$\csc x = \frac{1}{\sin x} \quad \sec x = \frac{1}{\cos x} \quad \cot x = \frac{1}{\tan x}$$

Quotient Identities

$$\tan x = \frac{\sin x}{\cos x} \quad \cot x = \frac{\cos x}{\sin x}$$

Pythagorean Identities

$$\sin^2 x + \cos^2 x = 1 \quad 1 + \tan^2 x = \sec^2 x \quad 1 + \cot^2 x = \csc^2 x$$

Even-Odd Identities

$$\sin(-x) = -\sin x \quad \cos(-x) = \cos x \quad \tan(-x) = -\tan x$$

$$\csc(-x) = -\csc x \quad \sec(-x) = \sec x \quad \cot(-x) = -\cot x$$

Study Tip

Memorize the identities in the box. You may need to use variations of these fundamental identities. For example, instead of

$$\sin^2 x + \cos^2 x = 1,$$

you might want to use

$$\sin^2 x = 1 - \cos^2 x$$

or

$$\cos^2 x = 1 - \sin^2 x.$$

Therefore, it is important to know each relationship well so that mental algebraic manipulation is possible.

1 Use the fundamental trigonometric identities to verify identities.

Using Fundamental Identities to Verify Other Identities

The fundamental trigonometric identities are used to establish other relationships among trigonometric functions. To **verify an identity**, we show that one side of the identity can be simplified so that it is identical to the other side. Each side of the equation is manipulated independently of the other side of the equation. Start with the side containing the more complicated expression. If you substitute one or more fundamental identities on the more complicated side, you will often be able to rewrite it in a form identical to that of the other side.

No one method can be used to verify every identity. Some identities can be verified by rewriting the more complicated side so that it contains only sines and cosines.

EXAMPLE 1 Changing to Sines and Cosines to Verify an Identity

Verify the identity: $\sec x \cot x = \csc x$.

Technology

You can use a graphing utility to provide evidence of an identity. For example, consider the equation in Example 1,

$$\sec x \cot x = \csc x.$$

Graph

$$y = \sec x \cot x$$

and

$$y = \csc x$$

in the same viewing rectangle. The screen shown indicates that the graphs appear to be the same. Example 1 shows the equivalence algebraically.

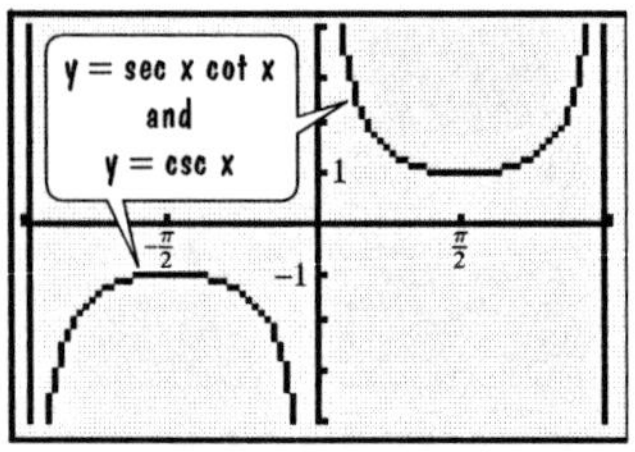

Solution The left side of the equation contains the more complicated expression. Thus, we work with the left side. Let us express this side of the identity in terms of sines and cosines. Perhaps this strategy will enable us to transform the left side into $\csc x$, the expression on the right.

$$\sec x \cot x = \frac{1}{\cos x} \cdot \frac{\cos x}{\sin x}$$ Apply a reciprocal identity: $\sec x = \frac{1}{\cos x}$ and a quotient identity: $\cot x = \frac{\cos x}{\sin x}$.

$$= \frac{1}{\cancel{\cos x}_1} \cdot \frac{\cancel{\cos x}^1}{\sin x}$$ Divide both the numerator and the denominator by $\cos x$, the common factor.

$$= \frac{1}{\sin x}$$ Multiply the remaining factors in the numerator and denominator.

$$= \csc x$$ Apply a reciprocal identity: $\csc x = \frac{1}{\sin x}$.

By working with the left side and simplifying it so that it is identical to the right side, we have verified the given identity.

Check Point 1 Verify the identity: $\csc x \tan x = \sec x$.

In verifying an identity, stay focused on your goal. When manipulating one side of the equation, continue to look at the other side to keep the desired form of the result in mind.

Study Tip

Verifying that an equation is an identity is different from solving an equation. You do not verify an identity by adding, subtracting, multiplying, or dividing each side by the same expression. If you do this, you have already assumed that the given statement is true. You do not know that it is true until after you have verified it.

EXAMPLE 2 Changing to Sines and Cosines to Verify an Identity

Verify the identity: $\sin x \tan x + \cos x = \sec x$.

Solution The left side is more complicated, so we start with it. Let us express this side of the identity so that it contains only sines and cosines. Thus, we apply a quotient identity and replace $\tan x$ by $\frac{\sin x}{\cos x}$. Perhaps this strategy will enable us to transform the left side into $\sec x$, the expression on the right.

$$\sin x \tan x + \cos x = \sin x\left(\frac{\sin x}{\cos x}\right) + \cos x$$
Apply a quotient identity: $\tan x = \frac{\sin x}{\cos x}$.

$$= \frac{\sin^2 x}{\cos x} + \cos x.$$
Multiply.

$$= \frac{\sin^2 x}{\cos x} + \cos x \cdot \frac{\cos x}{\cos x}$$
The least common denominator is $\cos x$. Write the second expression with a denominator of $\cos x$.

$$= \frac{\sin^2 x}{\cos x} + \frac{\cos^2 x}{\cos x}$$
Multiply.

$$= \frac{\sin^2 x + \cos^2 x}{\cos x}$$
Add numerators, putting this sum over the least common denominator.

$$= \frac{1}{\cos x}$$
Apply a Pythagorean identity: $\sin^2 x + \cos^2 x = 1$.

$$= \sec x$$
Apply a reciprocal identity: $\sec x = \frac{1}{\cos x}$.

By working with the left side and arriving at the right side, the identity is verified.

Check Point 2 Verify the identity: $\cos x \cot x + \sin x = \csc x$.

Some identities are verified using factoring to simplify a trigonometric expression.

EXAMPLE 3 Using Factoring to Verify an Identity

Verify the identity: $\cos x - \cos x \sin^2 x = \cos^3 x$.

Solution We start with the more complicated side, the left side. Factor out the greatest common factor, $\cos x$, from each of the two terms.

$$\cos x - \cos x \sin^2 x = \cos x(1 - \sin^2 x)$$
Factor $\cos x$ from the two terms.

$$= \cos x \cdot \cos^2 x$$
Use a variation of $\sin^2 x + \cos^2 x = 1$. Solving for $\cos^2 x$, we obtain $\cos^2 x = 1 - \sin^2 x$.

$$= \cos^3 x$$
Multiply.

We worked with the left side and arrived at the right side. Thus, the identity is verified.

Check Point 3 Verify the identity: $\sin x - \sin x \cos^2 x = \sin^3 x$.

How do we verify identities in which sums or differences of fractions with trigonometric functions appear on one side? Use the least common denominator and combine the fractions. This technique is especially useful when the other side of the identity contains only one term.

EXAMPLE 4 Combining Fractional Expressions to Verify an Identity

Verify the identity: $\dfrac{\cos x}{1 + \sin x} + \dfrac{1 + \sin x}{\cos x} = 2 \sec x$.

Solution We start with the more complicated side, the left side. The least common denominator of the fractions is $(1 + \sin x)(\cos x)$. We express each fraction in terms of this least common denominator by multiplying the numerator and denominator by the extra factor needed to form $(1 + \sin x)(\cos x)$.

Study Tip

Some students have difficulty verifying identities due to problems working with fractions. If this applies to you, review the material on rational expressions in Chapter P, Section P.6.

$$\frac{\cos x}{1 + \sin x} + \frac{1 + \sin x}{\cos x}$$

The least common denominator is $(1 + \sin x)(\cos x)$.

$$= \frac{\cos x(\cos x)}{(1 + \sin x)(\cos x)} + \frac{(1 + \sin x)(1 + \sin x)}{(1 + \sin x)(\cos x)}$$

Rewrite each fraction with the least common denominator.

$$= \frac{\cos^2 x}{(1 + \sin x)(\cos x)} + \frac{1 + 2\sin x + \sin^2 x}{(1 + \sin x)(\cos x)}$$

Use the FOIL method to multiply $(1 + \sin x)(1 + \sin x)$.

$$= \frac{\cos^2 x + 1 + 2\sin x + \sin^2 x}{(1 + \sin x)(\cos x)}$$

Add numerators. Put this sum over the least common denominator.

$$= \frac{(\sin^2 x + \cos^2 x) + 1 + 2\sin x}{(1 + \sin x)(\cos x)}$$

Regroup terms to apply a Pythagorean identity.

$$= \frac{1 + 1 + 2\sin x}{(1 + \sin x)(\cos x)}$$

Apply a Pythagorean identity: $\sin^2 x + \cos^2 x = 1$.

$$= \frac{2 + 2\sin x}{(1 + \sin x)(\cos x)}$$

Add constant terms in the numerator: $1 + 1 = 2$.

$$= \frac{2\cancel{(1 + \sin x)}}{\cancel{(1 + \sin x)}(\cos x)}$$

Factor and simplify.

$$= \frac{2}{\cos x}$$

$$= 2 \sec x$$

Apply a reciprocal identity: $\sec x = \dfrac{1}{\cos x}$.

We worked with the left side and arrived at the right side. Thus, the identity is verified.

Check Point 4 Verify the identity: $\dfrac{\sin x}{1 + \cos x} + \dfrac{1 + \cos x}{\sin x} = 2 \csc x.$

Some identities are verified using a technique that may remind you of rationalizing a denominator.

EXAMPLE 5 Multiplying the Numerator and Denominator by the Same Factor to Verify an Identity

Verify the identity: $\dfrac{\sin x}{1 + \cos x} = \dfrac{1 - \cos x}{\sin x}.$

Solution The suggestions given in the previous examples do not apply here. Everything is already expressed in terms of sines and cosines. Furthermore, there are no fractions to combine and neither side looks more complicated than the other. Let's solve the puzzle by working with the left side and making it look like the expression on the right. The expression on the right contains $1 - \cos x$ in the numerator. This suggests multiplying the numerator and denominator of the left side by $1 - \cos x$. By doing this, we obtain a factor of $1 - \cos x$ in the numerator, as in the numerator on the right.

Discovery

Verify the identity in Example 5 by making the right side look like the left side. Start with the expression on the right. Multiply the numerator and denominator by $1 + \cos x$.

$$\frac{\sin x}{1 + \cos x} = \frac{\sin x}{1 + \cos x} \cdot \frac{1 - \cos x}{1 - \cos x}$$

Multiply numerator and denominator by $1 - \cos x$.

$$= \frac{\sin x(1 - \cos x)}{1 - \cos^2 x}$$

Multiply. Use $(A + B)(A - B) = A^2 - B^2$, with $A = 1$ and $B = \cos x$, to multiply denominators.

$$= \frac{\sin x(1 - \cos x)}{\sin^2 x}$$

Use a variation of $\sin^2 x + \cos^2 x = 1$. Solving for $\sin^2 x$, we obtain $\sin^2 x = 1 - \cos^2 x$.

$$= \frac{1 - \cos x}{\sin x}$$

Simplify: $\dfrac{\sin x}{\sin^2 x} = \dfrac{\cancel{\sin x}}{\cancel{\sin x} \cdot \sin x} = \dfrac{1}{\sin x}$.

We worked with the left side and arrived at the right side. Thus, the identity is verified.

Check Point 5 Verify the identity: $\dfrac{\cos x}{1 + \sin x} = \dfrac{1 - \sin x}{\cos x}.$

EXAMPLE 6 Changing to Sines and Cosines to Verify an Identity

Verify the identity: $\frac{\tan x - \sin(-x)}{1 + \cos x} = \tan x$.

Solution We begin with the left side. Our goal is to obtain $\tan x$, the expression on the right.

$$\frac{\tan x - \sin(-x)}{1 + \cos x} = \frac{\tan x - (-\sin x)}{1 + \cos x}$$ The sine function is odd: $\sin(-x) = -\sin x$.

$$= \frac{\tan x + \sin x}{1 + \cos x}$$ Simplify.

$$= \frac{\frac{\sin x}{\cos x} + \sin x}{1 + \cos x}$$ Apply a quotient identity: $\tan x = \frac{\sin x}{\cos x}$.

$$= \frac{\frac{\sin x}{\cos x} + \frac{\sin x \cos x}{\cos x}}{1 + \cos x}$$ Express the terms in the numerator with the least common denominator, $\cos x$.

$$= \frac{\frac{\sin x + \sin x \cos x}{\cos x}}{1 + \cos x}$$ Add in the numerator.

$$= \frac{\sin x + \sin x \cos x}{\cos x} \div \frac{1 + \cos x}{1}$$ Rewrite the main fraction bar as ÷.

$$= \frac{\sin x + \sin x \cos x}{\cos x} \cdot \frac{1}{1 + \cos x}$$ Invert the divisor and multiply.

$$= \frac{\sin x\,\overset{1}{\cancel{(1 + \cos x)}}}{\cos x} \cdot \frac{1}{\underset{1}{\cancel{1 + \cos x}}}$$ Factor and simplify.

$$= \frac{\sin x}{\cos x}$$ Multiply the remaining factors in the numerator and the denominator.

$$= \tan x$$ Apply a quotient identity.

The left side simplifies to $\tan x$, the right side. Thus, the identity is verified.

Discovery

Try simplifying

$$\frac{\frac{\sin x}{\cos x} + \sin x}{1 + \cos x}$$

by multiplying the two terms in the numerator and the two terms in the denominator by $\cos x$. This method for simplifying the complex fraction involves multiplying the numerator and the denominator by the least common denominator of all fractions in the expression. Do you prefer this simplification procedure over the method used on the right?

Check Point 6 Verify the identity: $\frac{\sec x + \csc(-x)}{\sec x \csc x} = \sin x - \cos x$.

Is every identity verified by working with only one side? No. You can sometimes work with each side separately and show that both sides are equal to the same trigonometric expression. This is illustrated in Example 7.

EXAMPLE 7 Working with Both Sides Separately to Verify an Identity

Verify the identity: $\dfrac{1}{1+\cos\theta}+\dfrac{1}{1-\cos\theta}=2+2\cot^2\theta.$

Solution We begin by working with the left side.

$$\frac{1}{1+\cos\theta}+\frac{1}{1-\cos\theta}$$

The least common denominator is $(1+\cos\theta)(1-\cos\theta)$.

$$=\frac{1(1-\cos\theta)}{(1+\cos\theta)(1-\cos\theta)}+\frac{1(1+\cos\theta)}{(1+\cos\theta)(1-\cos\theta)}$$

Rewrite each fraction with the least common denominator.

$$=\frac{1-\cos\theta+1+\cos\theta}{(1+\cos\theta)(1-\cos\theta)}$$

Add numerators. Put this sum over the least common denominator.

$$=\frac{2}{(1+\cos\theta)(1-\cos\theta)}$$

Simplify the numerator: $-\cos\theta+\cos\theta=0$ and $1+1=2$.

$$=\frac{2}{1-\cos^2\theta}$$

Multiply the factors in the denominator.

Now we work with the right side. Our goal is to transform this side into the simplified form attained for the left side, $\dfrac{2}{1-\cos^2\theta}$.

$$2+2\cot^2\theta=2+2\left(\frac{\cos^2\theta}{\sin^2\theta}\right)$$

Use a quotient identity: $\cot\theta=\dfrac{\cos\theta}{\sin\theta}$.

$$=\frac{2\sin^2\theta}{\sin^2\theta}+\frac{2\cos^2\theta}{\sin^2\theta}$$

Rewrite each fraction with the least common denominator, $\sin^2\theta$.

$$=\frac{2\sin^2\theta+2\cos^2\theta}{\sin^2\theta}$$

Add numerators. Put this sum over the least common denominator.

$$=\frac{2(\sin^2\theta+\cos^2\theta)}{\sin^2\theta}$$

Factor out the greatest common factor, 2.

$$=\frac{2}{\sin^2\theta}$$

Apply a Pythagorean identity: $\sin^2\theta+\cos^2\theta=1$.

$$=\frac{2}{1-\cos^2\theta}$$

Use a variation of $\sin^2\theta+\cos^2\theta=1$ and solve for $\sin^2\theta$: $\sin^2\theta=1-\cos^2\theta$.

The identity is verified because both sides are equal to $\dfrac{2}{1-\cos^2\theta}$.

Check Point 7 Verify the identity: $\dfrac{1}{1 + \sin\theta} + \dfrac{1}{1 - \sin\theta} = 2 + 2\tan^2\theta.$

Guidelines for Verifying Trigonometric Identities

There is often more than one correct way to solve a puzzle, although one method may be shorter and more efficient than another. The same is true for verifying an identity. For example, how would you verify

$$\frac{\csc^2 x - 1}{\csc^2 x} = \cos^2 x?$$

One approach is to use a Pythagorean identity, $1 + \cot^2 x = \csc^2 x$, on the left side. Then change the resulting expression to sines and cosines.

$$\frac{\csc^2 x - 1}{\csc^2 x} = \frac{(1 + \cot^2 x) - 1}{\csc^2 x} = \frac{\cot^2 x}{\csc^2 x} = \frac{\frac{\cos^2 x}{\sin^2 x}}{\frac{1}{\sin^2 x}} = \frac{\cos^2 x}{\cancel{\sin^2 x}} \cdot \frac{\cancel{\sin^2 x}}{1} = \cos^2 x$$

Apply a Pythagorean identity: $1 + \cot^2 x = \csc^2 x$.

Use $\cot x = \frac{\cos x}{\sin x}$ and $\csc x = \frac{1}{\sin x}$ to change to sines and cosines.

Invert the divisor and multiply.

A more efficient strategy for verifying this identity may not be apparent at first glance. Work with the left side and divide each term in the numerator by the denominator, $\csc^2 x$.

$$\frac{\csc^2 x - 1}{\csc^2 x} = \frac{\csc^2 x}{\csc^2 x} - \frac{1}{\csc^2 x} = 1 - \sin^2 x = \cos^2 x$$

Apply a reciprocal identity: $\sin x = \frac{1}{\csc x}$.

Use $\sin^2 x + \cos^2 x = 1$ and solve for $\cos^2 x$.

With this strategy, we again obtain $\cos^2 x$, the expression on the right side, and it takes fewer steps than the first approach.

An even longer strategy, but one that works, is to replace each of the two occurrences of $\csc^2 x$ on the left side by $\dfrac{1}{\sin^2 x}$. This may be the approach that you first consider, particularly if you become accustomed to rewriting the more complicated side in terms of sines and cosines. The selection of an appropriate fundamental identity to solve the puzzle most efficiently is learned through lots of practice.

The more identities you prove, the more confident and efficient you will become. Although practice is the only way to learn how to verify identities, there are some guidelines developed throughout the section that should help you get started.

Guidelines for Verifying Trigonometric Identities

1. Work with each side of the equation independently of the other side. Start with the more complicated side and transform it in a step-by-step fashion until it looks exactly like the other side.
2. Analyze the identity and look for opportunities to apply the fundamental identities. Rewriting the more complicated side of the equation in terms of sines and cosines is often helpful.
3. If sums or differences of fractions appear on one side, use the least common denominator and combine the fractions.
4. Don't be afraid to stop and start over again if you are not getting anywhere. Creative puzzle solvers know that strategies leading to dead ends often provide good problem-solving ideas.

EXERCISE SET 6.1

Practice Exercises

In Exercises 1–60, verify each identity.

1. $\sin x \sec x = \tan x$

2. $\cos x \csc x = \cot x$

3. $\tan(-x)\cos x = -\sin x$

4. $\cot(-x)\sin x = -\cos x$

5. $\tan x \csc x \cos x = 1$

6. $\cot x \sec x \sin x = 1$

7. $\sec x - \sec x \sin^2 x = \cos x$

8. $\csc x - \csc x \cos^2 x = \sin x$

9. $\cos^2 x - \sin^2 x = 1 - 2\sin^2 x$

10. $\cos^2 x - \sin^2 x = 2\cos^2 x - 1$

11. $\csc\theta - \sin\theta = \cot\theta\cos\theta$

12. $\tan\theta + \cot\theta = \sec\theta\csc\theta$

13. $\dfrac{\tan\theta\cot\theta}{\csc\theta} = \sin\theta$

14. $\dfrac{\cos\theta\sec\theta}{\cot\theta} = \tan\theta$

15. $\sin^2\theta(1 + \cot^2\theta) = 1$

16. $\cos^2\theta(1 + \tan^2\theta) = 1$

17. $\sin t \tan t = \dfrac{1 - \cos^2 t}{\cos t}$

18. $\cos t \cot t = \dfrac{1 - \sin^2 t}{\sin t}$

19. $\dfrac{\csc^2 t}{\cot t} = \csc t \sec t$

20. $\dfrac{\sec^2 t}{\tan t} = \sec t \csc t$

21. $\dfrac{\tan^2 t}{\sec t} = \sec t - \cos t$

22. $\dfrac{\cot^2 t}{\csc t} = \csc t - \sin t$

23. $\dfrac{\sin t}{\csc t} + \dfrac{\cos t}{\sec t} = 1$

24. $\dfrac{\sin t}{\tan t} + \dfrac{\cos t}{\cot t} = \sin t + \cos t$

25. $\tan t + \dfrac{\cos t}{1 + \sin t} = \sec t$

26. $\cot t + \dfrac{\sin t}{1 + \cos t} = \csc t$

27. $1 - \dfrac{\sin^2 x}{1 + \cos x} = \cos x$

28. $1 - \dfrac{\cos^2 x}{1 + \sin x} = \sin x$

29. $\dfrac{\cos x}{1 - \sin x} + \dfrac{1 - \sin x}{\cos x} = 2\sec x$

30. $\dfrac{\sin x}{\cos x + 1} + \dfrac{\cos x - 1}{\sin x} = 0$

31. $\sec^2 x \csc^2 x = \sec^2 x + \csc^2 x$

32. $\csc^2 x \sec x = \sec x + \csc x \cot x$

33. $\dfrac{\sec x - \csc x}{\sec x + \csc x} = \dfrac{\tan x - 1}{\tan x + 1}$

34. $\dfrac{\csc x - \sec x}{\csc x + \sec x} = \dfrac{\cot x - 1}{\cot x + 1}$

35. $\dfrac{\sin^2 x - \cos^2 x}{\sin x + \cos x} = \sin x - \cos x$

36. $\dfrac{\tan^2 x - \cot^2 x}{\tan x + \cot x} = \tan x - \cot x$

37. $\tan^2 2x + \sin^2 2x + \cos^2 2x = \sec^2 2x$

38. $\cot^2 2x + \cos^2 2x + \sin^2 2x = \csc^2 2x$

39. $\dfrac{\tan 2\theta + \cot 2\theta}{\csc 2\theta} = \sec 2\theta$

40. $\dfrac{\tan 2\theta + \cot 2\theta}{\sec 2\theta} = \csc 2\theta$

41. $\dfrac{\tan x + \tan y}{1 - \tan x \tan y} = \dfrac{\sin x \cos y + \cos x \sin y}{\cos x \cos y - \sin x \sin y}$

42. $\dfrac{\cot x + \cot y}{1 - \cot x \cot y} = \dfrac{\cos x \sin y + \sin x \cos y}{\sin x \sin y - \cos x \cos y}$

43. $(\sec x - \tan x)^2 = \dfrac{1 - \sin x}{1 + \sin x}$

44. $(\csc x - \cot x)^2 = \dfrac{1 - \cos x}{1 + \cos x}$

45. $\dfrac{\sec t + 1}{\tan t} = \dfrac{\tan t}{\sec t - 1}$

46. $\dfrac{\csc t - 1}{\cot t} = \dfrac{\cot t}{\csc t + 1}$

47. $\dfrac{1 + \cos t}{1 - \cos t} = (\csc t + \cot t)^2$

48. $\dfrac{1 - \sin t}{1 + \sin t} = (\sec t - \tan t)^2$

49. $\cos^4 t - \sin^4 t = 1 - 2\sin^2 t$

50. $\sin^4 t - \cos^4 t = 1 - 2\cos^2 t$

51. $\dfrac{\sin\theta - \cos\theta}{\sin\theta} + \dfrac{\cos\theta - \sin\theta}{\cos\theta} = 2 - \sec\theta\csc\theta$

52. $\dfrac{\sin\theta}{1 - \cot\theta} - \dfrac{\cos\theta}{\tan\theta - 1} = \sin\theta + \cos\theta$

53. $(\tan^2\theta + 1)(\cos^2\theta + 1) = \tan^2\theta + 2$

54. $(\cot^2\theta + 1)(\sin^2\theta + 1) = \cot^2\theta + 2$

55. $(\cos\theta - \sin\theta)^2 + (\cos\theta + \sin\theta)^2 = 2$

56. $(3\cos\theta - 4\sin\theta)^2 + (4\cos\theta + 3\sin\theta)^2 = 25$

57. $\dfrac{\cos^2 x - \sin^2 x}{1 - \tan^2 x} = \cos^2 x$

58. $\dfrac{\sin x + \cos x}{\sin x} - \dfrac{\cos x - \sin x}{\cos x} = \sec x \csc x$

59. $(\sec x - \tan x)^2 = \dfrac{1 - \sin x}{1 + \sin x}$

60. $(\cot x - \csc x)^2 = \dfrac{1 - \cos x}{1 + \cos x}$

Writing in Mathematics

61. Explain how to verify an identity.

62. Describe two strategies that can be used to verify identities.

63. Describe how you feel when you successfully verify a difficult identity. What other activities do you engage in that evoke the same feelings?

64. A 10-point question on a quiz asks students to verify the identity

$$\frac{\sin^2 x - \cos^2 x}{\sin x + \cos x} = \sin x - \cos x.$$

One student begins with the left side and obtains the right side as shown in next column:

$$\frac{\sin^2 x - \cos^2 x}{\sin x + \cos x} = \frac{\sin^2 x}{\sin x} - \frac{\cos^2 x}{\cos x} = \sin x - \cos x.$$

How many points (out of 10) would you give this student? Explain your answer.

Technology Exercises

In Exercises 65–73, graph each side of the equation in the same viewing rectangle. If the graphs appear to coincide, verify that the equation is an identity. If the graphs do not appear to coincide, this indicates the equation is not an identity. In these exercises, find a value of x for which both sides are defined but not equal.

65. $\tan x = \sec x(\sin x - \cos x) + 1$

66. $\sin x = -\cos x \tan(-x)$

67. $\sin\left(x + \dfrac{\pi}{4}\right) = \sin x + \sin\dfrac{\pi}{4}$

68. $\cos\left(x + \dfrac{\pi}{4}\right) = \cos x + \cos\dfrac{\pi}{4}$

69. $\cos(x + \pi) = \cos x$

70. $\sin(x + \pi) = \sin x$

71. $\dfrac{\sin x}{1 - \cos^2 x} = \csc x$

72. $\sin x - \sin x \cos^2 x = \sin^3 x$

73. $\sqrt{\sin^2 x + \cos^2 x} = \sin x + \cos x$

Critical Thinking Exercises

In Exercises 74–76, verify each identity.

74. $\dfrac{\sin^3 x - \cos^3 x}{\sin x - \cos x} = 1 + \sin x \cos x$

75. $\dfrac{\sin x - \cos x + 1}{\sin x + \cos x - 1} = \dfrac{\sin x + 1}{\cos x}$

76. $\ln|\sec x| = -\ln|\cos x|$

77. Use one of the fundamental identities in the box on page 557 to create an original identity.

Group Exercise

78. Group members are to write a helpful list of items for a pamphlet called "The Underground Guide to Verifying Identities." The pamphlet will be used primarily by students who sit, stare, and freak out every time they are asked to verify an identity. List easy ways to remember the fundamental identities. What helpful guidelines can you offer from the perspective of a student that you probably won't find in math books? If you have your own strategies that work particularly well, include them in the pamphlet.

SECTION 6.2 Sum and Difference Formulas

Objectives

1. Use the formula for the cosine of the difference of two angles.
2. Use sum and difference formulas for cosines and sines.
3. Use sum and difference formulas for tangents.

Listen to the same note played on a piano and a violin. The notes have a different quality or "tone." Tone depends on the way an instrument vibrates. However, the less than 1% of the population with amusia, or true tone deafness, cannot tell the two sounds apart. Even simple, familiar tunes such as *Happy Birthday* and *Jingle Bells* are mystifying to amusics.

When a note is played, it vibrates at a specific fundamental frequency and has a particular amplitude. Amusics cannot tell the difference between two sounds from tuning forks modeled by $p = 3 \sin 2t$ and $p = 2 \sin(2t + \pi)$, respectively. However, they can recognize the difference between the two equations. Notice that the second equation contains the sine of the sum of two angles. In this section, we will be developing identities involving the sums or differences of two angles. These formulas are called the **sum and difference formulas.** We begin with $\cos(\alpha - \beta)$, the cosine of the difference of two angles.

The Cosine of the Difference of Two Angles

The Cosine of the Difference of Two Angles

$$\cos(\alpha - \beta) = \cos \alpha \cos \beta + \sin \alpha \sin \beta$$

The cosine of the difference of two angles equals the cosine of the first angle times the cosine of the second angle plus the sine of the first angle times the sine of the second angle.

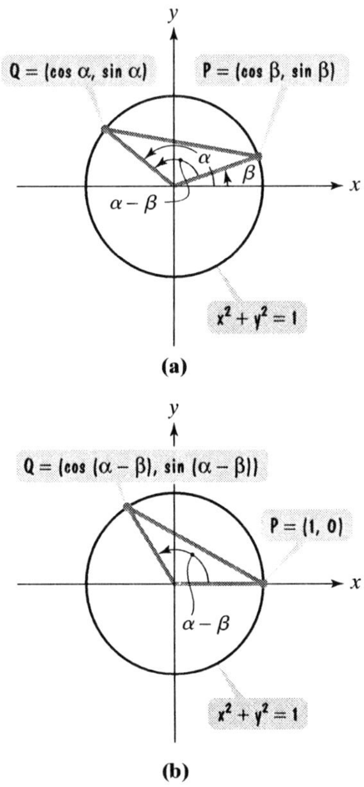

Figure 6.1 Using the unit circle and QP to develop a formula for $\cos(\alpha - \beta)$

We use Figure 6.1 to prove the identity in the box. The graph in Figure 6.1(a) shows a unit circle, $x^2 + y^2 = 1$. The figure uses the definitions of the cosine and sine functions as the x- and y-coordinates of points along the unit circle. For example, point P corresponds to angle β. By definition, the x-coordinate of P is $\cos \beta$ and the y-coordinate is $\sin \beta$. Similarly, point Q corresponds to angle α. By definition, the x-coordinate of Q is $\cos \alpha$ and the y-coordinate is $\sin \alpha$.

Note that if we draw a line segment between points P and Q, a triangle is formed. Angle $\alpha - \beta$ is one of the angles of this triangle. What happens if we rotate this triangle so that point P falls on the x-axis at $(1, 0)$? The result is shown in Figure 6.1(b). This rotation changes the coordinates of points P and Q. However, it has no effect on the length of line segment PQ.

We can use the distance formula, $d = \sqrt{(x_2 - x_1)^2 + (y_2 - y_1)^2}$, to find an expression for PQ in Figure 6.1(a) and in Figure 6.1(b). By equating the two expressions for PQ, we will obtain the identity for the cosine of the difference of two angles, $\alpha - \beta$. We first apply the distance formula in Figure 6.1(a).

$$PQ = \sqrt{(\cos\alpha - \cos\beta)^2 + (\sin\alpha - \sin\beta)^2}$$

Apply the distance formula, $d = \sqrt{(x_2 - x_1)^2 + (y_2 - y_1)^2}$, to find the distance between $(\cos\beta, \sin\beta)$ and $(\cos\alpha, \sin\alpha)$.

$$= \sqrt{\cos^2\alpha - 2\cos\alpha\cos\beta + \cos^2\beta + \sin^2\alpha - 2\sin\alpha\sin\beta + \sin^2\beta}$$

Square each expression using $(A - B)^2 = A^2 - 2AB + B^2$.

$$= \sqrt{(\sin^2\alpha + \cos^2\alpha) + (\sin^2\beta + \cos^2\beta) - 2\cos\alpha\cos\beta - 2\sin\alpha\sin\beta}$$

Regroup terms to apply a Pythagorean identity.

$$= \sqrt{1 + 1 - 2\cos\alpha\cos\beta - 2\sin\alpha\sin\beta}$$

Because $\sin^2 x + \cos^2 x = 1$, each expression in parentheses equals 1.

$$= \sqrt{2 - 2\cos\alpha\cos\beta - 2\sin\alpha\sin\beta}$$

Simplify.

Next, we apply the distance formula in Figure 6.1(b) to obtain a second expression for PQ. We let $(x_1, y_1) = (1, 0)$ and $(x_2, y_2) = (\cos(\alpha - \beta), \sin(\alpha - \beta))$.

$$PQ = \sqrt{[\cos(\alpha - \beta) - 1]^2 + [\sin(\alpha - \beta) - 0]^2}$$

Apply the distance formula to find the distance between $(1, 0)$ and $(\cos(\alpha - \beta), \sin(\alpha - \beta))$.

$$= \sqrt{\cos^2(\alpha - \beta) - 2\cos(\alpha - \beta) + 1 + \sin^2(\alpha - \beta)}$$

Square each expression.

Using a Pythagorean identity, $\sin^2(\alpha - \beta) + \cos^2(\alpha - \beta) = 1$

$$= \sqrt{1 - 2\cos(\alpha - \beta) + 1}$$

Use a Pythagorean identity.

$$= \sqrt{2 - 2\cos(\alpha - \beta)}$$

Simplify.

Now we equate the two expressions for PQ.

$$\sqrt{2 - 2\cos(\alpha - \beta)} = \sqrt{2 - 2\cos\alpha\cos\beta - 2\sin\alpha\sin\beta}$$

The rotation does not change the length of PQ.

$$2 - 2\cos(\alpha - \beta) = 2 - 2\cos\alpha\cos\beta - 2\sin\alpha\sin\beta$$

Square both sides to eliminate radicals.

$$-2\cos(\alpha - \beta) = -2\cos\alpha\cos\beta - 2\sin\alpha\sin\beta$$

Subtract 2 from both sides of the equation.

$$\cos(\alpha - \beta) = \cos\alpha\cos\beta + \sin\alpha\sin\beta$$

Divide both sides of the equation by -2.

This proves the identity for the cosine of the difference of two angles.

Now that we see where the identity for the cosine of the difference of two angles comes from, let's look at some applications of this result.

1 Use the formula for the cosine of the difference of two angles.

EXAMPLE 1 Using the Difference Formula for Cosines to Find Exact Values

Find the exact value of $\cos 15°$.

Solution We know exact values for trigonometric functions of 60° and 45°. Thus, we write 15° as $60° - 45°$ and use the difference formula for cosines.

$$\cos 15° = \cos(60° - 45°)$$

$$= \cos 60° \cos 45° + \sin 60° \sin 45° \quad \cos(\alpha - \beta) = \cos\alpha\cos\beta + \sin\alpha\sin\beta$$

$$= \frac{1}{2}\cdot\frac{\sqrt{2}}{2} + \frac{\sqrt{3}}{2}\cdot\frac{\sqrt{2}}{2} \quad \text{Substitute exact values from memory or use special right triangles.}$$

$$= \frac{\sqrt{2}}{4} + \frac{\sqrt{6}}{4} \quad \text{Multiply.}$$

$$= \frac{\sqrt{2} + \sqrt{6}}{4} \quad \text{Add.}$$

Check Point 1 We know that $\cos 30° = \frac{\sqrt{3}}{2}$. Obtain this exact value using $\cos 30° = \cos(90° - 60°)$ and the difference formula for cosines.

Sound Quality and Amusia

People with true tone deafness cannot hear the difference among tones produced by a tuning fork, a flute, an oboe, and a violin. They cannot dance or tell the difference between harmony and dissonance. People with amusia appear to have been born without the wiring necessary to process music. Intriguingly, they show no overt signs of brain damage and their brain scans appear normal. Thus, they can visually recognize the difference among sound waves that produce varying sound qualities.

Varying Sound Qualities

- Tuning fork: Sound waves are rounded and regular, giving a pure and gentle tone.

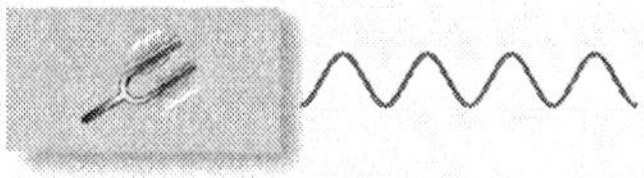

- Flute: Sound waves are smooth and give a fluid tone.

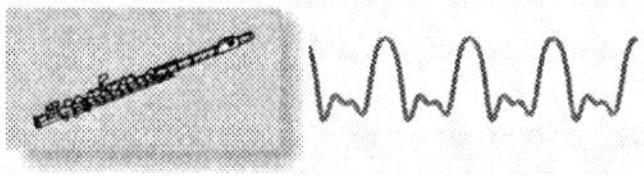

- Oboe: Rapid wave changes give a richer tone.

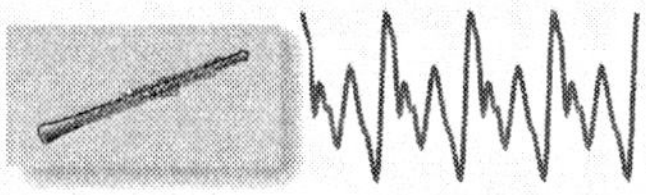

- Violin: Jagged waves give a brighter harsher tone.

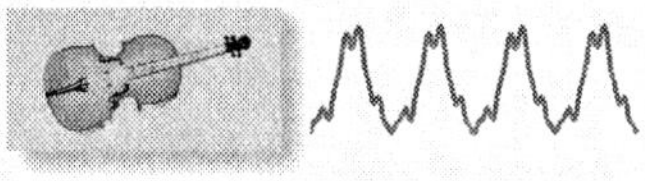

EXAMPLE 2 Using the Difference Formula for Cosines to Find Exact Values

Find the exact value of $\cos 80° \cos 20° + \sin 80° \sin 20°$.

Solution The given expression is the right side of the formula for $\cos(\alpha - \beta)$ with $\alpha = 80°$ and $\beta = 20°$.

$$\cos(\alpha - \beta) = \cos\alpha\cos\beta + \sin\alpha\sin\beta$$

$$\cos 80° \cos 20° + \sin 80° \sin 20° = \cos(80° - 20°) = \cos 60° = \frac{1}{2}$$

Check Point 2 Find the exact value of

$$\cos 70° \cos 40° + \sin 70° \sin 40°.$$

EXAMPLE 3 Verifying an Identity

Verify the identity: $\dfrac{\cos(\alpha - \beta)}{\sin\alpha\cos\beta} = \cot\alpha + \tan\beta.$

Solution We work with the left side.

$$\frac{\cos(\alpha - \beta)}{\sin\alpha\cos\beta} = \frac{\cos\alpha\cos\beta + \sin\alpha\sin\beta}{\sin\alpha\cos\beta}$$ Use the formula for cos $(\alpha - \beta)$.

$$= \frac{\cos\alpha}{\sin\alpha}\cdot\frac{\cos\beta}{\cos\beta} + \frac{\sin\alpha}{\sin\alpha}\cdot\frac{\sin\beta}{\cos\beta}$$ Divide each term in the numerator by $\sin\alpha\cos\beta$.

$$= \cot\alpha \cdot 1 + 1 \cdot \tan\beta$$ Use quotient identities.

$$= \cot\alpha + \tan\beta$$ Simplify.

We worked with the left side and arrived at the right side. Thus, the identity is verified.

Technology

The graphs of

$$y = \cos\left(\frac{\pi}{2} - x\right)$$

and

$$y = \sin x$$

are shown in the same viewing rectangle. The graphs are the same. The displayed math on the right shows the equivalence algebraically.

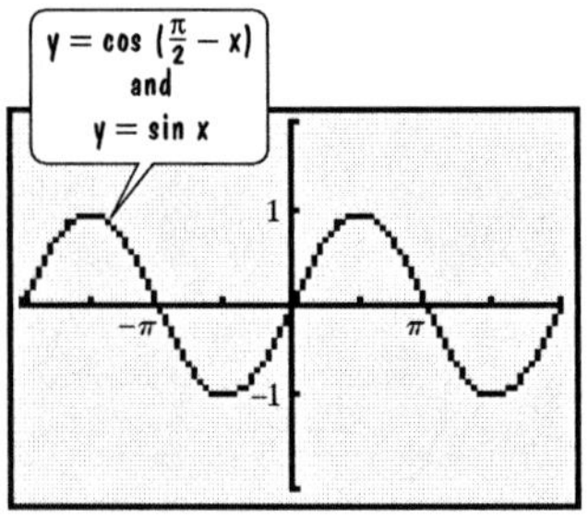

Check Point 3 Verify the identity: $\dfrac{\cos(\alpha - \beta)}{\cos\alpha\cos\beta} = 1 + \tan\alpha\tan\beta.$

The difference formula for cosines is used to establish other identities. For example, in our work with right triangles we noted that cofunctions of complements are equal. Thus, because $\frac{\pi}{2} - \theta$ and θ are complements,

$$\cos\left(\frac{\pi}{2} - \theta\right) = \sin\theta.$$

We can use the formula for cos $(\alpha - \beta)$ to prove this cofunction identity.

Apply cos$(\alpha - \beta)$ with $\alpha = 90°$ and $\theta = \beta$.
cos$(\alpha - \beta)$ = cos α cos β + sin α sin β

$$\cos\left(\frac{\pi}{2} - \theta\right) = \cos\frac{\pi}{2}\cos\theta + \sin\frac{\pi}{2}\sin\theta$$
$$= 0 \cdot \cos\theta + 1 \cdot \sin\theta$$
$$= \sin\theta$$

2 Use sum and difference formulas for cosines and sines.

Sum and Difference Formulas for Cosines and Sines

Our formula for cos $(\alpha - \beta)$ can be used to verify an identity for a sum involving cosines, as well as identities for a sum and a difference for sines.

Sum and Difference Formulas for Cosines and Sines

1. $\cos(\alpha + \beta) = \cos\alpha\cos\beta - \sin\alpha\sin\beta$
2. $\cos(\alpha - \beta) = \cos\alpha\cos\beta + \sin\alpha\sin\beta$
3. $\sin(\alpha + \beta) = \sin\alpha\cos\beta + \cos\alpha\sin\beta$
4. $\sin(\alpha - \beta) = \sin\alpha\cos\beta - \cos\alpha\sin\beta$

Up to now, we have concentrated on the second formula in the box. The first identity gives a formula for the cosine of the sum of two angles. It is proved as follows:

$$\cos(\alpha + \beta) = \cos[\alpha - (-\beta)]$$ Express addition as subtraction of an inverse.

$$= \cos \alpha \cos(-\beta) + \sin \alpha \sin(-\beta)$$ Use the difference formula for cosines.

$$= \cos \alpha \cos \beta + \sin \alpha (-\sin \beta)$$ Cosine is even: $\cos(-\beta) = \cos \beta$. Sine is odd: $\sin(-\beta) = -\sin \beta$.

$$= \cos \alpha \cos \beta - \sin \alpha \sin \beta.$$ Simplify.

Thus, the cosine of the sum of two angles equals the cosine of the first angle times the cosine of the second angle minus the sine of the first angle times the sine of the second angle.

The third identity in the box gives a formula for $\sin(\alpha + \beta)$, the sine of the sum of two angles. It is proved as follows:

$$\sin(\alpha + \beta) = \cos\left[\frac{\pi}{2} - (\alpha + \beta)\right]$$ Use a cofunction identity: $\sin \theta = \cos\left(\frac{\pi}{2} - \theta\right)$.

$$= \cos\left[\left(\frac{\pi}{2} - \alpha\right) - \beta\right]$$ Regroup.

$$= \cos\left(\frac{\pi}{2} - \alpha\right)\cos\beta + \sin\left(\frac{\pi}{2} - \alpha\right)\sin\beta$$ Use the difference formula for cosines.

$$= \sin\alpha\cos\beta + \cos\alpha\sin\beta.$$ Use cofunction identities.

Thus, the sine of the sum of two angles equals the sine of the first angle times the cosine of the second angle plus the cosine of the first angle times the sine of the second angle.

The final identity in the box gives a formula for $\sin(\alpha - \beta)$, the sine of the difference of two angles. It is proved by writing $\sin(\alpha - \beta)$ as $\sin[\alpha + (-\beta)]$ and then using the formula for the sine of a sum.

EXAMPLE 4 Using the Sine of a Sum to Find an Exact Value

Find the exact value of $\sin\frac{7\pi}{12}$ using the fact that $\frac{7\pi}{12} = \frac{\pi}{3} + \frac{\pi}{4}$.

Solution We apply the formula for the sine of a sum.

$$\sin\frac{7\pi}{12} = \sin\left(\frac{\pi}{3} + \frac{\pi}{4}\right)$$

$$= \sin\frac{\pi}{3}\cos\frac{\pi}{4} + \cos\frac{\pi}{3}\sin\frac{\pi}{4}$$ $\sin(\alpha + \beta) = \sin\alpha\cos\beta + \cos\alpha\sin\beta$

$$= \frac{\sqrt{3}}{2} \cdot \frac{\sqrt{2}}{2} + \frac{1}{2} \cdot \frac{\sqrt{2}}{2}$$ Substitute exact values.

$$= \frac{\sqrt{6} + \sqrt{2}}{4}$$ Simplify.

Check Point 4 Find the exact value of $\sin \frac{5\pi}{12}$ using the fact that

$$\frac{5\pi}{12} = \frac{\pi}{6} + \frac{\pi}{4}.$$

EXAMPLE 5 Finding Exact Values

Suppose that $\sin \alpha = \frac{12}{13}$ for a quadrant II angle α and $\sin \beta = \frac{3}{5}$ for a quadrant I angle β. Find the exact value of:

a. $\cos \alpha$ **b.** $\cos \beta$ **c.** $\cos(\alpha + \beta)$ **d.** $\sin(\alpha + \beta)$.

Solution

a. We find $\cos \alpha$ using a sketch that illustrates

$$\sin \alpha = \frac{12}{13} = \frac{y}{r}.$$

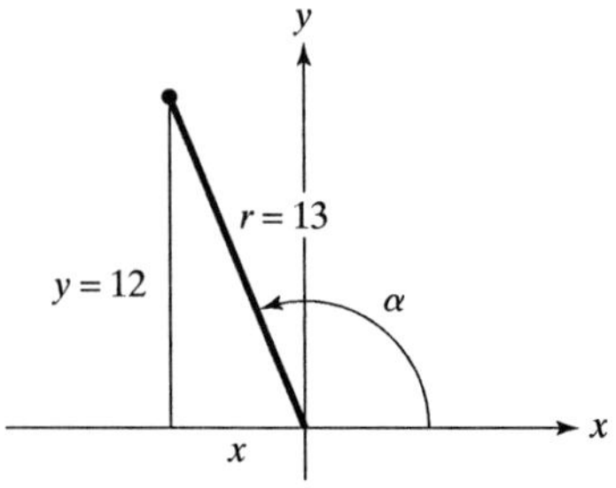

Figure 6.2 $\sin \alpha = \frac{12}{13}$: α lies in quadrant II

Figure 6.2 shows a quadrant II angle α with $\sin \alpha = \frac{12}{13}$. We find x using $x^2 + y^2 = r^2$. Because α lies in quadrant II, x is negative.

$$x^2 + 12^2 = 13^2 \quad x^2 + y^2 = r^2$$

$$x^2 + 144 = 169 \quad \text{Square 12 and 13, respectively.}$$

$$x^2 = 25 \quad \text{Subtract 144 from both sides.}$$

$$x = -\sqrt{25} = -5 \quad \text{In quadrant II, } x \text{ is negative.}$$

Thus,

$$\cos \alpha = \frac{x}{r} = \frac{-5}{13} = -\frac{5}{13}.$$

b. We find $\cos \beta$ using a sketch that illustrates

$$\sin \beta = \frac{3}{5} = \frac{y}{r}.$$

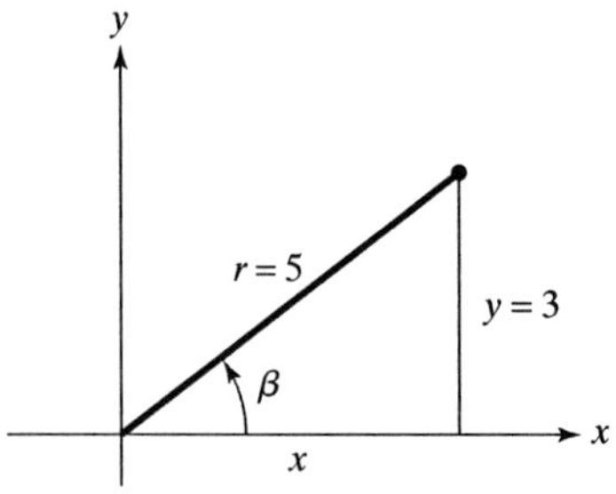

Figure 6.3 $\sin \beta = \frac{3}{5}$: β lies in quadrant I

Figure 6.3 shows a quadrant I angle β with $\sin \beta = \frac{3}{5}$. We find x using $x^2 + y^2 = r^2$.

$$x^2 + 3^2 = 5^2$$

$$x^2 = 25 - 9 = 16$$

$$x = \sqrt{16} = 4 \quad \text{In quadrant I, } x \text{ is positive.}$$

Thus,

$$\cos \beta = \frac{x}{r} = \frac{4}{5}.$$

We use the given values and the exact values that we determined to find exact values for $\cos(\alpha + \beta)$ and $\sin(\alpha + \beta)$.

These values are given. These are the values we found.

$$\sin \alpha = \frac{12}{13}, \ \sin \beta = \frac{3}{5} \qquad \cos \alpha = -\frac{5}{13}, \ \cos \beta = \frac{4}{5}$$

c. We use the formula for the cosine of a sum.

$$\cos(\alpha + \beta) = \cos\alpha\cos\beta - \sin\alpha\sin\beta$$
$$= -\frac{5}{13}\left(\frac{4}{5}\right) - \frac{12}{13}\left(\frac{3}{5}\right) = -\frac{56}{65}$$

d. We use the formula for the sine of a sum.

$$\sin(\alpha + \beta) = \sin\alpha\cos\beta + \cos\alpha\sin\beta$$
$$= \frac{12}{13}\cdot\frac{4}{5} + \left(-\frac{5}{13}\right)\cdot\frac{3}{5} = \frac{33}{65}$$

Check Point 5 Suppose that $\sin\alpha = \frac{4}{5}$ for a quadrant II angle α and $\sin\beta = \frac{1}{2}$ for a quadrant I angle β. Find the exact value of:

a. $\cos\alpha$ **b.** $\cos\beta$
c. $\cos(\alpha + \beta)$ **d.** $\sin(\alpha + \beta)$.

EXAMPLE 6 Verifying Observations on a Graphing Utility

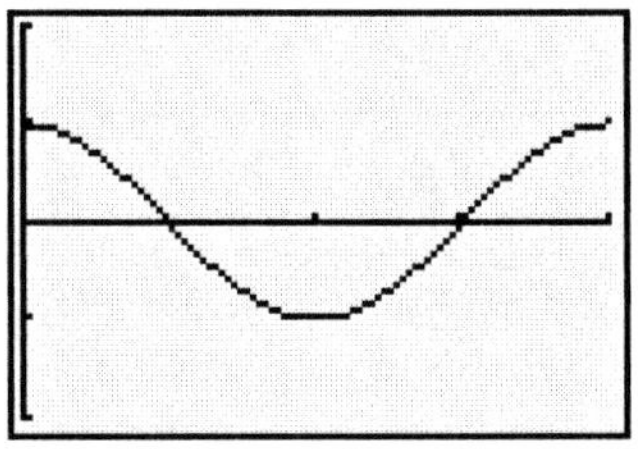

Figure 6.4 The graph of $y = \sin\left(x - \frac{3\pi}{2}\right)$ in a $\left[0, 2\pi, \frac{\pi}{2}\right]$ by $[-2, 2, 1]$ viewing rectangle

Figure 6.4 shows the graph of $y = \sin\left(x - \frac{3\pi}{2}\right)$ in a $\left[0, 2\pi, \frac{\pi}{2}\right]$ by $[-2, 2, 1]$ viewing rectangle.

a. Describe the graph using another equation.

b. Verify that the two equations are equivalent.

Solution

a. The graph appears to be the cosine curve $y = \cos x$. It cycles through maximum, intercept, minimum, intercept, and back to maximum. Thus, $y = \cos x$ also describes the graph.

b. We must show that

$$\sin\left(x - \frac{3\pi}{2}\right) = \cos x.$$

We apply the formula for the sine of a difference on the left side.

$$\sin\left(x - \frac{3\pi}{2}\right) = \sin x\cos\frac{3\pi}{2} - \cos x\sin\frac{3\pi}{2}$$ $\sin(\alpha - \beta) = \sin\alpha\cos\beta - \cos\alpha\sin\beta$

$$= \sin x \cdot 0 - \cos x(-1)$$ $\cos\frac{3\pi}{2} = 0$ and $\sin\frac{3\pi}{2} = -1$

$$= \cos x$$ Simplify.

This verifies our observation that $y = \sin\left(x - \frac{3\pi}{2}\right)$ and $y = \cos x$ describe the same graph.

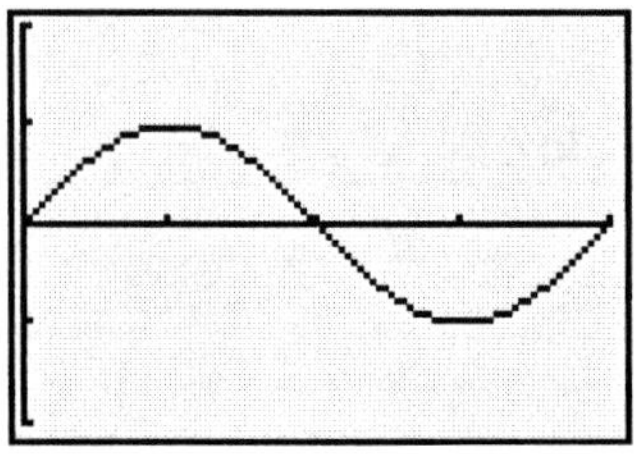

Figure 6.5

Check Point 6 Figure 6.5 shows the graph of $y = \cos\left(x + \frac{3\pi}{2}\right)$ in a $\left[0, 2\pi, \frac{\pi}{2}\right]$ by $[-2, 2, 1]$ viewing rectangle.

a. Describe the graph using another equation.

b. Verify that the two equations are equivalent.

3 Use sum and difference formulas for tangents.

Sum and Difference Formulas for Tangents

By writing $\tan(\alpha + \beta)$ as the quotient of $\sin(\alpha + \beta)$ and $\cos(\alpha + \beta)$ we can develop a formula for the tangent of a sum. Writing subtraction as addition of an inverse leads to a formula for the tangent of a difference.

Discovery

Derive the sum and difference formulas for tangents by working Exercises 55 and 56 in Exercise Set 6.2.

Sum and Difference Formulas for Tangents

$$\tan(\alpha + \beta) = \frac{\tan\alpha + \tan\beta}{1 - \tan\alpha\tan\beta}$$

The tangent of the sum of two angles equals the tangent of the first angle plus the tangent of the second angle divided by 1 minus their product.

$$\tan(\alpha - \beta) = \frac{\tan\alpha - \tan\beta}{1 + \tan\alpha\tan\beta}$$

The tangent of the difference of two angles equals the tangent of the first angle minus the tangent of the second angle divided by 1 plus their product.

EXAMPLE 7 Verifying an Identity

Verify the identity: $\tan\left(x - \frac{\pi}{4}\right) = \frac{\tan x - 1}{\tan x + 1}$.

Solution We work with the left side.

$$\tan\left(x - \frac{\pi}{4}\right) = \frac{\tan x - \tan\frac{\pi}{4}}{1 + \tan x \tan\frac{\pi}{4}} \qquad \tan(\alpha - \beta) = \frac{\tan\alpha - \tan\beta}{1 + \tan\alpha\tan\beta}$$

$$= \frac{\tan x - 1}{1 + \tan x \cdot 1} \qquad \tan\frac{\pi}{4} = 1$$

$$= \frac{\tan x - 1}{1 + \tan x}$$

Check Point 7 Verify the identity: $\tan(x + \pi) = \tan x$.

EXERCISE SET 6.2

Practice Exercises

Use the formula for the cosine of the difference of two angles to solve Exercises 1–12.

In Exercises 1–4, find the exact value of each expression.

1. $\cos(45° - 30°)$

2. $\cos(120° - 45°)$

3. $\cos\left(\frac{3\pi}{4} - \frac{\pi}{6}\right)$

4. $\cos\left(\frac{2\pi}{3} - \frac{\pi}{6}\right)$

In Exercises 5–8, each expression is the right side of the formula for $\cos(\alpha - \beta)$ *with particular values for* α *and* β.

a. *Identify* α *and* β *in each expression.*

b. *Write the expression as the cosine of an angle.*

c. *Find the exact value of the expression.*

5. $\cos 50° \cos 20° + \sin 50° \sin 20°$

6. $\cos 50° \cos 5° + \sin 50° \sin 5°$

7. $\cos\frac{5\pi}{12}\cos\frac{\pi}{12} + \sin\frac{5\pi}{12}\sin\frac{\pi}{12}$

8. $\cos\frac{5\pi}{18}\cos\frac{\pi}{9} + \sin\frac{5\pi}{18}\sin\frac{\pi}{9}$

In Exercises 9–12, verify each identity.

9. $\frac{\cos(\alpha - \beta)}{\cos\alpha\sin\beta} = \tan\alpha + \cot\beta$

10. $\frac{\cos(\alpha - \beta)}{\sin\alpha\sin\beta} = \cot\alpha\cot\beta + 1$

11. $\cos\left(x - \frac{\pi}{4}\right) = \frac{\sqrt{2}}{2}(\cos x + \sin x)$

12. $\cos\left(x - \frac{5\pi}{4}\right) = -\frac{\sqrt{2}}{2}(\cos x + \sin x)$

Use one or more of the six sum and difference identities to solve Exercises 13–54.

In Exercises 13–24, find the exact value of each expression.

13. $\sin(45° - 30°)$

14. $\sin(60° - 45°)$

15. $\sin 105°$

16. $\sin 75°$

17. $\tan(30° + 45°)$

18. $\tan(60° + 45°)$

19. $\tan(240° - 45°)$

20. $\tan(300° - 45°)$

21. $\cos\left(\frac{3\pi}{4} + \frac{\pi}{6}\right)$

22. $\cos\left(\frac{4\pi}{3} + \frac{\pi}{4}\right)$

23. $\cos\frac{5\pi}{12}$

24. $\cos\frac{7\pi}{12}$

In Exercises 25–32, write each expression as the sine, cosine, or tangent of an angle. Then find the exact value of the expression.

25. $\sin 25° \cos 5° + \cos 25° \sin 5°$

26. $\sin 40° \cos 20° + \cos 40° \sin 20°$

27. $\frac{\tan 10° + \tan 35°}{1 - \tan 10° \tan 35°}$

28. $\frac{\tan 50° - \tan 20°}{1 + \tan 50° \tan 20°}$

29. $\sin\frac{5\pi}{12}\cos\frac{\pi}{4} - \cos\frac{5\pi}{12}\sin\frac{\pi}{4}$

30. $\sin\frac{7\pi}{12}\cos\frac{\pi}{12} - \cos\frac{7\pi}{12}\sin\frac{\pi}{12}$

31. $\dfrac{\tan\frac{\pi}{5} - \tan\frac{\pi}{30}}{1 + \tan\frac{\pi}{5}\tan\frac{\pi}{30}}$

32. $\dfrac{\tan\frac{\pi}{5} + \tan\frac{4\pi}{5}}{1 - \tan\frac{\pi}{5}\tan\frac{4\pi}{5}}$

In Exercises 33–54, verify each identity.

33. $\sin\left(x + \frac{\pi}{2}\right) = \cos x$

34. $\sin\left(x + \frac{3\pi}{2}\right) = -\cos x$

35. $\cos\left(x - \frac{\pi}{2}\right) = \sin x$

36. $\cos(\pi - x) = -\cos x$

37. $\tan(2\pi - x) = -\tan x$

38. $\tan(\pi - x) = -\tan x$

39. $\sin(\alpha + \beta) + \sin(\alpha - \beta) = 2\sin\alpha\cos\beta$

40. $\cos(\alpha + \beta) + \cos(\alpha - \beta) = 2\cos\alpha\cos\beta$

41. $\frac{\sin(\alpha - \beta)}{\cos\alpha\cos\beta} = \tan\alpha - \tan\beta$

42. $\frac{\sin(\alpha + \beta)}{\cos\alpha\cos\beta} = \tan\alpha + \tan\beta$

43. $\tan\left(\theta + \frac{\pi}{4}\right) = \frac{\cos\theta + \sin\theta}{\cos\theta - \sin\theta}$

44. $\tan\left(\frac{\pi}{4} - \theta\right) = \frac{\cos\theta - \sin\theta}{\cos\theta - \sin\theta}$

45. $\cos(\alpha + \beta)\cos(\alpha - \beta) = \cos^2\beta - \sin^2\alpha$

46. $\sin(\alpha + \beta)\sin(\alpha - \beta) = \cos^2\beta - \cos^2\alpha$

47. $\frac{\sin(\alpha + \beta)}{\sin(\alpha - \beta)} = \frac{\tan\alpha + \tan\beta}{\tan\alpha - \tan\beta}$

48. $\frac{\cos(\alpha + \beta)}{\cos(\alpha - \beta)} = \frac{1 - \tan\alpha\tan\beta}{1 + \tan\alpha\tan\beta}$

49. $\frac{\cos(x + h) - \cos x}{h} = \cos x\frac{\cos h - 1}{h} - \sin x\frac{\sin h}{h}$

50. $\dfrac{\sin(x+h)-\sin x}{h} = \cos x \dfrac{\sin h}{h} + \sin x \dfrac{\cos h - 1}{h}$

51. $\sin 2\alpha = 2 \sin \alpha \cos \alpha$
Hint: Write $\sin 2\alpha$ as $\sin(\alpha + \alpha)$.

52. $\cos 2\alpha = \cos^2\alpha - \sin^2\alpha$
Hint: Write $\cos 2\alpha$ as $\cos(\alpha + \alpha)$.

53. $\tan 2\alpha = \dfrac{2\tan\alpha}{1-\tan^2\alpha}$
Hint: Write $\tan 2\alpha$ as $\tan(\alpha + \alpha)$.

54. $\tan\left(\dfrac{\pi}{4}+\alpha\right) - \tan\left(\dfrac{\pi}{4}-\alpha\right) = 2\tan 2\alpha$
Hint: Use the result in Exercise 53.

55. Derive the identity for $\tan(\alpha + \beta)$ using

$$\tan(\alpha+\beta) = \frac{\sin(\alpha+\beta)}{\cos(\alpha+\beta)}.$$

After applying the formulas for sums of sines and cosines, divide the numerator and denominator by $\cos\alpha\cos\beta$.

56. Derive the identity for $\tan(\alpha - \beta)$ using

$$\tan(\alpha-\beta) = \tan[\alpha + (-\beta)].$$

After applying the formula for the tangent of the sum of two angles, use the fact that the tangent is an odd function.

In Exercises 57–62, find the exact value of the following under the given conditions:

a. $\cos(\alpha+\beta)$ ***b.*** $\sin(\alpha+\beta)$ ***c.*** $\tan(\alpha+\beta)$

57. $\sin\alpha = \frac{3}{5}$, α lies in quadrant I, and $\sin\beta = \frac{5}{13}$, β lies in quadrant II.

58. $\sin\alpha = \frac{4}{5}$, α lies in quadrant I, and $\sin\beta = \frac{7}{25}$, β lies in quadrant II.

59. $\tan\alpha = -\frac{3}{4}$, α lies in quadrant II, and $\cos\beta = \frac{1}{3}$, β lies in quadrant I.

60. $\tan\alpha = -\frac{4}{3}$, α lies in quadrant II, and $\cos\beta = \frac{2}{3}$, β lies in quadrant I.

61. $\cos\alpha = \frac{8}{17}$, α lies in quadrant IV, and $\sin\beta = -\frac{1}{2}$, β lies in quadrant III.

62. $\cos\alpha = \frac{1}{2}$, α lies in quadrant IV, and $\sin\beta = -\frac{1}{3}$, β lies in quadrant III.

In Exercises 63–66, the graph with the given equation is shown in a $\left[0, 2\pi, \frac{\pi}{2}\right]$ *by* $[-2, 2, 1]$ *viewing rectangle.*

a. *Describe the graph using another equation.*

b. *Verify that the two equations are equivalent.*

63.

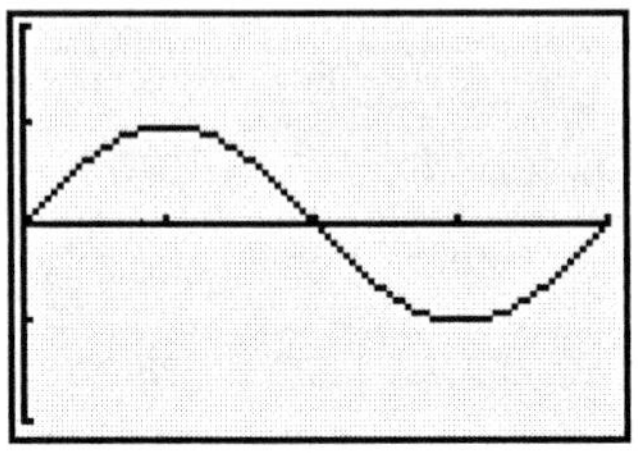

$y = \sin(\pi - x)$

64.

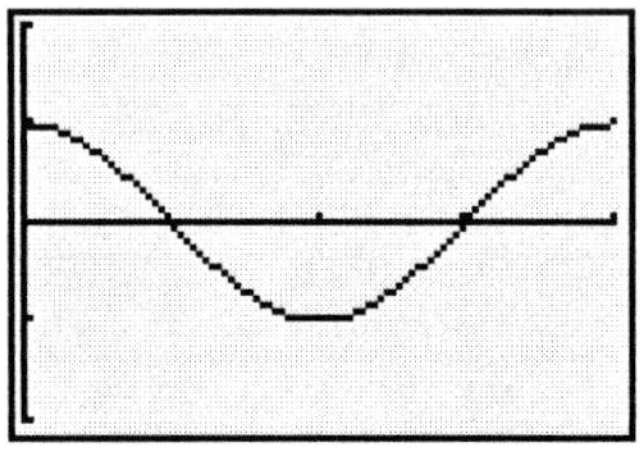

$y = \cos(x - 2\pi)$

65.

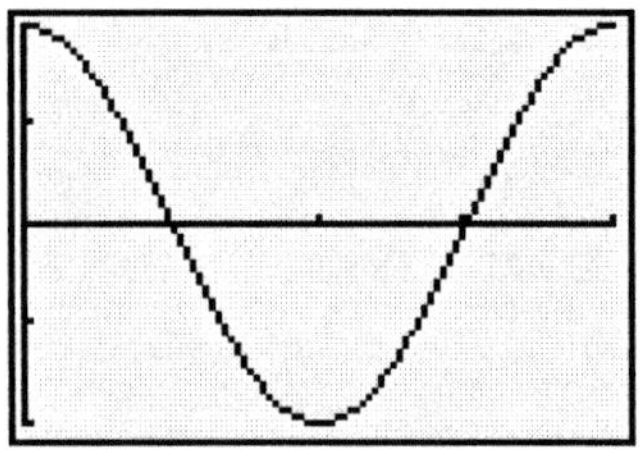

$y = \sin\left(x + \dfrac{\pi}{2}\right) + \sin\left(\dfrac{\pi}{2} - x\right)$

66.

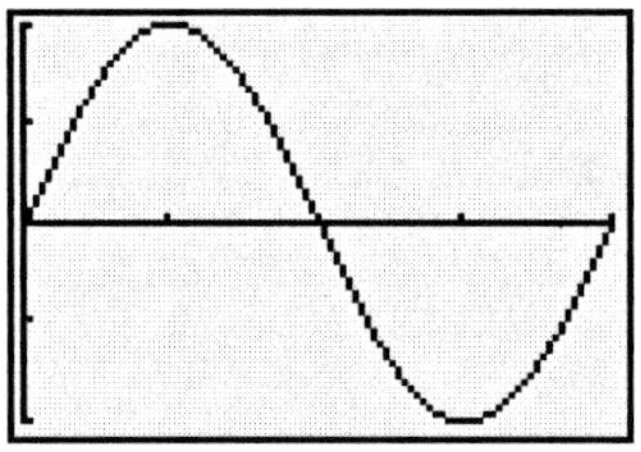

$y = \cos\left(x - \dfrac{\pi}{2}\right) - \cos\left(x + \dfrac{\pi}{2}\right)$

Application Exercises

67. A ball attached to a spring is raised 2 feet and released with an initial vertical velocity of 3 feet per second. The distance of the ball from its rest position after t seconds is given by $d = 2\cos t + 3\sin t$. Show that

$$2\cos t + 3\sin t = \sqrt{13}\cos(t - \theta)$$

where θ lies in quadrant I and $\tan\theta = \frac{3}{2}$. Use the identity to find the amplitude and the period of the ball's motion.

68. A tuning fork is held a certain distance from your ears and struck. Your eardrums' vibrations after t seconds are given by $p = 3\sin 2t$. When a second tuning fork is struck, the formula $p = 2\sin(2t + \pi)$ describes the effects of the sound on the eardrums' vibrations. The total vibrations are given by $p = 3\sin 2t + 2\sin(2t + \pi)$.

a. Express p using a single occurrence of the sine.

b. If the amplitude of p is zero, no sound is heard. Based on your equation in part (a), does this occur with the two tuning forks in this exercise? Explain your answer.

Writing in Mathematics

In Exercises 69–74, use words to describe the formula for:

69. the cosine of the difference of two angles.

70. the cosine of the sum of two angles.

71. the sine of the sum of two angles.

72. the sine of the difference of two angles.

73. the tangent of the difference of two angles.

74. the tangent of the sum of two angles.

75. The distance formula and the definitions for cosine and sine are used to prove the formula for the cosine of the difference of two angles. This formula logically leads the way to the other sum and difference identities. Using this development of ideas and formulas, describe a characteristic of mathematical logic.

Technology Exercises

In Exercises 76–81, graph each side of the equation in the same viewing rectangle. If the graphs appear to coincide, verify that the equation is an identity. If the graphs do not appear to coincide, this indicates that the equation is not an identity. In these exercises, find a value of x for which both sides are defined but not equal.

76. $\cos\left(\frac{3\pi}{2} - x\right) = -\sin x$

77. $\tan(\pi - x) = -\tan x$

78. $\sin\left(x + \frac{\pi}{2}\right) = \sin x + \sin\frac{\pi}{2}$

79. $\cos\left(x + \frac{\pi}{2}\right) = \cos x + \cos\frac{\pi}{2}$

80. $\cos 1.2x \cos 0.8x - \sin 1.2x \sin 0.8x = \cos 2x$

81. $\sin 1.2x \cos 0.8x + \cos 1.2x \sin 0.8x = \sin 2x$

Critical Thinking Exercises

82. Graph $y = \sin 5x \cos 3x - \cos 5x \sin 3x$ from 0 to 2π.

83. Verify the identity:

$$\frac{\sin(x-y)}{\cos x \cos y} + \frac{\sin(y-z)}{\cos y \cos z} + \frac{\sin(z-x)}{\cos z \cos x} = 0.$$

In Exercises 84–87, find the exact value of each expression. Do not use a calculator.

84. $\sin\left(\cos^{-1}\frac{1}{2} + \sin^{-1}\frac{3}{5}\right)$

85. $\sin\left[\sin^{-1}\frac{3}{5} - \cos^{-1}\left(-\frac{4}{5}\right)\right]$

86. $\cos\left(\tan^{-1}\frac{4}{3} + \cos^{-1}\frac{5}{13}\right)$

87. $\cos\left[\cos^{-1}\left(-\frac{\sqrt{3}}{2}\right) - \sin^{-1}\left(-\frac{1}{2}\right)\right]$

In Exercises 88–90, write each trigonometric expression as an algebraic expression as an algebraic expression (that is, without any trigonometric functions). Assume that x and y are positive and in the domain of the given inverse trigonometric function.

88. $\cos(\sin^{-1} x - \cos^{-1} y)$

89. $\sin(\tan^{-1} x - \sin^{-1} y)$

90. $\tan(\sin^{-1} x + \cos^{-1} y)$

Group Exercise

91. Remembering the six sum and difference identities can be difficult. Did you have problems with some exercises because the identity you were using in your head turned out to be an incorrect formula? Are there easy ways to remember the six new identities presented in this section? Group members should address this question, considering one identity at a time. For each formula, list ways to make it easier to remember.

SECTION 6.3 *Double-Angle and Half-Angle Formulas*

Objectives

1. Use the double-angle formulas.
2. Use the power-reducing formulas.
3. Use the half-angle formulas.

We have a long history of throwing things. Prior to 400 B.C., the Greeks competed in games that included discuss throwing. In the seventeenth century,

English soldiers organized cannonball-throwing competitions. In 1827, a Yale University student, disappointed over failing an exam, took out his frustrations at the passing of a collection plate in chapel. Seizing the monetary tray, he flung it in the direction of a large open space on campus. Yale students see this act of frustration as the origin of the Frisbee.

In this section, we develop other important classes of identities, called the double-angle and half-angle formulas. We will see how one of these formulas can be used by athletes to increase throwing distance.

1 Use the double-angle formulas.

Double-Angle Formulas

A number of basic identities follow from the sum formulas for sine, cosine, and tangent. The first category of identities involves **double-angle formulas.**

Study Tip

The 2 that appears in each of the double-angle expressions cannot be pulled to the front and written as a coefficient.

INCORRECT

~~$\sin 2\theta = 2\sin\theta$~~

~~$\cos 2\theta = 2\cos\theta$~~

~~$\tan 2\theta = 2\tan\theta$~~

The figure shows, in a $\left[0, 2\pi, \frac{\pi}{2}\right]$ by $[-3, 3, 1]$ viewing rectangle, that the graphs of

$$y = \sin 2x$$

and

$$y = 2\sin x$$

do not coincide.

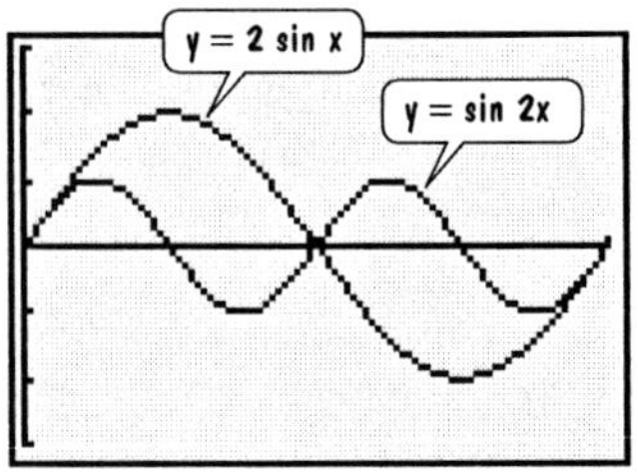

Double-Angle Formulas

$$\sin 2\theta = 2\sin\theta\cos\theta$$

$$\cos 2\theta = \cos^2\theta - \sin^2\theta$$

$$\tan 2\theta = \frac{2\tan\theta}{1 - \tan^2\theta}$$

To prove each of these formulas, we replace α and β by θ in the sum formulas for $\sin(\alpha + \beta)$, $\cos(\alpha + \beta)$, and $\tan(\alpha + \beta)$.

- $\sin 2\theta = \sin(\theta + \theta) = \sin\theta\cos\theta + \cos\theta\sin\theta = 2\sin\theta\cos\theta$

  We use $\sin(\alpha + \beta) = \sin\alpha\cos\beta + \cos\alpha\sin\beta$.

- $\cos 2\theta = \cos(\theta + \theta) = \cos\theta\cos\theta - \sin\theta\sin\theta = \cos^2\theta - \sin^2\theta$

  We use $\cos(\alpha + \beta) = \cos\alpha\cos\beta - \sin\alpha\sin\beta$.

- $\tan 2\theta = \tan(\theta + \theta) = \dfrac{\tan\theta + \tan\theta}{1 - \tan\theta\tan\theta} = \dfrac{2\tan\theta}{1 - \tan^2\theta}$

  We use $\tan(\alpha + \beta) = \dfrac{\tan\alpha + \tan\beta}{1 - \tan\alpha\tan\beta}$.

EXAMPLE 1 Using Double-Angle Formulas to Find Exact Values

If $\sin\theta = \frac{5}{13}$ and θ lies in quadrant II, find the exact value of:

a. $\sin 2\theta$ **b.** $\cos 2\theta$ **c.** $\tan 2\theta$.

Solution We begin with a sketch that illustrates

$$\sin\theta = \frac{5}{13} = \frac{y}{r}.$$

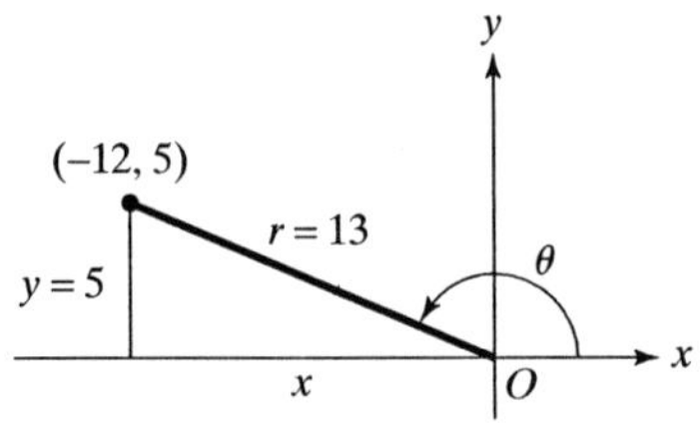

Figure 6.6 $\sin\theta = \frac{5}{13}$ and θ lies in quadrant II

Figure 6.6 shows a quadrant II angle θ for which $\sin\theta = \frac{5}{13}$. We find x using $x^2 + y^2 = r^2$. Because θ lies in quadrant II, x is negative.

$$x^2 + 5^2 = 13^2 \qquad x^2 + y^2 = r^2$$

$$x^2 = 13^2 - 5^2 = 144 \qquad \text{Solve for } x^2.$$

$$x = -\sqrt{144} = -12 \qquad \text{In quadrant II, } x \text{ is negative.}$$

Now we can use values for x, y, and r to find the required values. We will use $\cos\theta = \frac{x}{r} = -\frac{12}{13}$ and $\tan\theta = \frac{y}{x} = -\frac{5}{12}$. We were given $\sin\theta = \frac{5}{13}$.

a. $\sin 2\theta = 2\sin\theta\cos\theta = 2\left(\frac{5}{13}\right)\left(-\frac{12}{13}\right) = -\frac{120}{169}$

b. $\cos 2\theta = \cos^2\theta - \sin^2\theta = \left(-\frac{12}{13}\right)^2 - \left(\frac{5}{13}\right)^2 = \frac{144}{169} - \frac{25}{169} = \frac{119}{169}$

c. $\tan 2\theta = \dfrac{2\tan\theta}{1-\tan^2\theta} = \dfrac{2\left(-\frac{5}{12}\right)}{1-\left(-\frac{5}{12}\right)^2} = \dfrac{-\frac{5}{6}}{1-\frac{25}{144}} = \dfrac{-\frac{5}{6}}{\frac{119}{144}} = \left(-\frac{5}{6}\right)\left(\frac{144}{119}\right) = -\frac{120}{119}$

Check Point 1 If $\sin\theta = \frac{4}{5}$ and θ lies in quadrant II, find the exact value of:

a. $\sin 2\theta$ **b.** $\cos 2\theta$ **c.** $\tan 2\theta$.

EXAMPLE 2 Using the Double–Angle Formula for Tangents to Find an Exact Value

Find the exact value of $\dfrac{2\tan 15°}{1-\tan^2 15°}$.

Solution The given expression is the right side of the formula for $\tan 2\theta$ with $\theta = 15$.

$$\tan 2\theta = \frac{2\tan\theta}{1-\tan^2\theta}$$

$$\frac{2\tan 15°}{1-\tan^2 15°} = \tan(2 \cdot 15°) = \tan 30° = \frac{\sqrt{3}}{3}$$

Check Point 2 Find the exact value of $\cos^2 15° - \sin^2 15°$.

There are three forms of the double-angle formula for $\cos 2\theta$. The form we have seen involves both the cosine and the sine:

$$\cos 2\theta = \cos^2\theta - \sin^2\theta.$$

Using the Pythagorean identity $\sin^2\theta + \cos^2\theta = 1$, we can write this last formula in terms of the cosine only. We substitute $1 - \cos^2\theta$ for $\sin^2\theta$.

$$\cos 2\theta = \cos^2\theta - \sin^2\theta = \cos^2\theta - (1 - \cos^2\theta)$$
$$= \cos^2\theta - 1 + \cos^2\theta = 2\cos^2\theta - 1$$

We can also use a Pythagorean identity to write $\cos 2\theta$ in terms of sine only. We substitute $1 - \sin^2\theta$ for $\cos^2\theta$.

$$\cos 2\theta = \cos^2\theta - \sin^2\theta = 1 - \sin^2\theta - \sin^2\theta = 1 - 2\sin^2\theta$$

Three Forms of the Double-Angle Formula for cos 2θ

$$\cos 2\theta = \cos^2\theta - \sin^2\theta$$
$$\cos 2\theta = 2\cos^2\theta - 1$$
$$\cos 2\theta = 1 - 2\sin^2\theta$$

EXAMPLE 3 Verifying an Identity

Verify the identity: $\cos 3\theta = 4\cos^2\theta - 3\cos\theta$.

Solution We begin by working with the left side. In order to obtain an expression for $\cos 3\theta$, we use the sum formula and write 3θ as $2\theta + \theta$.

$\cos 3\theta = \cos(2\theta + \theta)$ — Write 3θ as $2\theta + \theta$.

$= \cos 2\theta \cos\theta - \sin 2\theta \sin\theta$ — $\cos(\alpha + \beta) = \cos\alpha\cos\beta - \sin\alpha\sin\beta$

($\cos 2\theta$ → $2\cos^2\theta - 1$; $\sin 2\theta$ → $2\sin\theta\cos\theta$)

$= (2\cos^2\theta - 1)\cos\theta - 2\sin\theta\cos\theta\sin\theta$ — Substitute double-angle formulas. Because the right side of the given equation involves cosines only, use this form for $\cos 2\theta$.

$= 2\cos^3\theta - \cos\theta - 2\sin^2\theta\cos\theta$ — Multiply.

($\sin^2\theta$ → $1 - \cos^2\theta$)

$= 2\cos^3\theta - \cos\theta - 2(1 - \cos^2\theta)\cos\theta$ — To get cosines only, use $\sin^2\theta + \cos^2\theta = 1$ and substitute $1 - \cos^2\theta$ for $\sin^2\theta$.

$= 2\cos^3\theta - \cos\theta - 2\cos\theta + 2\cos^3\theta$ — Multiply.

$= 4\cos^3\theta - 3\cos\theta$ — Simplify: $2\cos^3\theta + 2\cos^3\theta = 4\cos^3\theta$ and $-\cos\theta - 2\cos\theta = -3\cos\theta$.

By working with the left side and expressing it in a form identical to the right side, we have verified the identity.

Check Point 3 Verify the identity: $\sin 3\theta = 3\sin\theta - 4\sin^3\theta$.

2 Use the power-reducing formulas.

Power-Reducing Formulas

The double-angle formulas are used to derive the **power-reducing formulas:**

Power-Reducing Formulas

$$\sin^2\theta = \frac{1 - \cos 2\theta}{2} \qquad \cos^2\theta = \frac{1 + \cos 2\theta}{2} \qquad \tan^2\theta = \frac{1 - \cos 2\theta}{1 + \cos 2\theta}$$

We can prove the first two formulas in the box by working with two forms of the double-angle formula for $\cos 2\theta$.

This is the form with sine only. This is the form with cosine only.

$$\cos 2\theta = 1 - 2\sin^2\theta \qquad \cos 2\theta = 2\cos^2\theta - 1$$

Solve the formula on the left for $\sin^2\theta$. Solve the formula on the right for $\cos^2\theta$.

$$2\sin^2\theta = 1 - \cos 2\theta \qquad 2\cos^2\theta = 1 + \cos 2\theta$$

$$\sin^2\theta = \frac{1-\cos 2\theta}{2} \qquad \cos^2\theta = \frac{1+\cos 2\theta}{2}$$ Divide both sides of each equation by 2.

These are the first two formulas in the box. The third formula in the box is proved by writing the tangent as the quotient of the sine and the cosine.

$$\tan^2\theta = \frac{\sin^2\theta}{\cos^2\theta} = \frac{\frac{1-\cos 2\theta}{2}}{\frac{1+\cos 2\theta}{2}} = \frac{1-\cos 2\theta}{\cancel{2}_1} \cdot \frac{\cancel{2}^1}{1+\cos 2\theta} = \frac{1-\cos 2\theta}{1+\cos 2\theta}$$

Power-reducing formulas are quite useful in calculus. By reducing the power of trigonometric functions, calculus can better explore the relationship between a function and how it is changing at every single instant in time.

EXAMPLE 4 Reducing the Power of a Trigonometric Function

Write an equivalent expression for $\cos^4 x$ that does not contain powers of trigonometric functions greater than 1.

Solution We will apply the formula for $\cos^2\theta$ twice.

$$\cos^4 x = (\cos^2 x)^2$$

$$= \left(\frac{1+\cos 2x}{2}\right)^2$$ Use $\cos^2\theta = \frac{1+\cos 2\theta}{2}$ with $\theta = x$.

$$= \frac{1 + 2\cos 2x + \cos^2 2x}{4}$$ Square the numerator: $(A+B)^2 = A^2 + 2AB + B^2$. Square the denominator.

$$= \frac{1}{4} + \frac{1}{2}\cos 2x + \frac{1}{4}\cos^2 2x$$ Divide each term in the numerator by 4.

We can reduce the power of $\cos^2 2x$ using $\cos^2\theta = \frac{1+\cos 2\theta}{2}$ with $\theta = 2x$.

$$= \frac{1}{4} + \frac{1}{2}\cos 2x + \frac{1}{4}\left[\frac{1+\cos 2(2x)}{2}\right]$$ Use the power-reducing formula for $\cos^2\theta$ with $\theta = 2x$.

$$\cos^4 x = \frac{1}{4} + \frac{1}{2}\cos 2x + \frac{1}{4}\left[\frac{1 + \cos 2(2x)}{2}\right]$$ This is our expression for $\cos^4 x$ from the previous page.

$$= \frac{1}{4} + \frac{1}{2}\cos 2x + \frac{1}{8}(1 + \cos 4x)$$ Multiply.

$$= \frac{1}{4} + \frac{1}{2}\cos 2x + \frac{1}{8} + \frac{1}{8}\cos 4x$$ Distribute $\frac{1}{8}$ throughout parentheses.

$$= \frac{3}{8} + \frac{1}{2}\cos 2x + \frac{1}{8}\cos 4x$$ Simplify: $\frac{1}{4} + \frac{1}{8} = \frac{2}{8} + \frac{1}{8} = \frac{3}{8}$.

Thus, $\cos^4 x = \frac{3}{8} + \frac{1}{2}\cos 2x + \frac{1}{8}\cos 4x$. The expression for $\cos^4 x$ does not contain powers of trigonometric functions greater than 1.

Check Point 4 Write an equivalent expression for $\sin^4 x$ that does not contain powers of trigonometric functions greater than 1.

3 Use the half-angle formulas.

Half-Angle Formulas

Useful equivalent forms of the power-reducing formulas can be obtained by replacing θ by $\frac{\alpha}{2}$. Then solve for the trigonometric function on the left sides of the equations. The resulting identities are called the **half-angle formulas:**

Half-Angle Formulas

$$\sin\frac{\alpha}{2} = \pm\sqrt{\frac{1 - \cos\alpha}{2}}$$

$$\cos\frac{\alpha}{2} = \pm\sqrt{\frac{1 + \cos\alpha}{2}}$$

$$\tan\frac{\alpha}{2} = \pm\sqrt{\frac{1 - \cos\alpha}{1 + \cos\alpha}}$$

The + or − in each formula is determined by the quadrant in which $\frac{\alpha}{2}$ lies.

Study Tip

The $\frac{1}{2}$ that appears in each of the half-angle formulas cannot be pulled to the front and written as a coefficient.

INCORRECT

$$\sin\frac{\theta}{2} = \frac{1}{2}\sin\theta$$

$$\cos\frac{\theta}{2} = \frac{1}{2}\cos\theta$$

$$\tan\frac{\theta}{2} = \frac{1}{2}\tan\theta$$

The figure shows, in a $\left[0, 2\pi, \frac{\pi}{2}\right]$ by $[-2, 2, 1]$ viewing rectangle, that the graphs of

$$y = \sin\frac{x}{2}$$

and

$$y = \frac{1}{2}\sin x$$

do not coincide.

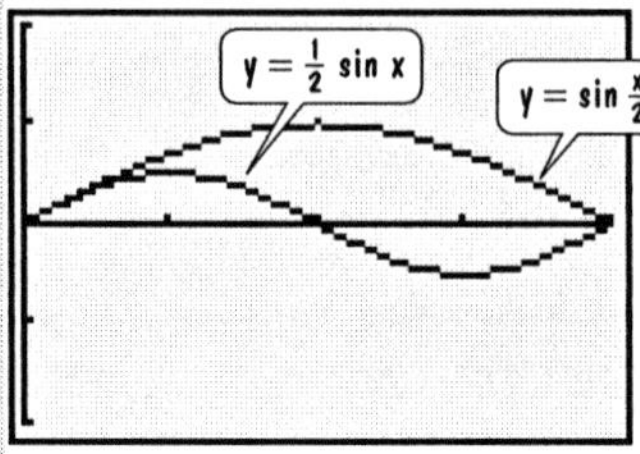

If we know the exact value for the sine, cosine, or tangent of an angle, we can use the half-angle formulas to find exact values for half that angle. For example, we know that $\cos 225° = -\frac{\sqrt{2}}{2}$. In the next example, we find the exact value of the cosine of half of 225°, or cos 112.5°.

EXAMPLE 5 Using a Half-Angle Formula to Find an Exact Value

Find the exact value of cos 112.5°.

Solution Because $112.5° = \frac{225°}{2}$, we use the half-angle formula for $\cos\frac{\alpha}{2}$ with $\alpha = 225°$. What sign should we use when we apply the formula? Because 112.5° lies in quadrant II, where only the sine and cosecant are positive, $\cos 112.5° < 0$. Thus, we use the − sign in the half-angle formula.

$$\cos 112.5° = \cos\frac{225°}{2}$$

$$= -\sqrt{\frac{1+\cos 225°}{2}}$$ Use $\cos\frac{\alpha}{2} = -\sqrt{\frac{1+\cos\alpha}{2}}$ with $\alpha = 225°$.

$$= -\sqrt{\frac{1+\left(-\frac{\sqrt{2}}{2}\right)}{2}}$$ $\cos 225° = -\frac{\sqrt{2}}{2}$

$$= -\sqrt{\frac{2-\sqrt{2}}{4}}$$ Multiply the radicand by $\frac{2}{2}$: $\frac{1+\left(-\frac{\sqrt{2}}{2}\right)}{2}\cdot\frac{2}{2} = \frac{2-\sqrt{2}}{4}$.

$$= -\frac{\sqrt{2-\sqrt{2}}}{2}$$ Simplify: $\sqrt{4} = 2$.

Discovery

Use your calculator to find approximations for

$$-\frac{\sqrt{2-\sqrt{2}}}{2}$$

and cos 112.5°. What do you observe?

Check Point 5 Use $\cos 210° = -\frac{\sqrt{3}}{2}$ to find the exact value of cos 105°.

There are alternate formulas for $\tan\frac{\alpha}{2}$ that do not require us to determine what sign to use when applying the formula. These formulas are logically connected to the identities in Example 6 and Check Point 6.

EXAMPLE 6 Verifying an Identity

Verify the identity: $\tan\theta = \frac{1-\cos 2\theta}{\sin 2\theta}$.

Solution We work with the right side.

$$\frac{1-\cos 2\theta}{\sin 2\theta} = \frac{1-(1-2\sin^2\theta)}{2\sin\theta\cos\theta}$$ The form $\cos 2\theta = 1 - 2\sin^2\theta$ is used because it produces only one term in the numerator. Use the double-angle formula for sine in the denominator.

$$= \frac{2\sin^2\theta}{2\sin\theta\cos\theta}$$ Simplify the numerator.

$$= \frac{\sin\theta}{\cos\theta}$$ Divide the numerator and denominator by $2\sin\theta$.

$$= \tan\theta$$ Use a quotient identity: $\tan\theta = \frac{\sin\theta}{\cos\theta}$.

The right side simplifies to $\tan\theta$, the expression on the left side. Thus, the identity is verified.

Check Point 6 Verify the identity: $\tan\theta = \dfrac{\sin 2\theta}{1 + \cos 2\theta}$.

$\tan\theta = \dfrac{1 - \cos 2\theta}{\sin 2\theta}$

Example 6 identity, repeated

Half-angle formulas for $\tan\dfrac{\alpha}{2}$ can be obtained using the identities in Example 6 and Check Point 6. Do you see how to do this? Replace each occurrence of θ with $\dfrac{\alpha}{2}$. This results in the following identities:

Half-Angle Formulas for tan $\dfrac{\alpha}{2}$

$$\tan\frac{\alpha}{2} = \frac{1 - \cos\alpha}{\sin\alpha}$$

$$\tan\frac{\alpha}{2} = \frac{\sin\alpha}{1 + \cos\alpha}$$

EXAMPLE 7 Verifying an Identity

Verify the identity: $\tan\dfrac{\alpha}{2} = \csc\alpha - \cot\alpha$.

Solution We begin with the right side.

$$\csc\alpha - \cot\alpha = \frac{1}{\sin\alpha} - \frac{\cos\alpha}{\sin\alpha} = \frac{1 - \cos\alpha}{\sin\alpha} = \tan\frac{\alpha}{2}$$

Express functions in terms of sines and cosines.

This is the first of the two half-angle formulas in the preceding box.

We worked with the right side and arrived at the left side. Thus, the identity is verified.

Check Point 7 Verify the identity: $\tan\dfrac{\alpha}{2} = \dfrac{\sec\alpha}{\sec\alpha\csc\alpha + \csc\alpha}$.

We conclude with a summary of the principal trigonometric identities developed in this section and the previous section. The fundamental identities can be found in the box on page 557.

Principal Trigonometric Identities

Sum and Difference Formulas

$$\sin(\alpha+\beta) = \sin\alpha\cos\beta + \cos\alpha\sin\beta \qquad \sin(\alpha-\beta) = \sin\alpha\cos\beta - \cos\alpha\sin\beta$$

$$\cos(\alpha+\beta) = \cos\alpha\cos\beta - \sin\alpha\sin\beta \qquad \cos(\alpha-\beta) = \cos\alpha\cos\beta + \sin\alpha\sin\beta$$

$$\tan(\alpha+\beta) = \frac{\tan\alpha + \tan\beta}{1 - \tan\alpha\tan\beta} \qquad \tan(\alpha-\beta) = \frac{\tan\alpha - \tan\beta}{1 + \tan\alpha\tan\beta}$$

Double-Angle Formulas

$$\sin 2\theta = 2\sin\theta\cos\theta$$

$$\cos 2\theta = \cos^2\theta - \sin^2\theta = 2\cos^2\theta - 1 = 1 - 2\sin^2\theta$$

$$\tan 2\theta = \frac{2\tan\theta}{1 - \tan^2\theta}$$

Power-Reducing Formulas

$$\sin^2\theta = \frac{1 - \cos 2\theta}{2} \qquad \cos^2\theta = \frac{1 + \cos 2\theta}{2} \qquad \tan^2\theta = \frac{1 - \cos 2\theta}{1 + \cos 2\theta}$$

Half-Angle Formulas

$$\sin\frac{\alpha}{2} = \pm\sqrt{\frac{1 - \cos\alpha}{2}} \qquad \cos\frac{\alpha}{2} = \pm\sqrt{\frac{1 + \cos\alpha}{2}}$$

$$\tan\frac{\alpha}{2} = \pm\sqrt{\frac{1 - \cos\alpha}{1 + \cos\alpha}} = \frac{1 - \cos\alpha}{\sin\alpha} = \frac{\sin\alpha}{1 + \cos\alpha}$$

Study Tip

To help remember the correct sign in the numerator in the first two power-reducing formulas and the first two half-angle formulas, remember *sinus-minus*–the sine is minus.

EXERCISE SET 6.3

Practice Exercises

In Exercises 1–6, use the figures to find the exact value of each trigonometric function.

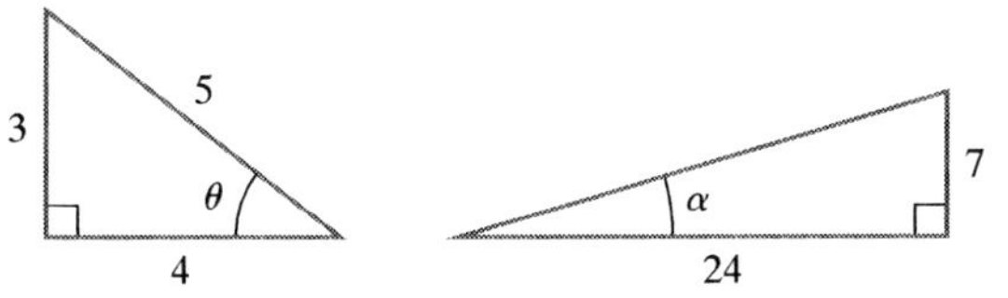

1. $\sin 2\theta$
2. $\cos 2\theta$
3. $\tan 2\theta$
4. $\sin 2\alpha$
5. $\cos 2\alpha$
6. $\tan 2\alpha$

In Exercises 7–14, use the given information to find the exact value of:

a. $\sin 2\theta$ b. $\cos 2\theta$ c. $\tan 2\theta$

7. $\sin\theta = \frac{15}{17}$, θ lies in quadrant II.
8. $\sin\theta = \frac{12}{13}$, θ lies in quadrant II.
9. $\cos\theta = \frac{24}{25}$, θ lies in quadrant IV.
10. $\cos\theta = \frac{40}{41}$, θ lies in quadrant IV.
11. $\cot\theta = 2$, θ lies in quadrant III.
12. $\cot\theta = 3$, θ lies in quadrant III.
13. $\sin\theta = -\frac{9}{41}$, θ lies in quadrant III.
14. $\sin\theta = -\frac{2}{3}$, θ lies in quadrant III.

In Exercises 15–22, write each expression as the sine, cosine, or tangent of a double angle. Then find the exact value of the expression.

15. $2\sin 15^\circ \cos 15^\circ$
16. $2\sin 22.5^\circ \cos 22.5^\circ$
17. $\cos^2 75^\circ - \sin^2 75^\circ$
18. $\cos^2 105^\circ - \sin^2 105^\circ$
19. $2\cos^2\frac{\pi}{8} - 1$
20. $1 - 2\sin^2\frac{\pi}{12}$
21. $\dfrac{2\tan\frac{\pi}{12}}{1 - \tan^2\frac{\pi}{12}}$
22. $\dfrac{2\tan\frac{\pi}{8}}{1 - \tan^2\frac{\pi}{8}}$

In Exercises 23–34, verify each identity.

23. $\sin 2\theta = \dfrac{2\tan\theta}{1 + \tan^2\theta}$
24. $\sin 2\theta = \dfrac{2\cot\theta}{1 + \cot^2\theta}$
25. $(\sin\theta + \cos\theta)^2 = 1 + \sin 2\theta$
26. $(\sin\theta - \cos\theta)^2 = 1 - \sin 2\theta$
27. $\sin^2 x + \cos 2x = \cos^2 x$

28. $1 - \tan^2 x = \dfrac{\cos 2x}{\cos^2 x}$

29. $\cot x = \dfrac{\sin 2x}{1 - \cos 2x}$

30. $\cot x = \dfrac{1 + \cos 2x}{\sin 2x}$

31. $\sin 2t - \tan t = \tan t \cos 2t$

32. $\sin 2t - \cot t = -\cot t \cos 2t$

33. $\sin 4t = 4 \sin t \cos^3 t - 4 \sin^3 t \cos t$

34. $\cos 4t = 8 \cos^4 t - 8 \cos^2 t + 1$

In Exercises 35–38, use the power-reducing formulas to rewrite each expression as an equivalent expression that does not contain powers of trigonometric functions greater than 1.

35. $6 \sin^4 x$

36. $10 \cos^4 x$

37. $\sin^2 x \cos^2 x$

38. $8 \sin^2 x \cos^2 x$

In Exercises 39–46, use a half-angle formula to find the exact value of each expression.

39. $\sin 15°$

40. $\cos 22.5°$

41. $\cos 157.5°$

42. $\sin 105°$

43. $\tan 75°$

44. $\tan 112.5°$

45. $\tan \dfrac{7\pi}{8}$

46. $\tan \dfrac{3\pi}{8}$

In Exercises 47–54, use the figures to find the exact value of each trigonometric function.

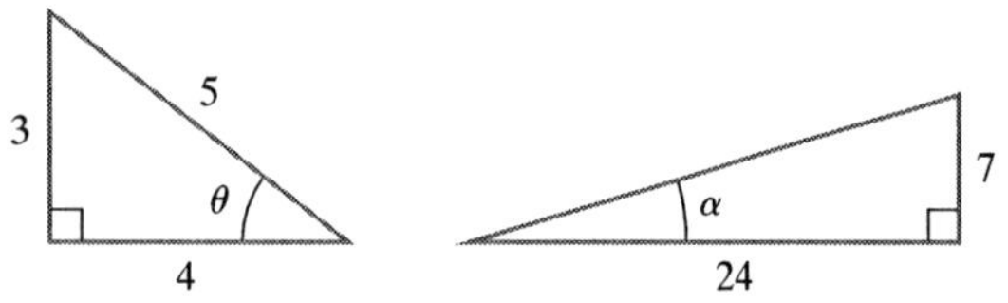

47. $\sin \dfrac{\theta}{2}$

48. $\cos \dfrac{\theta}{2}$

49. $\tan \dfrac{\theta}{2}$

50. $\sin \dfrac{\alpha}{2}$

51. $\cos \dfrac{\alpha}{2}$

52. $\tan \dfrac{\alpha}{2}$

53. $2 \sin \dfrac{\theta}{2} \cos \dfrac{\theta}{2}$

54. $2 \sin \dfrac{\alpha}{2} \cos \dfrac{\alpha}{2}$

In Exercises 55–58, use the given information to find the exact value of:

a. $\sin \dfrac{\alpha}{2}$ *b.* $\cos \dfrac{\alpha}{2}$ *c.* $\tan \dfrac{\alpha}{2}$.

55. $\tan \alpha = \frac{4}{3}$, α lies in quadrant III.

56. $\tan \alpha = \frac{8}{15}$, α lies in quadrant III.

57. $\sec \alpha = -\frac{13}{5}$, α lies in quadrant II.

58. $\sec \alpha = -3$, α lies in quadrant II.

In Exercises 59–68, verify each identity.

59. $\sin^2 \dfrac{\theta}{2} = \dfrac{\sec \theta - 1}{2 \sec \theta}$

60. $\sin^2 \dfrac{\theta}{2} = \dfrac{\csc \theta - \cot \theta}{2 \csc \theta}$

61. $\cos^2 \dfrac{\theta}{2} = \dfrac{\sin \theta + \tan \theta}{2 \tan \theta}$

62. $\cos^2 \dfrac{\theta}{2} = \dfrac{\sec \theta + 1}{2 \sec \theta}$

63. $\tan \dfrac{\alpha}{2} = \dfrac{\tan \alpha}{\sec \alpha + 1}$

64. $2 \tan \dfrac{\alpha}{2} = \dfrac{\sin^2 \alpha + 1 - \cos^2 \alpha}{\sin \alpha (1 + \cos \alpha)}$

65. $\cot \dfrac{x}{2} = \dfrac{\sin x}{1 - \cos x}$

66. $\cot \dfrac{x}{2} = \dfrac{1 + \cos x}{\sin x}$

67. $\tan \dfrac{x}{2} + \cot \dfrac{x}{2} = 2 \csc x$

68. $\tan \dfrac{x}{2} - \cot \dfrac{x}{2} = -2 \cot x$

Application Exercises

69. Throwing events in track and field include the shot put, the discuss throw, the hammer throw, and the javelin throw. The distance that the athlete can achieve depends on the initial speed of the object thrown and the angle above the horizontal at which the object leaves the hand. This angle is represented by θ in the figure shown. The distance, d, in feet, that the athlete throws is modeled by the formula

$$d = \frac{v_0^2}{16} \sin \theta \cos \theta$$

in which v_0 is the initial speed of the object thrown, in feet per second, and θ is the angle, in degrees, at which the object leaves the hand.

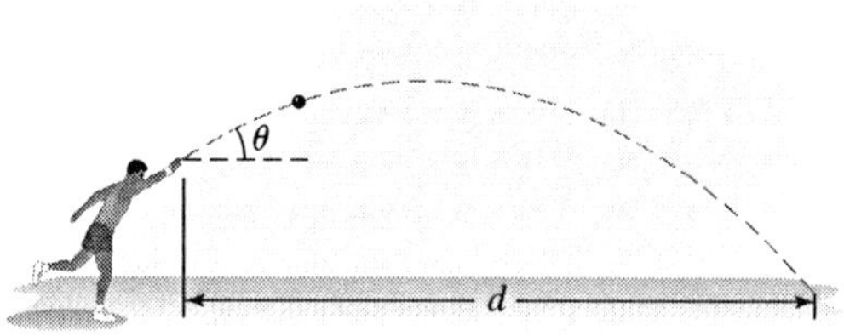

a. Use an identity to express the formula so that it contains the sine function only.

b. Use your formula from part (a) to find the angle, θ, that produces the maximum distance, d, for a given initial speed, v_0.

Use this information to solve Exercises 70–71.

The speed of a supersonic aircraft is usually represented by a Mach number, named after Austrian physicist Ernst Mach (1838–1916). A Mach number is the speed of the aircraft, in miles per hour, divided by the speed of sound, approximately 740 miles per hour. Thus, a plane flying at twice the speed of sound has a speed, M, of Mach 2. If an aircraft has a speed greater than Mach 1, a sonic boom is heard, created by sound waves that form a cone with a vertex angle θ, shown in the figure.

The relationship between the cone's vertex angle, θ, and the Mach speed, M, of an aircraft that is flying faster than the speed of sound is given by

$$\sin\frac{\theta}{2} = \frac{1}{M}.$$

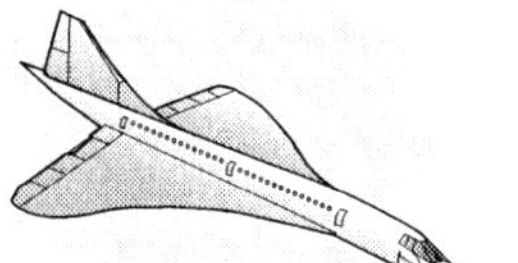

Concord
Mach 2.03

SR-71 Blackbird
Mach 3.3

70. If $\theta = \frac{\pi}{6}$, determine the Mach speed, M, of the aircraft. Express the speed as an exact value and as a decimal to the nearest tenth.

71. If $\theta = \frac{\pi}{4}$, determine the Mach speed, M, of the aircraft. Express the speed as an exact value and as a decimal to the nearest tenth.

Writing in Mathematics

In Exercises 72–79, use words to describe the formula for:

72. the sine of double an angle.

73. the cosine of double an angle. (Describe one of the three formulas.)

74. the tangent of double an angle.

75. the power-reducing formula for the sine squared of an angle.

76. the power-reducing formula for the cosine squared of an angle.

77. the sine of half an angle.

78. the cosine of half an angle.

79. the tangent of half an angle. (Describe one of the two formulas that does not involve a square root.)

80. Explain how the double-angle formulas are derived.

81. How can there be three forms of the double-angle formula for $\cos 2\theta$?

82. Without showing algebraic details, describe in words how to reduce the power of $\cos^4 x$.

83. Describe one or more of the techniques you use to help remember the identities in the box on page 585.

84. Your friend is about to compete as a shot-putter in a college field event. Using Exercise 69(b), write a short description to your friend on how to achieve the best distance possible in the throwing event.

Technology Exercises

In Exercises 85–88, graph each side of the equation in the same viewing rectangle. If the graphs appear to coincide, verify that the equation is an identity. If the graphs do not appear to coincide, find a value of x for which both sides are defined but not equal.

85. $3 - 6\sin^2 x = 3\cos 2x$

86. $4\cos^2\frac{x}{2} = 2 + 2\cos x$

87. $\sin\frac{x}{2} = \frac{1}{2}\sin x$

88. $\cos\frac{x}{2} = \frac{1}{2}\cos x$

In Exercises 89–91, graph each equation in a $\left[-2\pi, 2\pi, \frac{\pi}{2}\right]$ *by* $[-3, 3, 1]$ *viewing rectangle. Then* ***a.*** *Describe the graph using another equation, and* ***b.*** *Verify that the two equations are equivalent.*

89. $y = \dfrac{1 - 2\cos 2x}{2\sin x - 1}$

90. $y = \dfrac{2\tan\frac{x}{2}}{1 + \tan^2\frac{x}{2}}$

91. $y = \csc x - \cot x$

Critical Thinking Exercises

92. Verify the identity:

$$\sin^3 x + \cos^3 x = (\sin x + \cos x)\left(1 - \frac{\sin 2x}{2}\right).$$

93. Use the power-reducing formulas to rewrite $\sin^6 x$ as an equivalent expression that does not contain powers of trigonometric functions greater than 1.

94. Use a right triangle to write $\sin(2\sin^{-1} x)$ as an algebraic expression. Assume that x is positive and in the domain of the given inverse trigonometric function.

SECTION 6.4 Product-to-Sum and Sum-to-Product Formulas

Objectives

1. Use the product-to-sum formulas.
2. Use the sum-to-product formulas.

James K. Polk
Born November 2, 1795

Warren G. Harding
Born November 2, 1865

Of the 43 U.S. presidents, two share a birthday (same month and day). The probability of two or more people in a group sharing a birthday rises sharply as the group's size increases. Above 50 people, the probability approaches certainty. (You can verify the mathematics of this surprising result by studying Sections 11.6 and 11.7, and working Exercise 66 in Exercise Set 11.7.) So, come November 2, we salute Presidents Polk and Harding with

112, 163-, 112, 196-, 110, 8521-, 008, 121-.

Were you aware that each button on your touch-tone phone produces a unique sound? If we treat the commas as pauses and the hyphens as held notes, this sequence of numbers is *Happy Birthday* on a touch-tone phone.

Although *Happy Birthday* isn't Mozart or Sondheim, it is sinusoidal. Each of its touch-tone musical sounds can be described by the sum of two sine functions or the product of sines and cosines. In this section, we develop identities that enable us to use both descriptions. They are called the product-to-sum and sum-to-product formulas.

1 Use the product-to-sum formulas.

The Product-to-Sum Formulas

How do we write the products of sines and/or cosines as sums or differences? We use the following identities, which are called **product-to-sum formulas:**

Study Tip

You may not need to memorize the formulas in this section. When you need them, you can either refer to one of the two boxes in the section or perhaps even derive them using the methods shown.

Product-to-Sum Formulas

$$\sin\alpha\sin\beta = \tfrac{1}{2}[\cos(\alpha-\beta) - \cos(\alpha+\beta)]$$

$$\cos\alpha\cos\beta = \tfrac{1}{2}[\cos(\alpha-\beta) + \cos(\alpha+\beta)]$$

$$\sin\alpha\cos\beta = \tfrac{1}{2}[\sin(\alpha+\beta) + \sin(\alpha-\beta)]$$

$$\cos\alpha\sin\beta = \tfrac{1}{2}[\sin(\alpha+\beta) - \sin(\alpha-\beta)]$$

Although these formulas are difficult to remember, they are fairly easy to derive. For example, let's derive the first identity in the box,

$$\sin \alpha \sin \beta = \tfrac{1}{2}\left[\cos(\alpha - \beta) - \cos(\alpha + \beta)\right].$$

We begin with the difference and sum formulas for the cosine, and subtract the second identity from the first:

$$\begin{aligned} \cos(\alpha - \beta) &= \cos\alpha\cos\beta + \sin\alpha\sin\beta \\ -[\cos(\alpha + \beta) &= \cos\alpha\cos\beta - \sin\alpha\sin\beta] \\ \cos(\alpha - \beta) - \cos(\alpha + \beta) &= 0 + 2\sin\alpha\sin\beta\,. \end{aligned}$$

Subtract the identities.

Subtract terms on the left side.

Subtract terms on the right side: $\cos\alpha\cos\beta - \cos\alpha\cos\beta = 0$.

Subtract terms on the right side: $\sin\alpha\sin\beta - (-\sin\alpha\sin\beta) = 2\sin\alpha\sin\beta$.

Now we use this result to derive the product-to-sum formula for $\sin\alpha\sin\beta$.

$$2\sin\alpha\sin\beta = \cos(\alpha - \beta) - \cos(\alpha + \beta)$$

Reverse the sides in the preceding equation.

$$\sin\alpha\sin\beta = \tfrac{1}{2}\left[\cos(\alpha - \beta) - \cos(\alpha + \beta)\right]$$

Multiply each side by $\frac{1}{2}$.

This last equation is the desired formula. Likewise, we can derive the product-to-sum formula for cosine, $\cos\alpha\cos\beta = \frac{1}{2}\left[\cos(\alpha - \beta) + \cos(\alpha + \beta)\right]$. As we did for the previous derivation, begin with the difference and sum formulas for cosine. However, we *add* the formulas rather than subtracting them. Reversing both sides of this result and multiplying each side by $\frac{1}{2}$ produces the formula for $\cos\alpha\cos\beta$. The last two product-to-sum formulas, $\sin\alpha\cos\beta$ and $\cos\alpha\sin\beta$, are derived using the sum and difference formulas for sine in a similar manner.

EXAMPLE 1 Using the Product-to-Sum Formulas

Express each of the following products as a sum or difference:

a. $\sin 8x \sin 3x$ **b.** $\sin 4x \cos x$.

Solution The product-to-sum formula that we are using is shown in each of the voice balloons.

a. $\sin\alpha\sin\beta = \frac{1}{2}[\cos(\alpha-\beta) - \cos(\alpha+\beta)]$

$$\sin 8x \sin 3x = \tfrac{1}{2}\left[\cos(8x - 3x) - \cos(8x + 3x)\right] = \tfrac{1}{2}(\cos 5x - \cos 11x)$$

b. $\sin\alpha\cos\beta = \frac{1}{2}[\sin(\alpha+\beta) + \sin(\alpha-\beta)]$

$$\sin 4x \cos x = \tfrac{1}{2}\left[\sin(4x + x) + \sin(4x - x)\right] = \tfrac{1}{2}(\sin 5x + \sin 3x)$$

Check Point 1 Express each of the following products as a sum or difference:

a. $\sin 5x \sin 2x$ **b.** $\cos 7x \cos x$.

Technology

The graphs of

$$y = \sin 8x \sin 3x$$

and

$$y = \tfrac{1}{2}(\cos 5x - \cos 11x)$$

are shown in a $\left[-2\pi, 2\pi, \frac{\pi}{2}\right]$ by $[-1, 1, 1]$ viewing rectangle. The graphs coincide. This supports our algebraic work in Example 1(a).

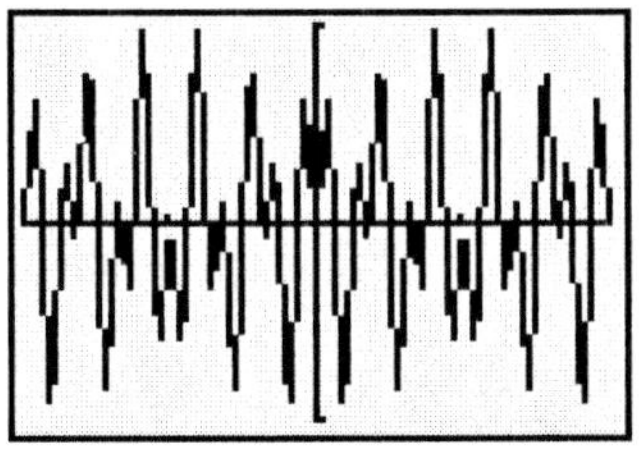

2 Use the sum-to-product formulas.

The Sum-to-Product Formulas

How do we write the sum or difference of sines and/or cosines as products? We use the following identities, which are called the **sum-to-product formulas:**

Sum-to-Product Formulas

$$\sin\alpha + \sin\beta = 2\sin\frac{\alpha+\beta}{2}\cos\frac{\alpha-\beta}{2}$$

$$\sin\alpha - \sin\beta = 2\sin\frac{\alpha-\beta}{2}\cos\frac{\alpha+\beta}{2}$$

$$\cos\alpha + \cos\beta = 2\cos\frac{\alpha+\beta}{2}\cos\frac{\alpha-\beta}{2}$$

$$\cos\alpha - \cos\beta = -2\sin\frac{\alpha+\beta}{2}\sin\frac{\alpha-\beta}{2}$$

We verify these formulas using the product-to-sum formulas. Let's verify the first sum-to-product formula

$$\sin\alpha + \sin\beta = 2\sin\frac{\alpha+\beta}{2}\cos\frac{\alpha-\beta}{2}.$$

We start with the right side of the formula, the side with the product. We can apply the product-to-sum formula for $\sin\alpha\cos\beta$ to this expression. By doing so, we obtain the left side of the formula, $\sin\alpha + \sin\beta$. Here's how:

$$\sin\alpha\cos\beta = \tfrac{1}{2}[\sin(\alpha+\beta) + \sin(\alpha-\beta)]$$

$$2\sin\frac{\alpha+\beta}{2}\cos\frac{\alpha-\beta}{2} = 2\cdot\frac{1}{2}\left[\sin\left(\frac{\alpha+\beta}{2}+\frac{\alpha-\beta}{2}\right) + \sin\left(\frac{\alpha+\beta}{2}-\frac{\alpha-\beta}{2}\right)\right]$$

$$= \sin\left(\frac{\alpha+\beta+\alpha-\beta}{2}\right) + \sin\left(\frac{\alpha+\beta-\alpha+\beta}{2}\right)$$

$$= \sin\frac{2\alpha}{2} + \sin\frac{2\beta}{2} = \sin\alpha + \sin\beta.$$

The three other sum-to-product formulas in the box are verified in a similar manner. Start with the right side and obtain the left side using an appropriate product-to-sum formula.

EXAMPLE 2 Using the Sum-to-Product Formulas

Express each sum or difference as a product:

a. $\sin 9x + \sin 5x$ **b.** $\cos 4x - \cos 3x$.

Solution The sum-to-product formula that we are using is shown in each of the voice balloons.

a. $\sin\alpha + \sin\beta = 2\sin\frac{\alpha+\beta}{2}\cos\frac{\alpha-\beta}{2}$

$$\sin 9x + \sin 5x = 2\sin\frac{9x+5x}{2}\cos\frac{9x-5x}{2}$$
$$= 2\sin\frac{14x}{2}\cos\frac{4x}{2}$$
$$= 2\sin 7x\cos 2x$$

b. $\cos\alpha - \cos\beta = -2\sin\frac{\alpha+\beta}{2}\sin\frac{\alpha-\beta}{2}$

$$\cos 4x - \cos 3x = -2\sin\frac{4x+3x}{2}\sin\frac{4x-3x}{2}$$
$$= -2\sin\frac{7x}{2}\sin\frac{x}{2}$$

Check Point 2 Express each sum or difference as a product:
a. $\sin 7x + \sin 3x$ **b.** $\cos 3x + \cos 2x$.

Some identities contain a fraction on one side with sums and differences of sines and/or cosines. Applying the sum-to-product formulas in the numerator and the denominator is often helpful in verifying these identities.

EXAMPLE 3 Using Sum-to-Product Formulas to Verify an Identity

Verify the identity: $\dfrac{\cos 3x - \cos 5x}{\sin 3x + \sin 5x} = \tan x$.

Solution Because the left side is more complicated, we will work with it. We use sum-to-product formulas for the numerator and the denominator of the fraction on this side.

$$\frac{\cos 3x - \cos 5x}{\sin 3x + \sin 5x}$$

$\cos\alpha - \cos\beta = -2\sin\frac{\alpha+\beta}{2}\sin\frac{\alpha-\beta}{2}$

$$= \frac{-2\sin\frac{3x+5x}{2}\sin\frac{3x-5x}{2}}{\sin 3x + \sin 5x}$$

$\sin\alpha + \sin\beta = 2\sin\frac{\alpha+\beta}{2}\cos\frac{\alpha-\beta}{2}$

$$= \frac{-2\sin\frac{3x+5x}{2}\sin\frac{3x-5x}{2}}{2\sin\frac{3x+5x}{2}\cos\frac{3x-5x}{2}}$$

$$\frac{\cos 3x - \cos 5x}{\sin 3x + \sin 5x} = \frac{-2\sin\frac{3x+5x}{2}\sin\frac{3x-5x}{2}}{2\sin\frac{3x+5x}{2}\cos\frac{3x-5x}{2}}$$

This is the expression from the previous page.

$$= \frac{-2\sin\frac{8x}{2}\sin\left(\frac{-2x}{2}\right)}{2\sin\frac{8x}{2}\cos\left(\frac{-2x}{2}\right)}$$

Perform the indicated additions.

$$= \frac{-\cancel{2}\,\cancel{\sin 4x}\,\sin(-x)}{\cancel{2}\,\cancel{\sin 4x}\,\cos(-x)}$$

Simplify.

$$= \frac{-(-\sin x)}{\cos x}$$

The sine function is odd: $\sin(-x) = -\sin x$. The cosine function is even: $\cos(-x) = \cos x$.

$$= \frac{\sin x}{\cos x}$$

Simplify.

$$= \tan x$$

Apply a quotient identity: $\tan x = \frac{\sin x}{\cos x}$.

We worked with the left side and arrived at the right side. Thus, the identity is verified.

Check Point 3 Verify the identity: $\frac{\cos 3x - \cos x}{\sin 3x + \sin x} = -\tan x$.

Sinusoidal Sounds

Music is all around us. A mere snippet of a song from the past can trigger vivid memories, inducing emotions ranging from unabashed joy to deep sorrow. Trigonometric functions can explain how sound travels from its source and describe its pitch, loudness, and quality. Still unexplained is the remarkable influence music has on the brain, including the deepest question of all: Why do we appreciate music?

When a note is played, it disturbs nearby air molecules, creating regions of higher-than-normal pressure and regions of lower-than-normal pressure. If we graph pressure, y, versus time, t, we get a sine wave that represents the note. The frequency of the sine wave is the number of high-low disturbances, or vibrations, per second. The greater the frequency, the higher the pitch; the lesser the frequency, the lower the pitch.

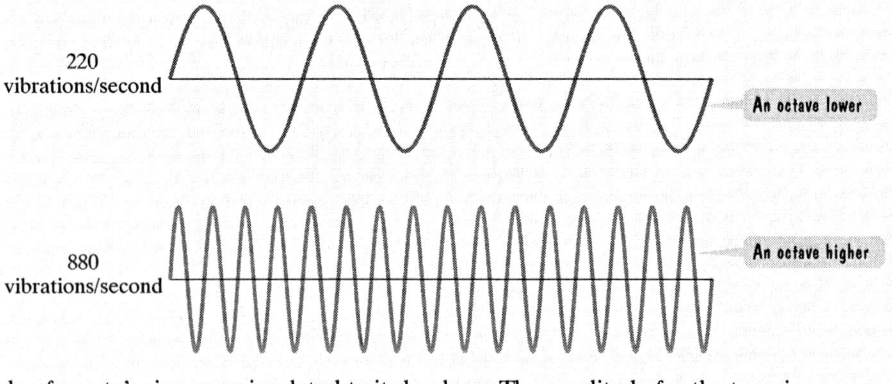

The amplitude of a note's sine wave is related to its loudness. The amplitude for the two sine waves shown above is the same. Thus, the notes have the same loudness, although they differ in pitch. The greater the amplitude, the louder the sound; the lesser the amplitude, the softer the sound. The amplitude and frequency are characteristic of every note—and thus of its graph—until the note dissipates.

EXERCISE SET 6.4

Practice Exercises

In Exercises 1–8, express each product as a sum or difference.

1. $\sin 6x \sin 2x$
2. $\sin 8x \sin 4x$
3. $\cos 7x \cos 3x$
4. $\cos 9x \cos 2x$
5. $\sin x \cos 2x$
6. $\sin 2x \cos 3x$
7. $\cos \frac{3x}{2} \sin \frac{x}{2}$
8. $\cos \frac{5x}{2} \sin \frac{x}{2}$

In Exercises 9–22, express each sum or difference as a product. If possible, find this product's exact value.

9. $\sin 6x + \sin 2x$
10. $\sin 8x + \sin 2x$
11. $\sin 7x - \sin 3x$
12. $\sin 11x - \sin 5x$
13. $\cos 4x + \cos 2x$
14. $\cos 9x - \cos 7x$
15. $\sin x + \sin 2x$
16. $\sin x - \sin 2x$
17. $\cos \frac{3x}{2} + \cos \frac{x}{2}$
18. $\sin \frac{3x}{2} + \sin \frac{x}{2}$
19. $\sin 75° + \sin 15°$
20. $\cos 75° - \cos 15°$
21. $\sin \frac{\pi}{12} - \sin \frac{5\pi}{12}$
22. $\cos \frac{\pi}{12} - \cos \frac{5\pi}{12}$

In Exercises 23–30, verify each identity.

23. $\frac{\sin 3x - \sin x}{\cos 3x - \cos x} = -\cot 2x$
24. $\frac{\sin x + \sin 3x}{\cos x + \cos 3x} = \tan 2x$
25. $\frac{\sin 2x + \sin 4x}{\cos 2x + \cos 4x} = \tan 3x$
26. $\frac{\cos 4x - \cos 2x}{\sin 2x - \sin 4x} = \tan 3x$
27. $\frac{\sin x - \sin y}{\sin x + \sin y} = \tan \frac{x - y}{2} \cot \frac{x + y}{2}$
28. $\frac{\sin x + \sin y}{\sin x - \sin y} = \tan \frac{x + y}{2} \cot \frac{x - y}{2}$
29. $\frac{\sin x + \sin y}{\cos x + \cos y} = \tan \frac{x + y}{2}$
30. $\frac{\sin x - \sin y}{\cos x - \cos y} = -\cot \frac{x + y}{2}$

In Exercises 31–34, the graph with the given equation is shown in a $\left[0, 2\pi, \frac{\pi}{2}\right]$ *by* $[-2, 2, 1]$ *viewing rectangle.*

a. Describe the graph using another equation.

b. Verify that the two equations are equivalent.

31.

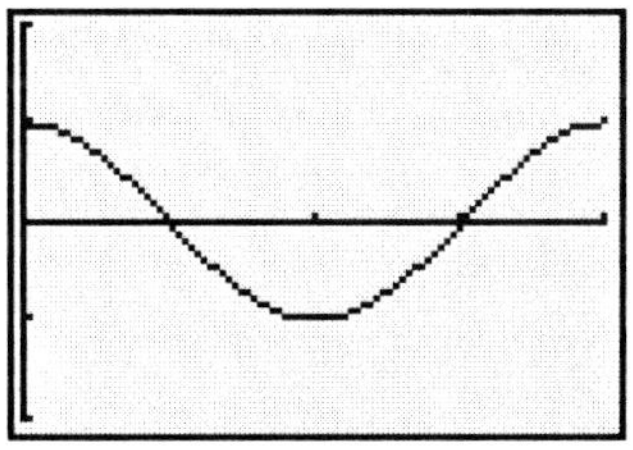

$$y = \frac{\sin x + \sin 3x}{2 \sin 2x}$$

32.

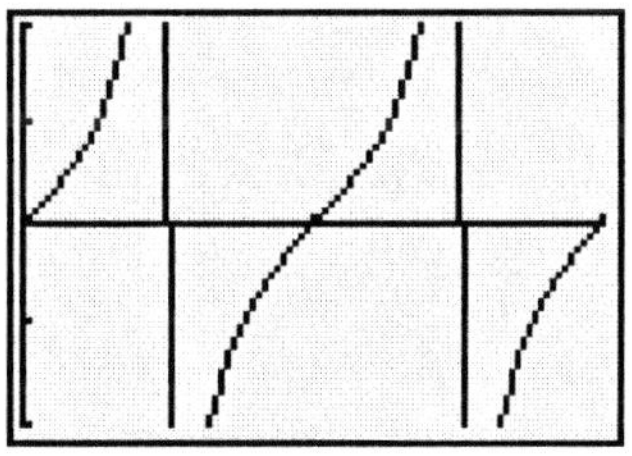

$$y = \frac{\cos x - \cos 3x}{\sin x + \sin 3x}$$

33.

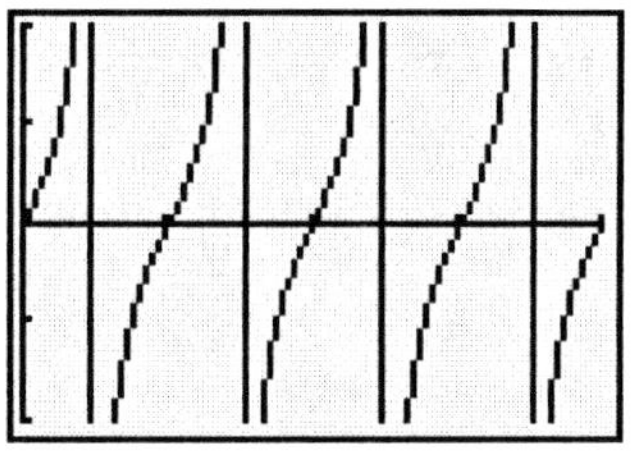

$$y = \frac{\cos x - \cos 5x}{\sin x + \sin 5x}$$

34.

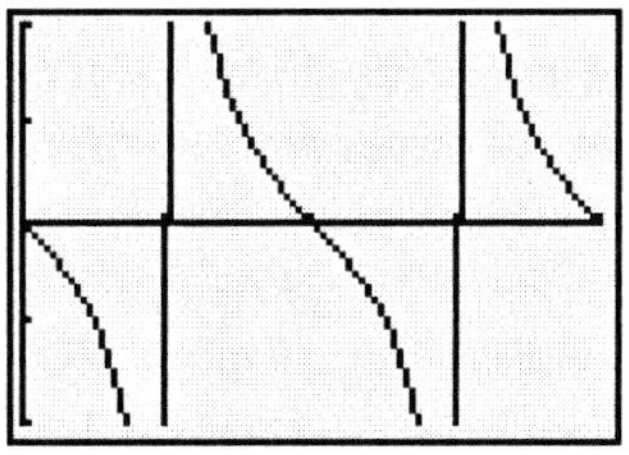

$$y = \frac{\cos 5x - \cos 3x}{\sin 5x + \sin 3x}$$

Application Exercises

Use this information to solve Exercises 35–36. The sound produced by touching each button on a touch-tone phone is described by

$$y = \sin 2\pi lt + \sin 2\pi ht,$$

where l and h are the low and high frequencies in the figure shown. For example, what sound is produced by touching 5? The low frequency is $l = 770$ cycles per second and the high frequency is $h = 1336$ cycles per second. The sound produced by touching 5 is described by

$$y = \sin 2\pi(770)t + \sin 2\pi(1336)t.$$

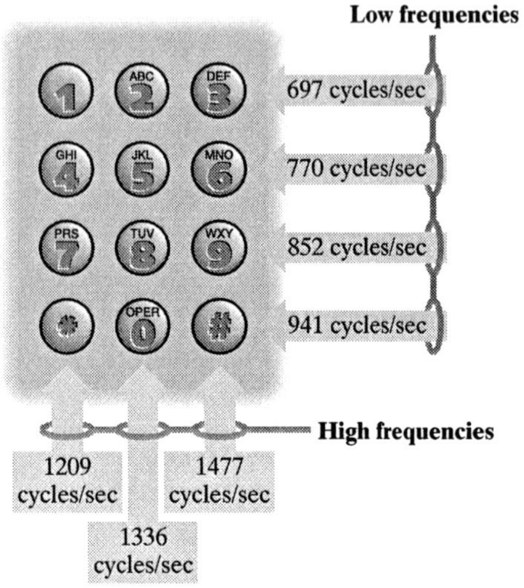

35. The touch-tone phone sequence for that most naive of melodies is given as follows:

Mary Had A Little Lamb

3212333,222,399,3212333322321.

a. Many numbers do not appear in this sequence, including 7. If you accidently touch 7 for one of the notes, describe this sound as the sum of sines.
b. Describe this accidental sound as a product of sines and cosines.

36. The touch-tone phone sequence for *Jingle Bells* is given as follows:

Jingle Bells

333,333,39123,666-663333322329,333,333,39123,666-6633,399621.

a. The first six notes of the song are produced by repeatedly touching 3. Describe this repeated sound as the sum of sines.
b. Describe the repeated sound as a product of sines and cosines.

Writing in Mathematics

In Exercises 37–40, use words to describe the given formula.

37. $\sin\alpha \sin\beta = \frac{1}{2}[\cos(\alpha - \beta) - \cos(\alpha + \beta)]$

38. $\cos\alpha \cos\beta = \frac{1}{2}[\cos(\alpha - \beta) + \cos(\alpha + \beta)]$

39. $\sin\alpha + \sin\beta = 2\sin\dfrac{\alpha + \beta}{2}\cos\dfrac{\alpha - \beta}{2}$

40. $\cos\alpha + \cos\beta = 2\cos\dfrac{\alpha + \beta}{2}\cos\dfrac{\alpha - \beta}{2}$

41. Describe identities that can be verified using the sum-to-product formulas.

42. Why do the sounds produced by touching each button on a touch-tone phone have the same loudness? Answer the question using the equation described for Exercises 35 and 36, $y = \sin 2\pi lt + \sin 2\pi ht$, and determine the maximum value of y for each sound.

Technology Exercises

In Exercises 43–46, graph each side of the equation in the same viewing rectangle. If the graphs appear to coincide, verify that the equation is an identity. If the graphs do not appear to coincide, find a value of x for which both sides are defined but not equal.

43. $\sin x + \sin 2x = \sin 3x$

44. $\cos x + \cos 2x = \cos 3x$

45. $\sin x + \sin 3x = 2\sin 2x \cos x$

46. $\cos x + \cos 3x = 2\cos 2x \cos x$

47. In Exercise 35(a), you wrote an equation for the sound produced by touching 7 on a touch-tone phone. Graph the equation in a $[0, 0.01, 0.001]$ by $[-2, 2, 1]$ viewing rectangle.

48. In Exercise 36(a), you wrote an equation for the sound produced by touching 3 on a touch-tone phone. Graph the equation in a $[0, 0.01, 0.001]$ by $[-2, 2, 1]$ viewing rectangle.

49. In this section, we saw how sums could be expressed as products. Sums of trigonometric functions can also be used to describe functions that are not trigonometric. French mathematician Jean Fourier (1768–1830) showed that *any function* can be described by a series of trigonometric functions. For example, the basic linear function $f(x) = x$ can also be represented by

$$f(x) = 2\left(\frac{\sin x}{1} - \frac{\sin 2x}{2} + \frac{\sin 3x}{3} - \frac{\sin 4x}{4} + \cdots\right).$$

a. Graph

$$y = 2\left(\frac{\sin x}{1}\right),$$

$$y = 2\left(\frac{\sin x}{1} - \frac{\sin 2x}{2}\right),$$

$$y = 2\left(\frac{\sin x}{1} - \frac{\sin 2x}{2} + \frac{\sin 3x}{3}\right)$$

and

$$y = 2\left(\frac{\sin x}{1} - \frac{\sin 2x}{2} + \frac{\sin 3x}{3} - \frac{\sin 4x}{4}\right)$$

in a $\left[-\pi, \pi, \frac{\pi}{2}\right]$ by $[-3, 3, 1]$ viewing rectangle. What patterns do you observe?

b. Graph

$$y = 2\left(\frac{\sin x}{1} - \frac{\sin 2x}{2} + \frac{\sin 3x}{3} - \frac{\sin 4x}{4} + \frac{\sin 5x}{5} - \frac{\sin 6x}{6} + \frac{\sin 7x}{7} - \frac{\sin 8x}{8} + \frac{\sin 9x}{9} - \frac{\sin 10x}{10}\right)$$

in a $\left[-\pi, \pi, \frac{\pi}{2}\right]$ by $[-3, 3, 1]$ viewing rectangle. Is a portion of the graph beginning to look like the graph of $f(x) = x$? Obtain a better approximation for the line by graphing functions that contain more and more terms involving sines of multiple angles.

c. Use

$$x = 2\left(\frac{\sin x}{1} - \frac{\sin 2x}{2} + \frac{\sin 3x}{3} - \frac{\sin 4x}{4} + \cdots\right)$$

and substitute $\frac{\pi}{2}$ for x to obtain a formula for $\frac{\pi}{2}$. Show at least four nonzero terms. Then multiply both sides of your formula by 2 to write a nonending series of subtractions and additions that approaches π. Use this series to obtain an approximation for π that is more accurate than the one given by your graphing utility.

Critical Thinking Exercises

Use the identities for sin $(\alpha + \beta)$ and sin $(\alpha - \beta)$ to solve Exercises 50–51.

50. Add the left and right sides of the identities and derive the product-to-sum formula for $\sin\alpha\cos\beta$.

51. Subtract the left and right sides of the identities and derive the product-to-sum formula for $\cos\alpha\sin\beta$.

In Exercises 52–53, verify the given sum-to-product formula. Start with the right side and obtain the expression on the left side by using an appropriate product-to-sum formula.

52. $\sin\alpha - \sin\beta = 2\sin\frac{\alpha - \beta}{2}\cos\frac{\alpha + \beta}{2}$

53. $\cos\alpha + \cos\beta = 2\cos\frac{\alpha + \beta}{2}\cos\frac{\alpha - \beta}{2}$

In Exercises 54–55, verify each identity.

54. $\dfrac{\sin 2x + (\sin 3x + \sin x)}{\cos 2x + (\cos 3x + \cos x)} = \tan 2x$

55. $4\cos x\cos 2x\sin 3x = \sin 2x + \sin 4x + \sin 6x$

Group Exercise

56. This activity should result in an unusual group display entitled "*Frere Jacques*, a New Perspective." Here is the touch-tone phone sequence:

Frere Jacques

4564,4564,69#,69#,#*#964,#*#964,414,414.

Group members should write every sound in the sequence as both the sum of sines and the product of sines and cosines. Use the sum of sines form and a graphing utility with a $[0, 0.01, 0.001]$ by $[-2, 2, 1]$ viewing rectangle to obtain a graph for every sound. Download these graphs. Use the graphs and equations to create your display in such a way that adults find the trigonometry of this naive melody interesting.

SECTION 6.5 Trigonometric Equations

Objectives

1. Find all solutions of a trigonometric equation.
2. Solve equations with multiple angles.
3. Solve trigonometric equations quadratic in form.
4. Use factoring to separate different functions in trigonometric equations.
5. Use identities to solve trigonometric equations.

Exponential functions display the manic energies of uncontrolled growth. By contrast, trigonometric functions repeat their behavior. Do they embody in their regularity some basic rhythm of the universe? The cycles of periodic phenomena provide events that we can comfortably count on. When will the moon look just as it does at this moment? When can I count on 13.5 hours of daylight? When will my breathing be exactly as it is right now? Models with trigonometric functions embrace the periodic rhythms of our world. Equations containing trigonometric functions are used to answer questions about these models.

1 Find all solutions of a trigonometric equation.

Trigonometric Equations and Their Solutions

A **trigonometric equation** is an equation that contains a trigonometric expression with a variable, such as $\sin x$. We have seen that some trigonometric equations are identities, such as $\sin^2 x + \cos^2 x = 1$. These equations are true for every value of the variable for which the expressions are defined. In this section, we consider trigonometric equations that are true for only some values of the variable. The values that satisfy the equation are its **solutions.** (There are trigonometric equations that have no solution.)

An example of a trigonometric equation is

$$\sin x = \tfrac{1}{2}.$$

A solution of this equation is $\frac{\pi}{6}$ because $\sin\frac{\pi}{6} = \frac{1}{2}$. By contrast, π is not a solution because $\sin \pi = 0 \neq \frac{1}{2}$.

Is $\frac{\pi}{6}$ the only solution of $\sin x = \frac{1}{2}$? The answer is no. Because of the periodic nature of the sine function, there are infinitely many values of x for which $\sin x = \frac{1}{2}$. Figure 6.7 shows five of the solutions, including $\frac{\pi}{6}$, for $-\frac{3\pi}{2} \leq x \leq \frac{7\pi}{2}$. Notice that the x-coordinates of the points where the graph of $y = \sin x$ intersects the line $y = \frac{1}{2}$ are the solutions of the equation $\sin x = \frac{1}{2}$.

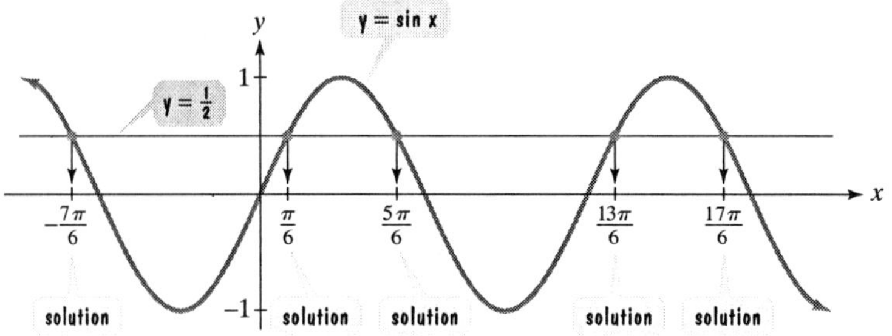

Figure 6.7 The equation $\sin x = \frac{1}{2}$ has five solutions when x is restricted to the interval $\left[-\frac{3\pi}{2}, \frac{7\pi}{2}\right]$

How do we represent all solutions of $\sin x = \frac{1}{2}$? Because the period of the sine function is 2π, first find all solutions in $[0, 2\pi)$. The solutions are

$$x = \frac{\pi}{6} \quad \text{and} \quad x = \pi - \frac{\pi}{6} = \frac{5\pi}{6}.$$

The sine is positive in quadrants I and II.

Any multiple of 2π can be added to these values and the sine is still $\frac{1}{2}$. Thus, all solutions of $\sin x = \frac{1}{2}$ are given by

$$x = \frac{\pi}{6} + 2n\pi \quad \text{or} \quad x = \frac{5\pi}{6} + 2n\pi$$

where n is any integer. By choosing any two integers, such as $n = 0$ and $n = 1$, we can find some solutions of $\sin x = \frac{1}{2}$. Thus, four of the solutions are:

Let $n = 0$.

$$x = \frac{\pi}{6} + 2 \cdot 0\pi = \frac{\pi}{6} \qquad x = \frac{5\pi}{6} + 2 \cdot 0\pi = \frac{5\pi}{6}$$

Let $n = 1$.

$$x = \frac{\pi}{6} + 2 \cdot 1\pi = \frac{\pi}{6} + 2\pi = \frac{\pi}{6} + \frac{12\pi}{6} = \frac{13\pi}{6} \qquad x = \frac{5\pi}{6} + 2 \cdot 1\pi = \frac{5\pi}{6} + 2\pi = \frac{5\pi}{6} + \frac{12\pi}{6} = \frac{17\pi}{6}.$$

These four solutions are shown among the five solutions in Figure 6.7.

Equations Involving a Single Trigonometric Function

To solve an equation containing a single trigonometric function:

- Isolate the function on one side of the equation.
- Solve for the variable.

EXAMPLE 1 Finding All Solutions of a Trigonometric Equation

Solve the equation: $3 \sin x - 2 = 5 \sin x - 1$.

Solution The equation contains a single trigonometric function, $\sin x$.

Step 1 Isolate the function on one side of the equation. We can solve for $\sin x$ by collecting terms with $\sin x$ on the left side and constant terms on the right side.

$3 \sin x - 2 = 5 \sin x - 1$	This is the given equation.
$3 \sin x - 5 \sin x - 2 = 5 \sin x - 5 \sin x - 1$	Subtract $5 \sin x$ from both sides.
$-2 \sin x - 2 = -1$	Simplify.
$-2 \sin x = 1$	Add 2 to both sides.
$\sin x = -\frac{1}{2}$	Divide both sides by -2 and solve for $\sin x$.

Urban Canyons

A city's tall buildings and narrow streets reduce the amount of sunlight. If h is the average height of the buildings and w is the width of the street, the angle of elevation from the street to the top of the buildings is given by the trigonometric equation

$$\tan\theta = \frac{h}{w}.$$

A value of $\theta = 63°$ can result in an 85% loss of illumination.

Step 2 Solve for the variable. We must solve for x in $\sin x = -\frac{1}{2}$. Because $\sin\frac{\pi}{6} = \frac{1}{2}$, the solutions of $\sin x = -\frac{1}{2}$ in $[0, 2\pi)$ are

$$x = \pi + \frac{\pi}{6} = \frac{6\pi}{6} + \frac{\pi}{6} = \frac{7\pi}{6} \qquad x = 2\pi - \frac{\pi}{6} = \frac{12\pi}{6} - \frac{\pi}{6} = \frac{11\pi}{6}.$$

The sine is negative in quadrant III.

The sine is negative in quadrant IV.

Because the period of the sine function is 2π, the solutions of the equation are given by

$$x = \frac{7\pi}{6} + 2n\pi \quad \text{and} \quad x = \frac{11\pi}{6} + 2n\pi$$

where n is any integer.

Check Point 1 Solve the equation: $5\sin x = 3\sin x + \sqrt{3}$.

Now we will concentrate on finding solutions of trigonometric equations for $0 \le x < 2\pi$. You can use a graphing utility to check the solutions of these equations. Graph the left side and graph the right side. The solutions are the x-coordinates of the points where the graphs intersect.

2 Solve equations with multiple angles.

Equations Involving Multiple Angles

Here are examples of two equations that include multiple angles:

$$\tan 3x = 1 \qquad \sin\frac{x}{2} = \frac{\sqrt{3}}{2}.$$

This angle is a multiple of 3.

This angle is a multiple of $\frac{1}{2}$.

We will solve each equation for $0 \le x < 2\pi$. The period of the function plays an important role in ensuring that we do not leave out any solutions.

EXAMPLE 2 Solving an Equation with a Multiple Angle

Solve the equation: $\tan 3x = 1, \quad 0 \le x < 2\pi$.

Solution The period of the tangent function is π. In the interval $[0, \pi)$, the only value for which the tangent function is 1 is $\frac{\pi}{4}$. This means that $3x = \frac{\pi}{4}$. Because the period is π, all the solutions to $\tan 3x = 1$ are given by

$$3x = \frac{\pi}{4} + n\pi \qquad n \text{ is any integer.}$$

$$x = \frac{\pi}{12} + \frac{n\pi}{3} \qquad \text{Divide both sides by 3 and solve for } x.$$

Technology

Shown below are the graphs of

$$y = \tan 3x$$

and

$$y = 1$$

in a $\left[0, 2\pi, \frac{\pi}{2}\right]$ by $[-3, 3, 1]$ viewing rectangle.
The solutions of

$$\tan 3x = 1$$

in $[0, 2\pi)$ are shown by the x-coordinates of the six intersection points.

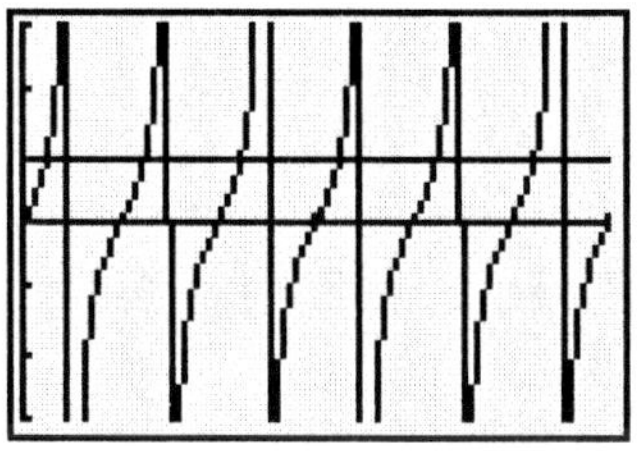

In the interval $[0, 2\pi)$, we obtain the solutions of $\tan 3x = 1$ as follows:

Let $n = 0$.

$$x = \frac{\pi}{12} + \frac{0\pi}{3} = \frac{\pi}{12}$$

Let $n = 1$.

$$x = \frac{\pi}{12} + \frac{1\pi}{3} = \frac{\pi}{12} + \frac{4\pi}{12} = \frac{5\pi}{12}$$

Let $n = 2$.

$$x = \frac{\pi}{12} + \frac{2\pi}{3} = \frac{\pi}{12} + \frac{8\pi}{12} = \frac{9\pi}{12} = \frac{3\pi}{4}$$

Let $n = 3$.

$$x = \frac{\pi}{12} + \frac{3\pi}{3} = \frac{\pi}{12} + \frac{12\pi}{12} = \frac{13\pi}{12}$$

Let $n = 4$.

$$x = \frac{\pi}{12} + \frac{4\pi}{3} = \frac{\pi}{12} + \frac{16\pi}{12} = \frac{17\pi}{12}$$

Let $n = 5$.

$$x = \frac{\pi}{12} + \frac{5\pi}{3} = \frac{\pi}{12} + \frac{20\pi}{12} = \frac{21\pi}{12} = \frac{7\pi}{4}.$$

If you let $n = 6$, you will obtain $x = \frac{25\pi}{12}$. This value exceeds 2π. In the interval $[0, 2\pi)$, the solutions of $\tan 3x = 1$ are $\frac{\pi}{12}, \frac{5\pi}{12}, \frac{3\pi}{4}, \frac{13\pi}{12}, \frac{17\pi}{12}$, and $\frac{7\pi}{4}$. These solutions are illustrated by the six intersection points in the technology box.

Check Point 2 Solve the equation: $\tan 2x = \sqrt{3}, 0 \le x < 2\pi$.

EXAMPLE 3 Solving an Equation with a Multiple Angle

Solve the equation: $\sin \frac{x}{2} = \frac{\sqrt{3}}{2}, \quad 0 \le x < 2\pi$.

Solution The period of the sine function is 2π. In the interval $[0, 2\pi)$, there are two values at which the sine function is $\frac{\sqrt{3}}{2}$. One of these values is $\frac{\pi}{3}$. The sine is positive in quadrant II; thus, the other value is $\pi - \frac{\pi}{3}$, or $\frac{2\pi}{3}$. This means that $\frac{x}{2} = \frac{\pi}{3}$ or $\frac{x}{2} = \frac{2\pi}{3}$. Because the period is 2π, all the solutions of $\sin \frac{x}{2} = \frac{\sqrt{3}}{2}$ are given by

$$\frac{x}{2} = \frac{\pi}{3} + 2n\pi \quad \text{or} \quad \frac{x}{2} = \frac{2\pi}{3} + 2n\pi. \quad n \text{ is any integer.}$$

$$x = \frac{2\pi}{3} + 4n\pi \qquad x = \frac{4\pi}{3} + 4n\pi. \quad \text{Multiply both sides by 2 and solve for } x.$$

If $n = 0$, we obtain $x = \frac{2\pi}{3}$ from the equation on the left and $x = \frac{4\pi}{3}$ from the equation on the right. If we let $n = 1$, we are adding $4 \cdot 1 \cdot \pi$, or 4π, to each of these expressions. These values of x exceed 2π. Thus, in the interval $[0, 2\pi)$, the only solutions of $\sin \frac{x}{2} = \frac{\sqrt{3}}{2}$ are $\frac{2\pi}{3}$ and $\frac{4\pi}{3}$.

Check Point 3 Solve the equation: $\sin\frac{x}{3} = \frac{1}{2}, 0 \le x < 2\pi$.

3 Solve trigonometric equations quadratic in form.

Trigonometric Equations Quadratic in Form

Some trigonometric equations are in the form of a quadratic equation $at^2 + bt + c = 0$, where t is a trigonometric function. Here are two examples of trigonometric equations that are quadratic in form:

$$2\cos^2 x + \cos x - 1 = 0 \qquad 2\sin^2 x - 3\sin x + 1 = 0.$$

The form of this equation is $2t^2 + t - 1 = 0$ with $t = \cos x$.

The form of this equation is $2t^2 - 3t + 1 = 0$ with $t = \sin x$.

To solve this kind of equation, try using factoring. If the trigonometric expression does not factor, use the quadratic formula.

EXAMPLE 4 Solving a Trigonometric Equation Quadratic in Form

Solve the equation: $2\cos^2 x + \cos x - 1 = 0, \quad 0 \le x < 2\pi$.

Solution The given equation is in quadratic form $2t^2 + t - 1 = 0$ with $t = \cos x$. Let us attempt to solve the equation by factoring.

$$2\cos^2 x + \cos x - 1 = 0$$ This is the given equation.

$$(2\cos x - 1)(\cos x + 1) = 0$$ Factor. Notice that $2t^2 + t - 1$ factors as $(2t - 1)(t + 1)$.

$$2\cos x - 1 = 0 \quad \text{or} \quad \cos x + 1 = 0$$ Set each factor equal to 0.

$$2\cos x = 1 \qquad \cos x = -1$$ Solve for cos x.

$$\cos x = \tfrac{1}{2}$$

$$x = \frac{\pi}{3} \qquad x = 2\pi - \frac{\pi}{3} = \frac{5\pi}{3} \qquad x = \pi$$ Solve each equation for x, $0 \le x < 2\pi$.

The cosine is positive in quadrants I and IV.

The solutions in the interval $[0, 2\pi)$ are $\frac{\pi}{3}$, π, and $\frac{5\pi}{3}$.

Technology

The graph of

$$y = 2\cos^2 x + \cos x - 1$$

is shown in a

$$\left[0, 2\pi, \frac{\pi}{2}\right] \text{ by } [-3, 3, 1]$$

viewing rectangle. The x-intercepts,

$$\frac{\pi}{3}, \pi, \text{and } \frac{5\pi}{3},$$

verify the three solutions of

$$2\cos^2 x + \cos x - 1 = 0$$

in $[0, 2\pi)$.

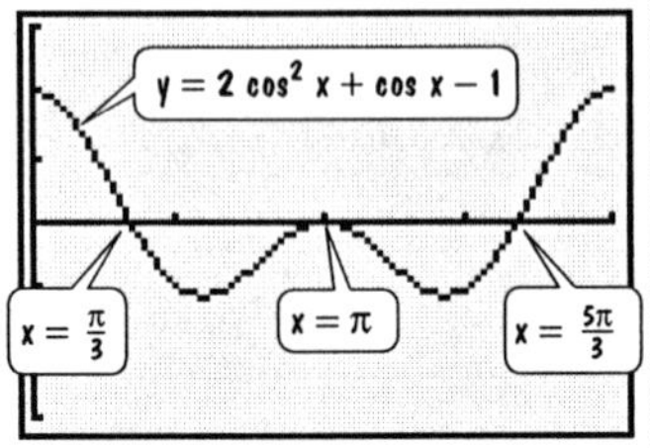

Check Point 4 Solve the equation: $2\sin^2 x - 3\sin x + 1 = 0, 0 \le x < 2\pi$.

4 Use factoring to separate different functions in trigonometric equations.

Using Factoring to Separate Two Different Trigonometric Functions in an Equation

We have seen that factoring is used to solve some trigonometric equations that are quadratic in form. Factoring can also be used to solve some trigonometric equations that contain two different functions such as

$$\tan x \sin^2 x = 3 \tan x.$$

In such a case, move all terms to one side and obtain zero on the other side. Then try to use factoring to separate the different functions. Example 5 shows how this is done.

Study Tip

In solving

$$\tan x \sin^2 x = 3 \tan x,$$

do not begin by dividing both sides by $\tan x$. Division by zero is undefined. If you divide by $\tan x$, you lose the two solutions for which $\tan x = 0$, namely 0 and π.

EXAMPLE 5 Using Factoring to Separate Different Functions

Solve the equation: $\tan x \sin^2 x = 3 \tan x, \quad 0 \le x < 2\pi$.

Solution

Move all terms to one side and obtain zero on the other side.

$\tan x \sin^2 x = 3 \tan x$ — This is the given equation.

$\tan x \sin^2 x - 3 \tan x = 0$ — Subtract $3 \tan x$ from both sides.

Use factoring to separate the two functions.

$\tan x(\sin^2 x - 3) = 0$ — Factor out $\tan x$ from the two terms on the left side.

$\tan x = 0 \quad \text{or} \quad \sin^2 x - 3 = 0$ — Set each factor equal to 0.

$x = 0 \quad x = \pi \qquad \sin^2 x = 3$ — Solve for x.

$\sin x = \pm\sqrt{3}$

This equation has no solution because sin x cannot be greater than 1 or less than −1.

The solutions in the interval $[0, 2\pi)$ are 0 and π.

Check Point 5 Solve the equation: $\sin x \tan x = \sin x, \; 0 \le x < 2\pi$.

5 Use identities to solve trigonometric equations.

Using Identities to Solve Trigonometric Equations

Some trigonometric equations contain more than one function on the same side and these functions cannot be separated by factoring. For example, consider the equation

$$2 \sin^2 x - \cos x - 1 = 0.$$

How can we obtain an equivalent equation that has only one trigonometric function? We use the identity $\sin^2 x + \cos^2 x = 1$ and substitute $1 - \cos^2$ for $\sin^2 x$.

$2 \sin^2 x - \cos x - 1 = 0$ — This is the given equation.

$2(1 - \cos^2 x) - \cos x - 1 = 0$ — $\sin^2 x = 1 - \cos^2 x$

$2 - 2\cos^2 x - \cos x - 1 = 0$ — Use the distributive property.

$-2\cos^2 x - \cos x + 1 = 0$ — Combine like terms.

Multiplying both sides of $-2\cos^2 x - \cos x + 1 = 0$ by -1, we obtain

$$2\cos^2 x + \cos x - 1 = 0.$$

This equivalent equation contains only the cosine function. This is the equation that we solved in Example 4.

EXAMPLE 6 Using an Identity to Solve a Trigonometric Equation

Solve the equation: $\cos 2x + 3\sin x - 2 = 0,\ 0 \le x < 2\pi.$

Solution The given equation contains a cosine function and a sine function. The cosine is a function of $2x$ and the sine is a function of x. We want one trigonometric function of the same angle. This can be accomplished by using the double-angle identity $\cos 2x = 1 - 2\sin^2 x$ to obtain an equivalent equation involving $\sin x$ only.

$$\cos 2x + 3\sin x - 2 = 0 \quad \text{This is the given equation.}$$
$$1 - 2\sin^2 x + 3\sin x - 2 = 0 \quad \cos 2x = 1 - 2\sin^2 x$$
$$-2\sin^2 x + 3\sin x - 1 = 0 \quad \text{Combine like terms.}$$
$$2\sin^2 x - 3\sin x + 1 = 0 \quad \text{Multiply both sides by } -1.$$

The equation is now in quadratic form $2t^2 - 3t + 1 = 0$ with $t = \sin x$. We solve using factoring.

$$(2\sin x - 1)(\sin x - 1) = 0 \quad \text{Factor. Notice that } 2t^2 - 3t + 1 \text{ factors as } (2t - 1)(t - 1).$$

$$2\sin x - 1 = 0 \quad \text{or} \quad \sin x - 1 = 0 \quad \text{Set each factor equal to 0.}$$
$$\sin x = \tfrac{1}{2} \qquad \sin x = 1 \quad \text{Solve for } \sin x.$$
$$x = \frac{\pi}{6} \quad x = \pi - \frac{\pi}{6} = \frac{5\pi}{6} \qquad x = \frac{\pi}{2} \quad \text{Solve each equation for } x,\ 0 \le x < 2\pi.$$

The sine is positive in quadrants I and II.

The solutions in the interval $[0, 2\pi)$ are $\frac{\pi}{6}, \frac{\pi}{2}$, and $\frac{5\pi}{6}$.

Check Point 6 Solve the equation: $\cos 2x + \sin x = 0,\ 0 \le x < 2\pi.$

Sometimes it is necessary to do something to both sides of a trigonometric equation to substitute an identity. For example, consider the equation

$$\sin x \cos x = \tfrac{1}{2}.$$

This equation contains both a sine and a cosine function. How can we obtain a single function? Multiply both sides by 2. In this way, we can use the double-angle identity $\sin 2x = 2\sin x \cos x$ and obtain $\sin 2x$, a single function, on the left side.

Technology

Shown below are the graphs of

$$y = \sin x \cos x$$

and

$$y = \tfrac{1}{2}$$

in a $\left[0, 2\pi, \frac{\pi}{2}\right]$ by $[-1, 1, 1]$ viewing rectangle.

The solutions of

$$\sin x \cos x = \tfrac{1}{2}$$

are shown by the x-coordinates of the two intersection points.

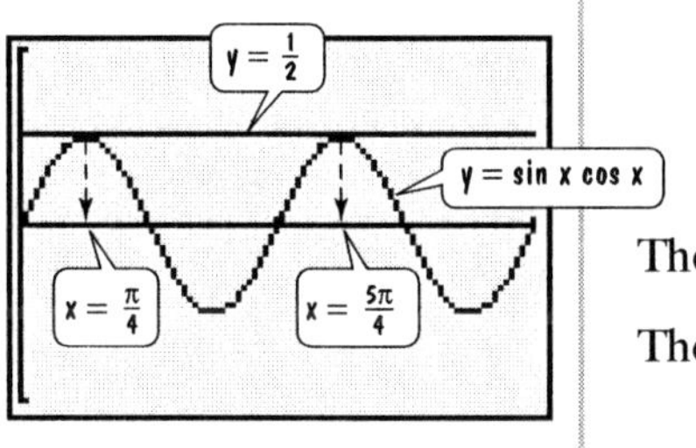

EXAMPLE 7 Using an Identity to Solve a Trigonometric Equation

Solve the equation: $\sin x \cos x = \frac{1}{2}$, $0 \le x < 2\pi$.

Solution

$\sin x \cos x = \frac{1}{2}$ — *This is the given equation.*

$2 \sin x \cos x = 1$ — *Multiply both sides by 2 in anticipation of using $\sin 2x = 2 \sin x \cos x$.*

$\sin 2x = 1$ — *Use a double-angle identity.*

Notice that we have an equation with $2x$, a multiple angle. The period of the sine function is 2π. In the interval $[0, 2\pi)$, the only value for which the sine function is 1 is $\frac{\pi}{2}$. This means that $2x = \frac{\pi}{2}$. Because the period is 2π, all the solutions of $\sin 2x = 1$ are given by

$2x = \frac{\pi}{2} + 2n\pi$ — *n is any integer.*

$x = \frac{\pi}{4} + n\pi$ — *Divide both sides by 2 and solve for x.*

The solutions in the interval $[0, 2\pi)$ are obtained by letting $n = 0$ and $n = 1$. The solutions are $\frac{\pi}{4}$ and $\frac{5\pi}{4}$.

Check Point 7 Solve the equation: $\sin x \cos x = -\frac{1}{2}$, $0 \le x < 2\pi$.

Let's look at another equation that contains two different functions, $\sin x - \cos x = 1$. Can you think of an identity that can be used to produce only one function? Perhaps $\sin^2 x + \cos^2 x = 1$ might be helpful. The next example shows how we use this identity by squaring both sides of the given equation. Remember that if we raise both sides of an equation to an even power, we have the possibility of introducing extraneous solutions. Thus, we must check each proposed solution in the given equation. Alternatively, we can use a graphing utility to verify actual solutions.

EXAMPLE 8 Using an Identity to Solve a Trigonometric Equation

Solve the equation: $\sin x - \cos x = 1$, $0 \le x < 2\pi$.

Solution We square both sides of the equation in anticipation of using $\sin^2 x + \cos^2 x = 1$.

$\sin x - \cos x = 1$ — *This is the given equation.*

$(\sin x - \cos x)^2 = 1^2$ — *Square both sides.*

$\sin^2 x - 2 \sin x \cos x + \cos^2 x = 1$ — *Square the left side using $(A - B)^2 = A^2 - 2AB + B^2$.*

Technology

A graphing utility can be used instead of the algebraic check on the right. Shown are the graphs of

$$y = \sin x - \cos x$$

and

$$y = 1$$

in a $\left[0, 2\pi, \frac{\pi}{2}\right]$ by $[-2, 2, 1]$ viewing rectangle. The actual solutions of

$$\sin x - \cos x = 1$$

are shown by the x-coordinates of the two intersection points, $\frac{\pi}{2}$ and π.

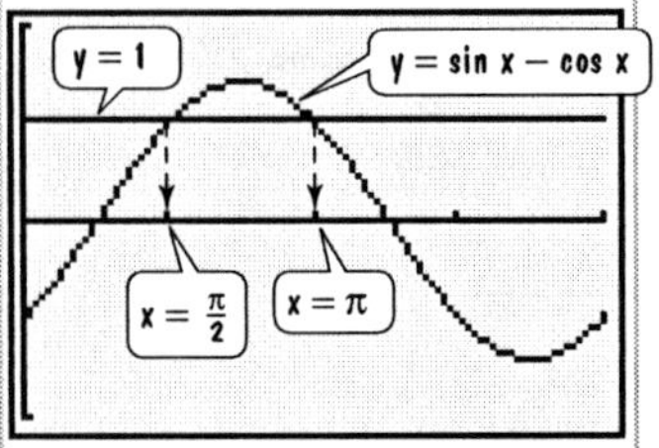

$$\sin^2 x - 2\sin x\cos x + \cos^2 x = 1$$ We have repeated the equation from the previous page.

$$\sin^2 x + \cos^2 x - 2\sin x\cos x = 1$$ Rearrange terms.

$$1 - 2\sin x\cos x = 1$$ Apply a Pythagorean identity: $\sin^2 x + \cos^2 x = 1$.

$$-2\sin x\cos x = 0$$ Subtract 1 from both sides of the equation.

$$\sin x\cos x = 0$$ Divide both sides of the equation by −2.

$$\sin x = 0 \quad \text{or} \quad \cos x = 0$$ Set each factor equal to 0.

$$x = 0 \quad x = \pi \qquad x = \frac{\pi}{2} \quad x = \frac{3\pi}{2}$$ Solve for x in $[0, 2\pi)$.

We check these proposed solutions to see if any are extraneous.

Check 0:	**Check $\frac{\pi}{2}$:**	**Check π:**	**Check $\frac{3\pi}{2}$:**
$\sin x - \cos x = 1$	$\sin x - \cos x = 1$	$\sin x - \cos x = 1$	$\sin x - \cos x = 1$
$\sin 0 - \cos 0 \stackrel{?}{=} 1$	$\sin\frac{\pi}{2} - \cos\frac{\pi}{2} \stackrel{?}{=} 1$	$\sin\pi - \cos\pi \stackrel{?}{=} 1$	$\sin\frac{3\pi}{2} - \cos\frac{3\pi}{2} \stackrel{?}{=} 1$
$0 - 1 \stackrel{?}{=} 1$	$1 - 0 \stackrel{?}{=} 1$	$0 - (-1) \stackrel{?}{=} 1$	$-1 - 0 \stackrel{?}{=} 1$
$-1 = 1$, False	$1 = 1$, True	$1 = 1$, True	$-1 = 1$, False

0 and $\frac{3\pi}{2}$ are extraneous.

The actual solutions in the interval $[0, 2\pi)$ are $\frac{\pi}{2}$ and π.

Check Point 8 Solve the equation: $\cos x - \sin x = -1, \ 0 \le x < 2\pi$.

EXERCISE SET 6.5

Practice Exercises

In Exercises 1–10, use substitution to determine whether the given x-value is a solution of the equation.

1. $\cos x = \frac{\sqrt{2}}{2}, \quad x = \frac{\pi}{4}$

2. $\tan x = \sqrt{3}, \quad x = \frac{\pi}{3}$

3. $\sin x = \frac{\sqrt{3}}{2}, \quad x = \frac{\pi}{6}$

4. $\sin x = \frac{\sqrt{2}}{2}, \quad x = \frac{\pi}{3}$

5. $\cos x = -\frac{1}{2}, \quad x = \frac{2\pi}{3}$

6. $\cos x = -\frac{1}{2}, \quad x = \frac{4\pi}{3}$

7. $\tan 2x = -\frac{\sqrt{3}}{3}, \quad x = \frac{5\pi}{12}$

8. $\cos\frac{2x}{3} = -\frac{1}{2}, \quad x = \pi$

9. $\cos x = \sin 2x, \ x = \frac{\pi}{3}$

10. $\cos x + 2 = \sqrt{3}\sin x, \quad x = \frac{\pi}{6}$

In Exercises 11–24, find all solutions of each equation.

11. $\sin x = \frac{\sqrt{3}}{2}$

12. $\cos x = \frac{\sqrt{3}}{2}$

13. $\tan x = 1$

14. $\tan x = \sqrt{3}$

15. $\cos x = -\frac{1}{2}$

16. $\sin x = -\frac{\sqrt{2}}{2}$

17. $\tan x = 0$

18. $\sin x = 0$

19. $2\cos x + \sqrt{3} = 0$

20. $2\sin x + \sqrt{3} = 0$

21. $4\sin\theta - 1 = 2\sin\theta$

22. $5\sin\theta + 1 = 3\sin\theta$

23. $3\sin\theta + 5 = -2\sin\theta$

24. $7\cos\theta + 9 = -2\cos\theta$

Exercises 25–38 involve equations with multiple angles. Solve each equation on the interval $[0, 2\pi)$.

25. $\sin 2x = \frac{\sqrt{3}}{2}$

26. $\cos 2x = \frac{\sqrt{2}}{2}$

27. $\cos 4x = -\frac{\sqrt{3}}{2}$

28. $\sin 4x = -\frac{\sqrt{2}}{2}$

29. $\tan 3x = \frac{\sqrt{3}}{3}$

30. $\tan 3x = \sqrt{3}$

31. $\tan \frac{x}{2} = \sqrt{3}$

32. $\tan \frac{x}{2} = \frac{\sqrt{3}}{3}$

33. $\sin \frac{2\theta}{3} = -1$

34. $\cos \frac{2\theta}{3} = -1$

35. $\sec \frac{3\theta}{2} = -2$

36. $\cot \frac{3\theta}{2} = -\sqrt{3}$

37. $\sin\left(2x + \frac{\pi}{6}\right) = \frac{1}{2}$

38. $\sin\left(2x - \frac{\pi}{4}\right) = \frac{\sqrt{2}}{2}$

Exercises 39–46 involve trigonometric equations quadratic in form. Solve each equation on the interval $[0, 2\pi)$.

39. $2\sin^2 x - \sin x - 1 = 0$

40. $2\sin^2 x + \sin x - 1 = 0$

41. $2\cos^2 x + 3\cos x + 1 = 0$

42. $\cos^2 x + 2\cos x - 3 = 0$

43. $2\sin^2 x = \sin x + 3$

44. $2\sin^2 x = 4\sin x + 6$

45. $\sin^2 \theta - 1 = 0$

46. $\cos^2 \theta - 1 = 0$

In Exercises 47–56, solve each equation on the interval $[0, 2\pi)$.

47. $(\tan x - 1)(\cos x + 1) = 0$

48. $(\tan x + 1)(\sin x - 1) = 0$

49. $(2\cos x + \sqrt{3})(2\sin x + 1) = 0$

50. $(2\cos x - \sqrt{3})(2\sin x - 1) = 0$

51. $\cot x(\tan x - 1) = 0$

52. $\cot x(\tan x + 1) = 0$

53. $\sin x + 2\sin x\cos x = 0$

54. $\cos x - 2\sin x\cos x = 0$

55. $\tan^2 x\cos x = \tan^2 x$

56. $\cot^2 x\sin x = \cot^2 x$

In Exercises 57–78, use an identity to solve each equation on the interval $[0, 2\pi)$.

57. $2\cos^2 x + \sin x - 1 = 0$

58. $2\cos^2 x - \sin x - 1 = 0$

59. $\sin^2 x - 2\cos x - 2 = 0$

60. $4\sin^2 x + 4\cos x - 5 = 0$

61. $4\cos^2 x = 5 - 4\sin x$

62. $3\cos^2 x = \sin^2 x$

63. $\sin 2x = \cos x$

64. $\sin 2x = \sin x$

65. $\cos 2x = \cos x$

66. $\cos 2x = \sin x$

67. $\cos 2x + 5\cos x + 3 = 0$

68. $\cos 2x + \cos x + 1 = 0$

69. $\sin x\cos x = \frac{\sqrt{2}}{4}$

70. $\sin x\cos x = \frac{\sqrt{3}}{4}$

71. $\sin x + \cos x = 1$

72. $\sin x + \cos x = -1$

73. $\sin\left(x + \frac{\pi}{4}\right) + \sin\left(x - \frac{\pi}{4}\right) = 1$

74. $\sin\left(x + \frac{\pi}{3}\right) + \sin\left(x - \frac{\pi}{3}\right) = 1$

75. $\sin 2x\cos x + \cos 2x\sin x = \frac{\sqrt{2}}{2}$

76. $\sin 3x\cos 2x + \cos 3x\sin 2x = 1$

77. $\tan x + \sec x = 1$

78. $\tan x - \sec x = 1$

Application Exercises

Use this information to solve Exercises 79–80. Our cycle of normal breathing takes place every 5 seconds. Velocity of air flow, y, measured in liters per second, after x seconds is modeled by

$$y = 0.6\sin\frac{2\pi}{5}x.$$

Velocity of air flow is positive when we inhale and negative when we exhale.

79. Within each breathing cycle, when are we inhaling at 0.3 liter per second? Round to the nearest tenth of a second.

80. Within each breathing cycle, when are we exhaling at 0.3 liter per second? Round to the nearest tenth of a second.

Use this information to solve Exercises 81–82. The number of hours of daylight in Boston is given by

$$y = 3\sin\left[\frac{2\pi}{365}(x - 79)\right] + 12$$

where x is the number of days after January 1.

81. Within a year, when does Boston have 10.5 hours of daylight? Give your answer in days after January 1 and round to the nearest day.

82. Within a year, when does Boston have 13.5 hours of daylight? Give your answer in days after January 1 and round to the nearest day.

Use this information to solve Exercises 83–84. A ball on a spring is pulled 4 inches below its rest position and then released. After t seconds, the ball's distance, d, in inches from its rest position is given by

$$d = -4\cos\frac{\pi}{3}t.$$

83. Find all values of t for which the ball is 2 inches above its rest position.

84. Find all values of t for which the ball is 2 inches below its rest position.

Use this information to solve Exercises 85–86. When throwing an object, the distance achieved depends on its initial velocity, v_0, and the angle above the horizontal at which the object is thrown, θ. The distance, d, in feet, that describes the range covered is given by

$$d = \frac{v_0^2}{16}\sin\theta\cos\theta$$

where v_0 is measured in feet per second.

85. You and your friend are throwing a baseball back and forth. If you throw the ball with an initial velocity of $v_0 = 90$ feet per second, at what angle of elevation, θ, to the nearest degree, should you direct your throw so that it can be easily caught by your friend located 170 feet away?

86. In Exercise 85, you increase the distance between you and your friend to 200 feet. With this increase, at what angle of elevation, θ, to the nearest degree, should you direct your throw?

Writing in Mathematics

87. What are the solutions of a trigonometric equation?

88. Describe the difference between verifying a trigonometric identity and solving a trigonometric equation.

89. Without actually solving the equation, describe how to solve

$$3\tan x - 2 = 5\tan x - 1.$$

90. In the interval $[0, 2\pi)$, the solutions of $\sin x = \cos 2x$ are $\frac{\pi}{6}, \frac{5\pi}{6}$, and $\frac{3\pi}{2}$. Explain how to use graphs generated by a graphing utility to check these solutions.

91. Suppose you are solving equations in the interval $[0, 2\pi)$. Without actually solving equations, what is the difference between the number of solutions of $\sin x = \frac{1}{2}$ and $\sin 2x = \frac{1}{2}$? How do you account for this difference?

In Exercises 92–93, describe a general strategy for solving each equation. Do not solve the equation.

92. $2\sin^2 x + 5\sin x + 3 = 0$

93. $\sin 2x = \sin x$

94. Describe a natural periodic phenomenon. Give an example of a question that can be answered by a trigonometric equation in the study of this phenomenon.

95. Some people experience depression with loss of sunlight. Use the essay on page 597 to determine whether such a person should live on a city street that is 80 feet wide with buildings whose heights average 400 feet. Explain your answer and include θ, to the nearest degree, in your argument.

Technology Exercises

96. Use a graphing utility to verify the solutions of any five equations that you solved in Exercises 57–78.

In Exercises 97–101, use a graphing utility to approximate the solutions of each equation in the interval $[0, 2\pi)$. Round to the nearest hundredth of a radian.

97. $15\cos^2 x + 7\cos x - 2 = 0$

98. $\cos x = x$

99. $2\sin^2 x = 1 - 2\sin x$

100. $\sin 2x = 2 - x^2$

101. $\sin x + \sin 2x + \sin 3x = 0$

Critical Thinking Exercises

102. Which one of the following is true?

a. The equation $(\sin x - 3)(\cos x + 2) = 0$ has no solution.

b. The equation $\tan x = \frac{\pi}{2}$ has no solution.

c. A trigonometric equation with an infinite number of solutions is an identity.

d. The equations $\sin 2x = 1$ and $\sin 2x = \frac{1}{2}$ have the same number of solutions on the interval $[0, 2\pi)$.

In Exercises 103–105, solve each equation on the interval $[0, 2\pi)$. Do not use a calculator.

103. $2\cos x - 1 + 3\sec x = 0$

104. $\sin 3x + \sin x + \cos x = 0$

105. $\sin x + 2\sin\frac{x}{2} = \cos\frac{x}{2} + 1$

CHAPTER SUMMARY, REVIEW, AND TEST

Summary

DEFINITIONS AND CONCEPTS	EXAMPLES
6.1 Verifying Trigonometric Identities	
a. Identities are trigonometric equations that are true for all values of the variable for which the expressions are defined.	Ex. 1, p. 558; Exs. 2 & 3, p. 559
b. Fundamental trigonometric identities are given in the box on page 557.	Ex. 4, p. 560
c. Guidelines for verifying trigonometric identities are given in the box on page 565.	Ex. 5, p. 561; Ex. 6, p. 562; Ex. 7, p. 563

DEFINITIONS AND CONCEPTS	EXAMPLES
6.2 and 6.3 Sum, Difference, Double-Angle, and Half-Angle Formulas	
a. Sum and difference formulas, double-angle formulas, power-reducing formulas, and half-angle formulas are given in the box on page 585.	Exs. 1-7, pp. 569–574; Exs. 1-7, pp. 578–584
6.4 Product-to-Sum and Sum-to-Product Formulas	
a. The product-to-sum formulas are given in the box on page 588.	Ex. 1, p. 589
b. The sum-to-product formulas are given in the box on page 590. These formulas are useful to verify identities with fractions that contain sums and differences of sines and/or cosines.	Ex. 2, p. 590; Ex. 3, p. 591
6.5 Trigonometric Equations	
a. The values that satisfy a trigonometric equation are its solutions.	Exs. 1-8, pp. 597–603
b. Algebraic techniques such as isolating an expression on one side of the equation and factoring are useful in solving trigonometric equations. Identities are also used to solve some trigonometric equations.	

Review Exercises

6.1

In Exercises 1–12, verify each identity.

1. $\sec x - \cos x = \tan x \sin x$
2. $\cos x + \sin x \tan x = \sec x$
3. $\sin^2\theta(1 + \cot^2\theta) = 1$
4. $(\sec\theta - 1)(\sec\theta + 1) = \tan^2\theta$
5. $\dfrac{1}{\sin t - 1} + \dfrac{1}{\sin t + 1} = -2\tan t \sec t$
6. $\dfrac{1 + \sin t}{\cos^2 t} = \tan^2 t + 1 + \tan t \sec t$
7. $\dfrac{\cos x}{1 - \sin x} = \dfrac{1 + \sin x}{\cos x}$
8. $1 - \dfrac{\cos^2 x}{1 + \sin x} = \sin x$
9. $(\tan\theta + \cot\theta)^2 = \sec^2\theta + \csc^2\theta$
10. $\dfrac{1}{\sin\theta + \cos\theta} + \dfrac{1}{\sin\theta - \cos\theta} = \dfrac{2\sin\theta}{\sin^4\theta - \cos^4\theta}$
11. $\dfrac{\cos t}{\cot t - 5\cos t} = \dfrac{1}{\csc t - 5}$
12. $\dfrac{1 - \cos t}{1 + \cos t} = (\csc t - \cot t)^2$

6.2 and 6.3

In Exercises 13–18, use a sum or difference formula to find the exact value of each expression.

13. $\cos(45° + 30°)$
14. $\sin 195°$
15. $\tan\left(\dfrac{4\pi}{3} - \dfrac{\pi}{4}\right)$
16. $\tan\dfrac{5\pi}{12}$
17. $\cos 65° \cos 5° + \sin 65° \sin 5°$
18. $\sin 80° \cos 50° - \cos 80° \sin 50°$

In Exercises 19–30, verify each identity.

19. $\sin\left(x + \dfrac{\pi}{6}\right) - \cos\left(x + \dfrac{\pi}{3}\right) = \sqrt{3}\sin x$
20. $\tan\left(x + \dfrac{3\pi}{4}\right) = \dfrac{\tan x - 1}{1 + \tan x}$
21. $\sec(\alpha + \beta) = \dfrac{\sec\alpha \sec\beta}{1 - \tan\alpha\tan\beta}$
22. $\dfrac{\cos(\alpha - \beta)}{\cos\alpha\cos\beta} = 1 + \tan\alpha\tan\beta$
23. $\cos^4 t - \sin^4 t = \cos 2t$
24. $\sin t - \cos 2t = (2\sin t - 1)(\sin t + 1)$
25. $\dfrac{\sin 2\theta - \sin\theta}{\cos 2\theta + \cos\theta} = \dfrac{1 - \cos\theta}{\sin\theta}$
26. $\dfrac{\sin 2\theta}{1 - \sin^2\theta} = 2\tan\theta$
27. $\tan 2t = 2\sin t \cos t \sec 2t$
28. $\cos 4t = 1 - 8\sin^2 t \cos^2 t$
29. $\tan\dfrac{x}{2}(1 + \cos x) = \sin x$
30. $\tan\dfrac{x}{2} = \dfrac{\sec x - 1}{\tan x}$

In Exercises 31–33, the graph with the given equation is shown on the next page in a $\left[0, 2\pi, \frac{\pi}{2}\right]$ by $[-2, 2, 1]$ viewing rectangle.

a. Describe the graph using another equation.

b. Verify that the two equations are equivalent.

31. 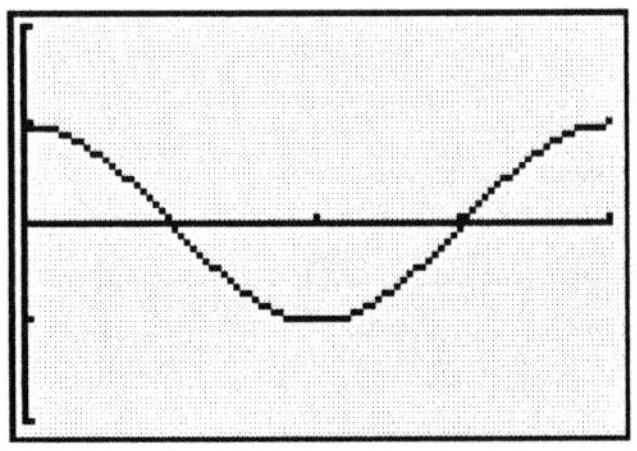

$y = \sin\left(x - \frac{3\pi}{2}\right)$

32. 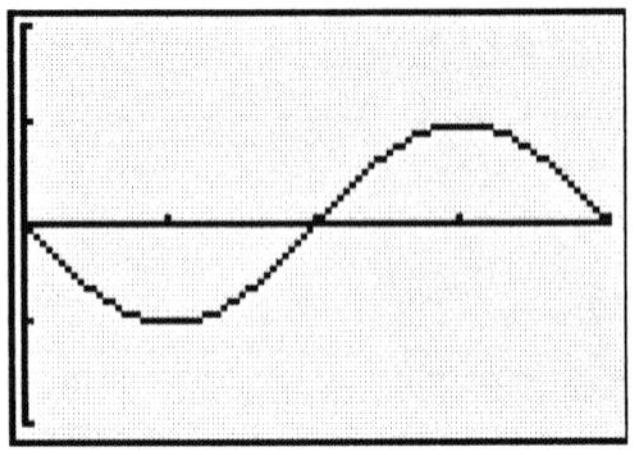

$y = \cos\left(x + \frac{\pi}{2}\right)$

33. 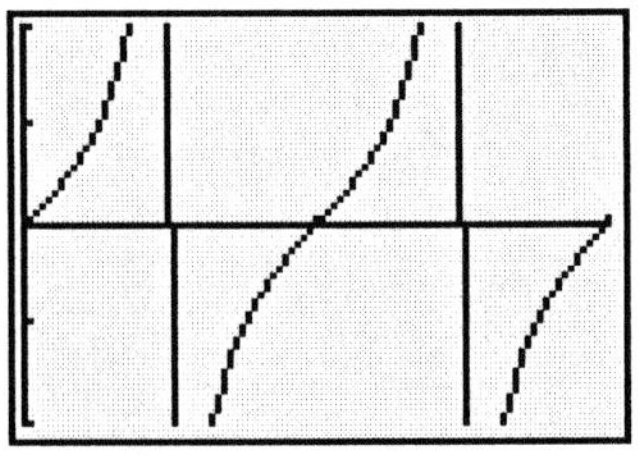

$y = \frac{\tan x - 1}{1 - \cot x}$

In Exercises 34–37, find the exact value of the following under the given conditions:

a. $\sin(\alpha + \beta)$ *b.* $\cos(\alpha - \beta)$ *c.* $\tan(\alpha + \beta)$

d. $\sin 2\alpha$ *e.* $\cos\frac{\beta}{2}$

34. $\sin\alpha = \frac{3}{5}$, α lies in quadrant I, and $\sin\beta = \frac{12}{13}$, β lies in quadrant II.

35. $\tan\alpha = \frac{4}{3}$, α lies in quadrant III, and $\tan\beta = \frac{5}{12}$, β lies in quadrant I.

36. $\tan\alpha = -3$, α lies in quadrant II, and $\cot\beta = -3$, β lies in quadrant IV.

37. $\sin\alpha = -\frac{1}{3}$, α lies in quadrant III, and $\cos\beta = -\frac{1}{3}$, β lies in quadrant III.

In Exercises 38–41, use double- and half-angle formulas to find the exact value of each expression.

38. $\cos^2 15° - \sin^2 15°$

39. $\dfrac{2\tan\frac{5\pi}{12}}{1 - \tan^2\frac{5\pi}{12}}$

40. $\sin 22.5°$

41. $\tan\frac{\pi}{12}$

6.4

In Exercises 42–43, express each product as a sum or difference.

42. $\sin 6x \sin 4x$

43. $\sin 7x \cos 3x$

In Exercises 44–45, express each sum or difference as a product. If possible, find this product's exact value.

44. $\sin 2x - \sin 4x$

45. $\cos 75° + \cos 15°$

In Exercises 46–47, verify each identity.

46. $\dfrac{\cos 3x + \cos 5x}{\cos 3x - \cos 5x} = \cot x \cot 4x$

47. $\dfrac{\sin 2x + \sin 6x}{\sin 2x - \sin 6x} = -\tan 4x \cot 2x$

48. The graph with the given equation is shown in a $\left[0, 2\pi, \frac{\pi}{2}\right]$ by $[-2, 2, 1]$ viewing rectangle.

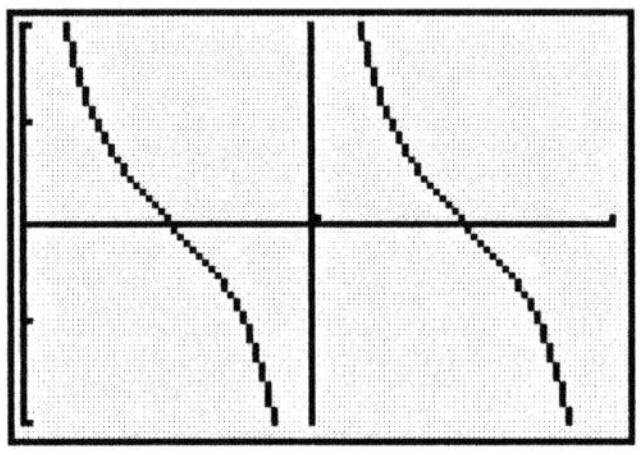

$y = \dfrac{\cos 3x + \cos x}{\sin 3x - \sin x}$

a. Describe the graph using another equation.
b. Verify that the two equations are equivalent.

6.5

In Exercises 49–52, find all solutions of each equation.

49. $\cos x = -\frac{1}{2}$

50. $\sin x = \frac{\sqrt{2}}{2}$

51. $2\sin x + 1 = 0$

52. $\sqrt{3}\tan x - 1 = 0$

In Exercises 53–62, solve each equation on the interval $[0, 2\pi)$.

53. $\cos 2x = -1$

54. $\sin 3x = 1$

55. $\tan\frac{x}{2} = -1$

56. $\tan x = 2\cos x \tan x$

57. $\cos^2 x - 2\cos x = 3$

58. $2\cos^2 x - \sin x = 1$

59. $4\sin^2 x = 1$

60. $\cos 2x - \sin x = 1$

61. $\sin 2x = \sqrt{3}\sin x$

62. $\sin x = \tan x$

63. A ball on a spring is pulled 6 inches below its rest position and then released. After t seconds, the ball's distance, d, in inches from its rest position is given by

$$d = -6\cos\frac{\pi}{2}t.$$

Find all values of t for which the ball is 3 inches below its rest position.

64. You are playing catch with a friend located 100 feet away. If you throw the ball with an initial velocity of $v_0 = 90$ feet per second, at what angle of elevation, θ, to the nearest degree should you direct your throw so that it can be caught easily?

Use the formula

$$d = \frac{v_0^2}{16}\sin\theta\cos\theta.$$

Chapter 6 Test

Use the following conditions to solve Exercises 1–4:

$\sin\alpha = \frac{4}{5}$, α lies in quadrant II

$\cos\beta = \frac{5}{13}$, β lies in quadrant I.

Find the exact value of each of the following.

1. $\cos(\alpha + \beta)$ **2.** $\tan(\alpha - \beta)$

3. $\sin 2\alpha$ **4.** $\cos\frac{\beta}{2}$

5. Use $105° = 135° - 30°$ to find the exact value of $\sin 105°$.

In Exercises 6–11, verify each identity.

6. $\cos x \csc x = \cot x$ **7.** $\dfrac{\sec x}{\cot x + \tan x} = \sin x$

8. $1 - \dfrac{\cos^2 x}{1 + \sin x} = \sin x$ **9.** $\cos\left(\theta + \frac{\pi}{2}\right) = -\sin\theta$

10. $\dfrac{\sin(\alpha - \beta)}{\sin\alpha\cos\beta} = 1 - \cot\alpha\tan\beta$

11. $\sin t\cos t(\tan t + \cot t) = 1$

In Exercises 12–15, solve each equation on the interval $[0, 2\pi)$.

12. $\sin 3x = -\frac{1}{2}$ **13.** $\sin 2x + \cos x = 0$

14. $2\cos^2 x - 3\cos x + 1 = 0$

15. $2\sin^2 x + \cos x = 1$

Cumulative Review Exercises (Chapters 1–6)

Solve each equation or inequality in Exercises 1–4.

1. $x^3 + x^2 - x + 15 = 0$ **2.** $11^{x-1} = 125$

3. $x^2 + 2x - 8 > 0$

4. $\cos 2x + 3 = 5\cos x, \quad 0 \le x < 2\pi$

In Exercises 5–10, graph each equation.

5. $y = \sqrt{x + 2} - 1$; Use transformations of the graph of $y = \sqrt{x}$.

6. $(x - 1)^2 + (y + 2)^2 = 9$

7. $y + 2 = \frac{1}{3}(x - 1)$

8. $y = 3\cos 2x, \quad -2\pi \le x \le 2\pi$

9. $y = 2\sin\frac{x}{2} + 1, \quad -2\pi \le x \le 2\pi$

10. $f(x) = (x - 1)^2(x - 3)$

11. If $f(x) = x^2 + 3x - 1$, find $\dfrac{f(a + h) - f(a)}{h}$.

12. Find the exact value of $\sin 225°$.

13. Verify the identity: $\sec^4 x - \sec^2 x = \tan^4 x + \tan^2 x$.

14. Convert 320° to radians.

15. How long would it take for any amount of money, compounded continuously at 5.75% per year, to triple? Round to the nearest tenth of a year.

16. If $f(x) = \dfrac{2x + 1}{x - 3}$, find $f^{-1}(x)$.

17. If C is a right angle in triangle ABC with $A = 23°$ and $a = 12$, solve the triangle.

18. A formula for calculating an infant's dosage for medication is

$$\text{Infant's dose} = \frac{\text{age of infant in months}}{150} \times \text{adult dose}.$$

If a 12-month-old infant is to receive 8.5 mg of medication, find the equivalent adult dose to the nearest milligram.

19. From a point on the ground 12 feet from the base of a flagpole, the angle of elevation to the top of the pole is 53°. Approximate the height of the flagpole to the nearest tenth of a foot.

20. In *A Tour of the Calculus*, David Berlinski describes trigonometric identities in the following way: "An invisible inner connection exists among the trigonometric functions, one revealed in various identities, strange places where the trigonometric functions appear fluidly to exchange identities or to resolve themselves into unlikely numbers." What does Berlinski mean by this? Use two fundamental trigonometric identities to illustrate your answer.

CHAPTER 5

Section 5.1

Check Point Exercises

1. a.

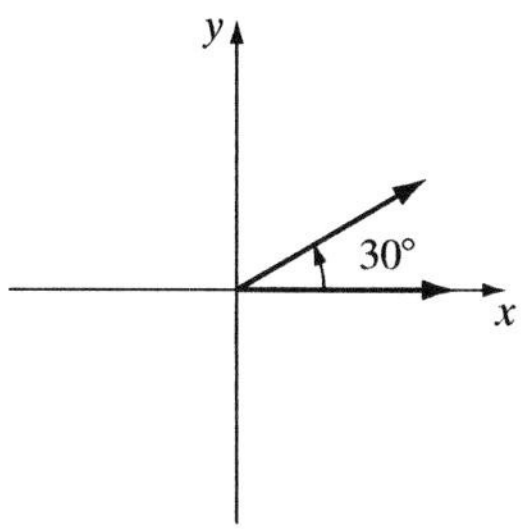

b.

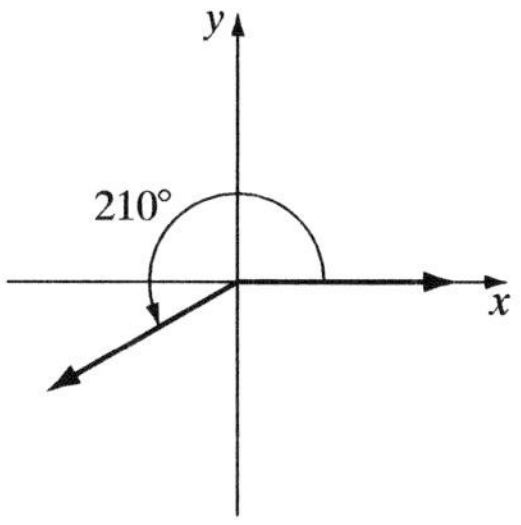

c.

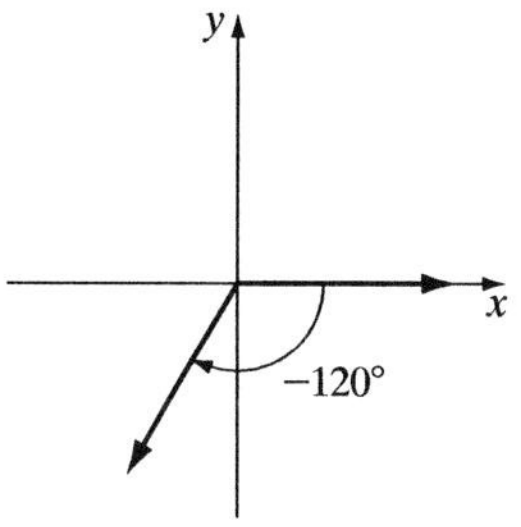

d.

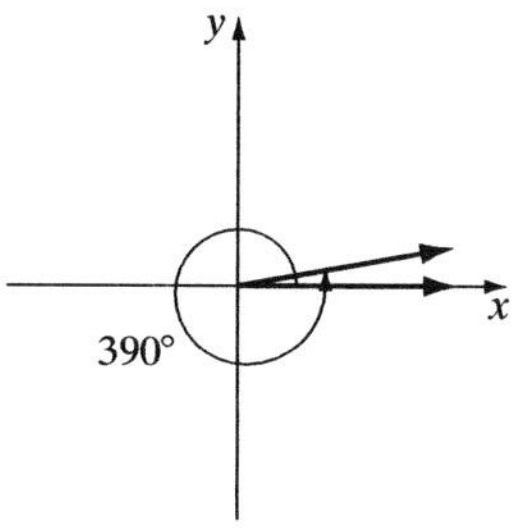

2. a. 40° **b.** 225° **3. a.** 12°; 102° **b.** no complementary angle; 30° **4.** 3.5 radians **5. a.** $\frac{\pi}{3}$ radians **b.** $\frac{3\pi}{2}$ radians **c.** $-\frac{5\pi}{3}$ radians **6. a.** 45° **b.** −240° **c.** 343.8° **7.** $\frac{3\pi}{2}$ in. ≈ 4.71 in. **8.** 135π in./min ≈ 424 in./min

Exercise Set 5.1

1. quadrant II **3.** quadrant III **5.** quadrant I **7.** obtuse **9.** straight

11.

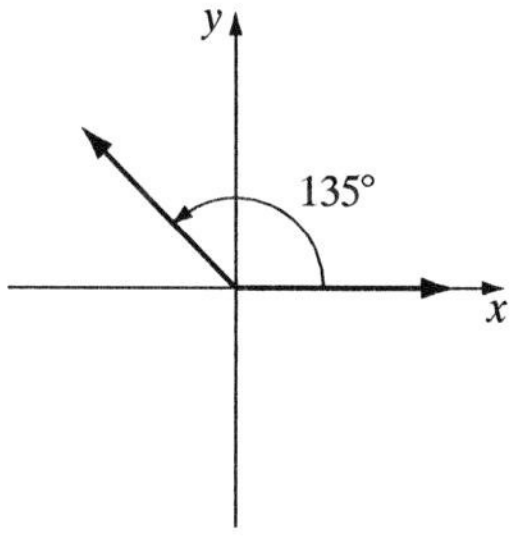

13.

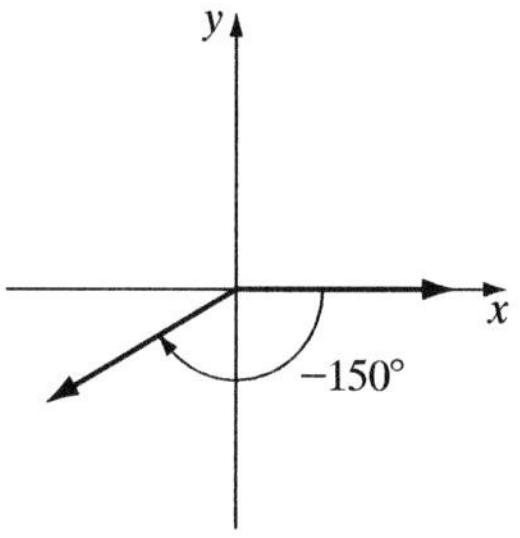

15.

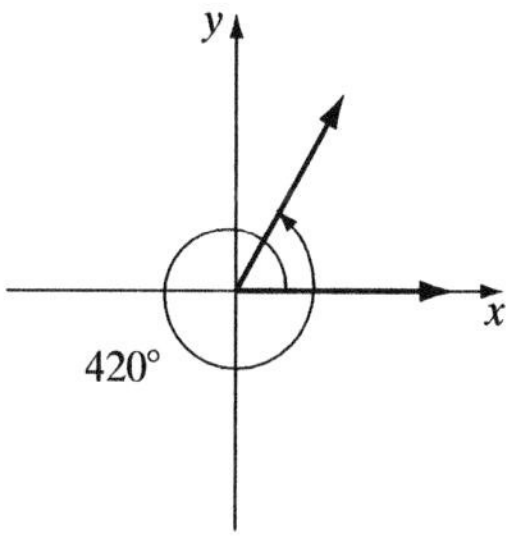

17.

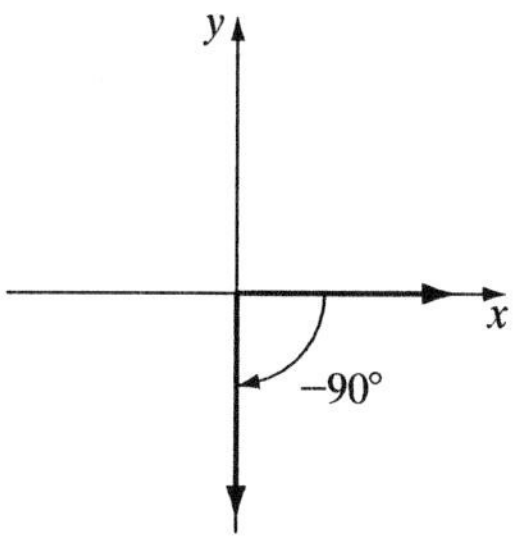

19. 35° **21.** 210° **23.** 315° **25.** 38°; 128° **27.** 52.6°; 142.6° **29.** no complement; 69° **31.** 4 radians **33.** $\frac{4}{3}$ radians **35.** 4 radians **37.** $\frac{\pi}{4}$ radians **39.** $\frac{3\pi}{4}$ radians **41.** $\frac{5\pi}{3}$ radians **43.** $-\frac{5\pi}{4}$ radians **45.** 90° **47.** 120° **49.** 210° **51.** −540° **53.** 0.31 radians **55.** −0.70 radians **57.** 3.49 radians **59.** 114.59° **61.** 13.85° **63.** −275.02° **65.** 3π in. ≈ 9.42 in. **67.** 10π ft ≈ 31.42 ft. **69.** $\frac{12\pi \text{ radians}}{\text{second}}$ **71.** 60°; $\frac{\pi}{3}$ radians **73.** $\frac{8\pi}{3}$ in. ≈ 8.38 in. **75.** 12π in. ≈ 37.70 in. **77.** 2 radians; 114.59° **79.** 2094 mi **81.** 1047 mph **83.** 1508 ft/min **97.** 30.25° **99.** 30°25′12″ **101.** smaller than a right angle **103.** 1815 mi

Section 5.2

Check Point Exercises

1. $\sin\theta = \frac{3}{5}$; $\cos\theta = \frac{4}{5}$; $\tan\theta = \frac{3}{4}$; $\csc\theta = \frac{5}{3}$; $\sec\theta = \frac{5}{4}$; $\cot\theta = \frac{4}{3}$ **2.** $\sqrt{2}$; $\sqrt{2}$; 1 **3.** $\sqrt{3}$; $\frac{\sqrt{3}}{3}$ **4.** $\tan\theta = \frac{2\sqrt{5}}{5}$; $\csc\theta = \frac{3}{2}$; $\sec\theta = \frac{3\sqrt{5}}{5}$; $\cot\theta = \frac{\sqrt{5}}{2}$ **5.** $\frac{\sqrt{3}}{2}$ **6. a.** cos 44° **b.** $\tan\frac{5\pi}{12}$ **7. a.** 0.9553 **b.** 1.0025 **8.** 333.9 yd **9.** 54°

Exercise Set 5.2

1. 15; $\sin\theta = \frac{3}{5}$; $\cos\theta = \frac{4}{5}$; $\tan\theta = \frac{3}{4}$; $\csc\theta = \frac{5}{3}$; $\sec\theta = \frac{5}{4}$; $\cot\theta = \frac{4}{3}$ **3.** 20; $\sin\theta = \frac{20}{29}$; $\cos\theta = \frac{21}{29}$; $\tan\theta = \frac{20}{21}$; $\csc\theta = \frac{29}{20}$; $\sec\theta = \frac{29}{21}$; $\cot\theta = \frac{21}{20}$ **5.** 24; $\sin\theta = \frac{5}{13}$; $\cos\theta = \frac{12}{13}$; $\tan\theta = \frac{5}{12}$; $\csc\theta = \frac{13}{5}$; $\sec\theta = \frac{13}{12}$; $\cot\theta = \frac{12}{5}$ **7.** 28; $\sin\theta = \frac{4}{5}$; $\cos\theta = \frac{3}{5}$; $\tan\theta = \frac{4}{3}$; $\csc\theta = \frac{5}{4}$; $\sec\theta = \frac{5}{3}$; $\cot\theta = \frac{3}{4}$ **9.** $\frac{\sqrt{3}}{2}$ **11.** $\sqrt{2}$ **13.** $\sqrt{3}$ **15.** 0

17. $\tan\theta = \frac{8}{15}$; $\csc\theta = \frac{17}{8}$; $\sec\theta = \frac{17}{15}$; $\cot\theta = \frac{15}{8}$ **19.** $\tan\theta = \frac{\sqrt{2}}{4}$; $\csc\theta = 3$; $\sec\theta = \frac{3\sqrt{2}}{4}$; $\cot\theta = 2\sqrt{2}$ **21.** $\frac{\sqrt{13}}{7}$

23. $\frac{5}{8}$ **25.** 1 **27.** 1 **29.** 1 **31.** cos 83° **33.** sec 65° **35.** $\cot\frac{7\pi}{18}$ **37.** $\sin\frac{\pi}{10}$ **39.** 0.6157 **41.** 0.6420 **43.** 3.4203

45. 0.9511 **47.** 3.7321 **49.** 188 cm **51.** 182 in. **53.** 41 m **55.** 17° **57.** 78° **59.** 1.147 radians **61.** 0.3950 radians

63. 529 yd **65.** 36° **67.** 2879 ft **69.** 37°

83. 0.92106, –0.19735; 0.95534, –0.148878; 0.98007, –0.099667; 0.99500, –0.04996; 0.99995, –0.005; 0.9999995, –0.0005; 0.999999995, –0.00005; 0.99999999995, –0.000005; $\frac{\cos\theta - 1}{\theta}$ approaches 0 as θ approaches 0.

85. In a right triangle, the hypotenuse is greater than either other side. Therefore, both $\frac{\text{opposite}}{\text{hypotenuse}}$ and $\frac{\text{adjacent}}{\text{hypotenuse}}$ must be less than 1 for an acute angle in a right triangle.

87. a. 357 ft **b.** 394 ft

Section 5.3

Check Point Exercises

1. $\sin\theta = -\frac{3}{5}$; $\cos\theta = \frac{4}{5}$; $\tan\theta = -\frac{3}{4}$; $\csc\theta = -\frac{5}{3}$; $\sec\theta = \frac{5}{4}$; $\cot\theta = -\frac{4}{3}$ **2. a.** 1; undefined **b.** 0; 1 **c.** –1; undefined **d.** 0; –1 **3.** quadrant III **4.** $\frac{\sqrt{10}}{10}$; $-\frac{\sqrt{10}}{3}$ **5. a.** 30° **b.** $\frac{\pi}{4}$ **c.** 60° **d.** 0.46 **6. a.** $-\frac{\sqrt{3}}{2}$ **b.** 1 **c.** $\frac{2\sqrt{3}}{3}$

Exercise Set 5.3

1. $\sin\theta = \frac{3}{5}$; $\cos\theta = -\frac{4}{5}$; $\tan\theta = -\frac{3}{4}$; $\csc\theta = \frac{5}{3}$; $\sec\theta = -\frac{5}{4}$; $\cot\theta = -\frac{4}{3}$ **3.** $\sin\theta = \frac{3\sqrt{13}}{13}$; $\cos\theta = \frac{2\sqrt{13}}{13}$; $\tan\theta = \frac{3}{2}$; $\csc\theta = \frac{\sqrt{13}}{3}$; $\sec\theta = \frac{\sqrt{13}}{2}$; $\cot\theta = \frac{2}{3}$ **5.** $\sin\theta = -\frac{\sqrt{2}}{2}$; $\cos\theta = \frac{\sqrt{2}}{2}$; $\tan\theta = -1$; $\csc\theta = -\sqrt{2}$; $\sec\theta = \sqrt{2}$; $\cot\theta = -1$ **7.** $\sin\theta = -\frac{5\sqrt{29}}{29}$; $\cos\theta = -\frac{2\sqrt{29}}{29}$; $\tan\theta = \frac{5}{2}$; $\csc\theta = -\frac{\sqrt{29}}{5}$; $\sec\theta = -\frac{\sqrt{29}}{2}$; $\cot\theta = \frac{2}{5}$ **9.** –1 **11.** –1 **13.** undefined **15.** 0 **17.** quadrant I

19. quadrant III **21.** quadrant II **23.** $\sin\theta = -\frac{4}{5}$; $\tan\theta = \frac{4}{3}$; $\csc\theta = -\frac{5}{4}$; $\sec\theta = -\frac{5}{3}$; $\cot\theta = \frac{3}{4}$ **25.** $\cos\theta = -\frac{12}{13}$; $\tan\theta = -\frac{5}{12}$; $\csc\theta = \frac{13}{5}$; $\sec\theta = -\frac{13}{12}$; $\cot\theta = -\frac{12}{5}$ **27.** $\sin\theta = -\frac{15}{17}$; $\tan\theta = -\frac{15}{8}$; $\csc\theta = -\frac{17}{15}$; $\sec\theta = \frac{17}{8}$; $\cot\theta = -\frac{8}{15}$ **29.** $\sin\theta = \frac{2\sqrt{13}}{13}$; $\cos\theta = -\frac{3\sqrt{13}}{3}$; $\csc\theta = \frac{\sqrt{13}}{2}$; $\sec\theta = -\frac{\sqrt{13}}{3}$; $\cot\theta = -\frac{3}{2}$ **31.** $\sin\theta = -\frac{4}{5}$; $\cos\theta = -\frac{3}{5}$; $\csc\theta = -\frac{5}{4}$; $\sec\theta = -\frac{5}{3}$; $\cot\theta = \frac{3}{4}$

33. $\sin\theta = -\frac{2\sqrt{2}}{3}$; $\cos\theta = -\frac{1}{3}$; $\tan\theta = 2\sqrt{2}$; $\csc\theta = -\frac{3\sqrt{2}}{4}$; $\cot\theta = \frac{\sqrt{2}}{4}$ **35.** 20° **37.** 25° **39.** 5° **41.** $\frac{\pi}{4}$ **43.** $\frac{\pi}{6}$ **45.** 30°

47. 25° **49.** 1.56 **51.** $-\frac{\sqrt{2}}{2}$ **53.** $\frac{\sqrt{3}}{3}$ **55.** $\sqrt{3}$ **57.** $\frac{\sqrt{3}}{2}$ **59.** –2 **61.** 1 **63.** $\frac{\sqrt{3}}{2}$ **65.** –1

Section 5.4

Check Point Exercises

1. $\sin\pi = 0$; $\cos\pi = -1$; $\tan\pi = 0$; $\csc\pi$ is undefined; $\sec\pi = -1$; $\cot\pi$ is undefined **2. a.** $\frac{1}{2}$ **b.** $-\frac{\sqrt{3}}{3}$ **3. a.** $\frac{\sqrt{2}}{2}$ **b.** $\sqrt{3}$

Exercise Set 5.4

1. $\sin t = \frac{8}{17}$; $\cos t = -\frac{15}{17}$; $\tan t = -\frac{8}{15}$; $\csc t = \frac{17}{8}$; $\sec t = -\frac{17}{15}$; $\cot t = -\frac{15}{8}$ **3.** $\sin\left(-\frac{\pi}{4}\right) = -\frac{\sqrt{2}}{2}$; $\cos\left(-\frac{\pi}{4}\right) = \frac{\sqrt{2}}{2}$; $\tan\left(-\frac{\pi}{4}\right) = -1$; $\csc\left(-\frac{\pi}{4}\right) = -\sqrt{2}$; $\sec\left(-\frac{\pi}{4}\right) = \sqrt{2}$; $\cot\left(-\frac{\pi}{4}\right) = -1$ **5.** $-\frac{\sqrt{2}}{2}$ **7.** 2 **9.** $-\frac{\sqrt{2}}{2}$ **11.** $\frac{\sqrt{3}}{3}$ **13. a.** 12 hr **b.** 20.3 hr **c.** 3.7 hr **15. a.** 1; 0; –1; 0; 1 **b.** 28 days **23.** 0 **25.** $-\frac{1}{4}$ **c.** 9.6hr

Section 5.5

Check Point Exercises

1.

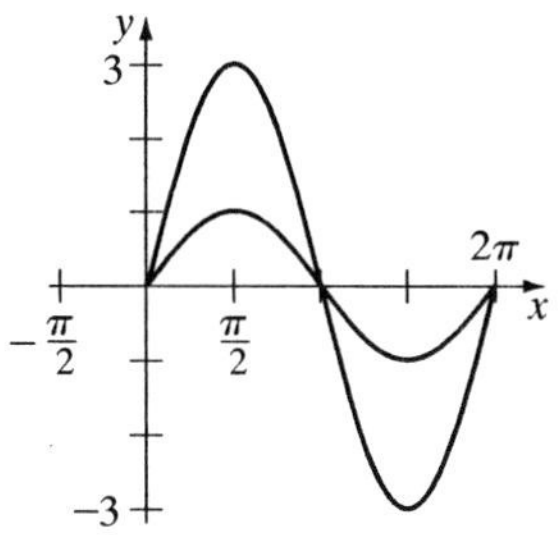

2.

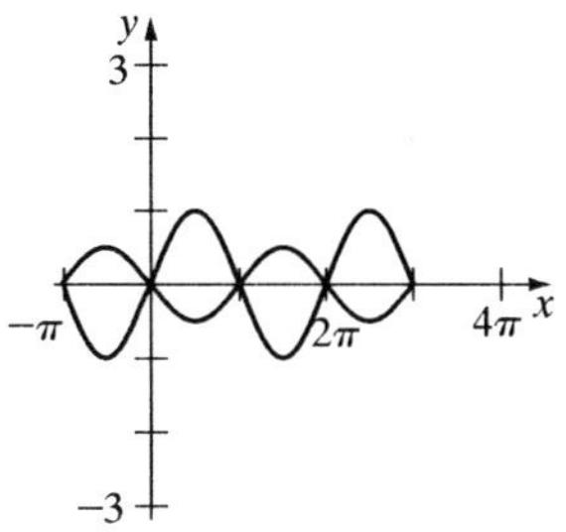

3.

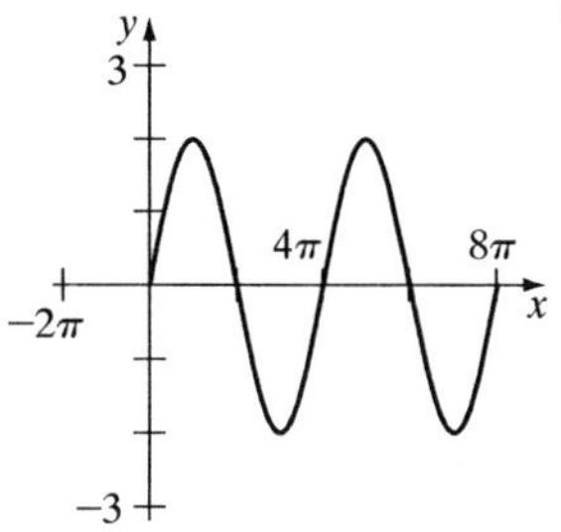

4.

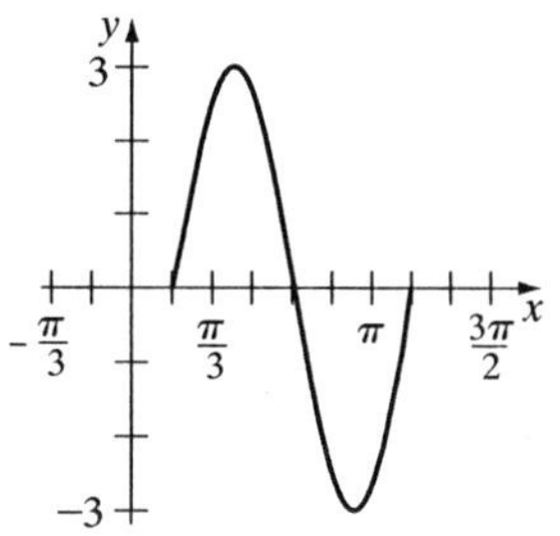

5.

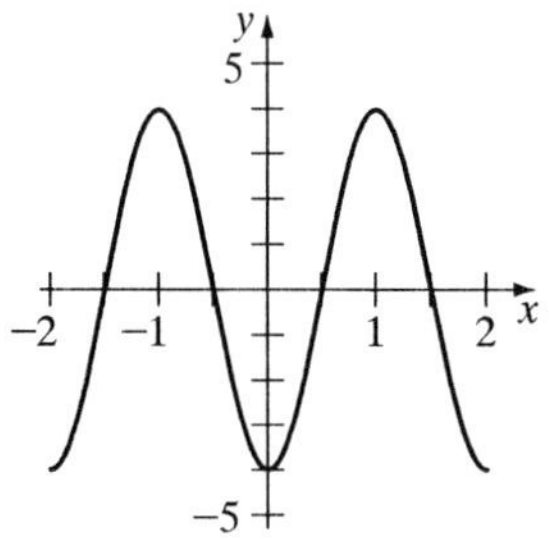

6.

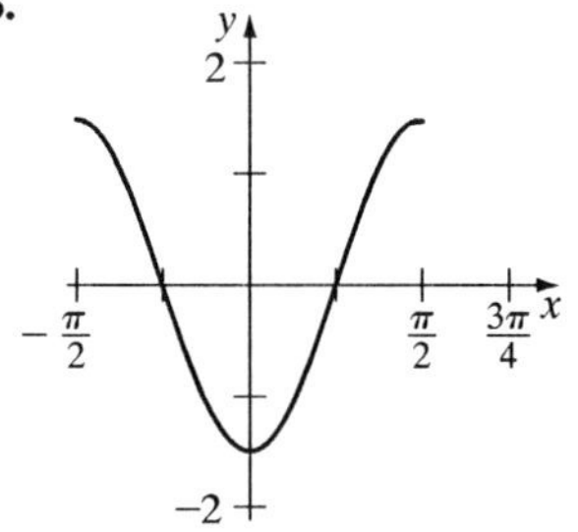

7.

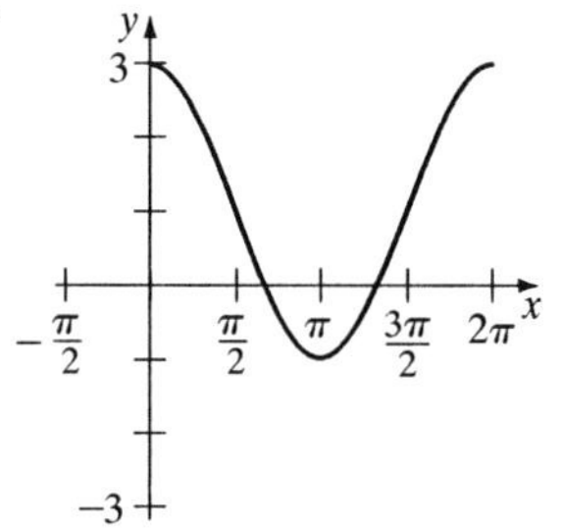

8. $y = 4 \sin 4x$

9. $y = 2 \sin\left(\frac{\pi}{6}x - \frac{\pi}{2}\right) + 12$

Exercise Set 5.5

1. 4

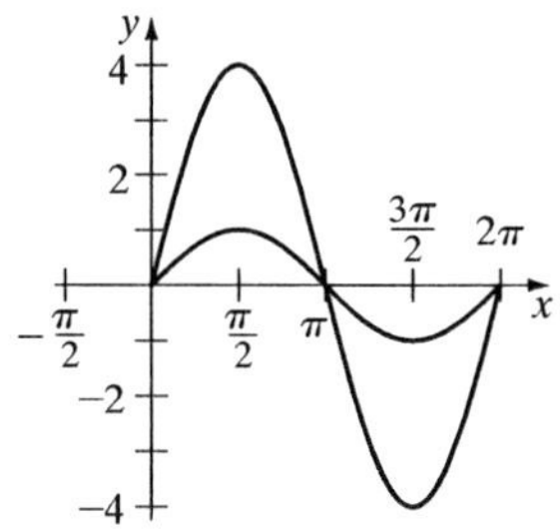

3. $\frac{1}{3}$

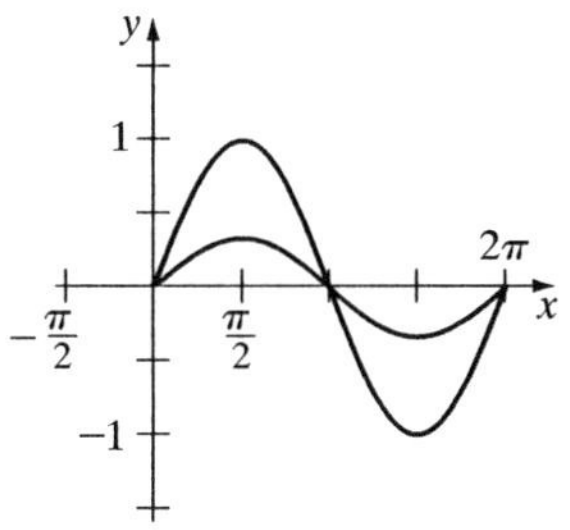

5. 3

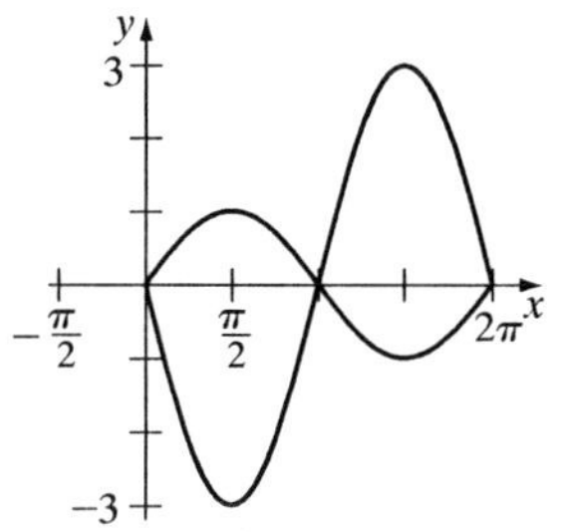

7. 1; π

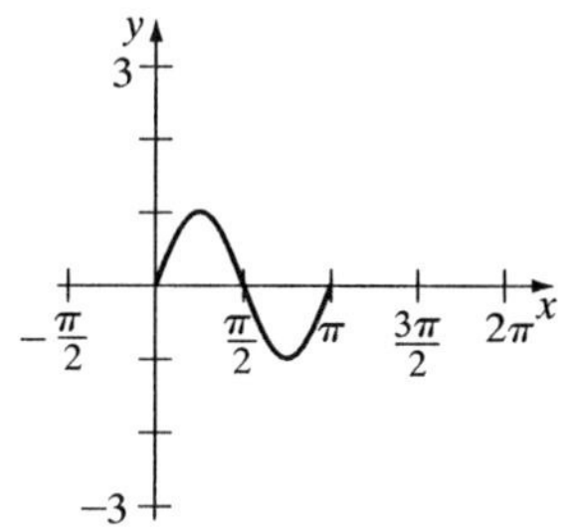

9. 3; 4π

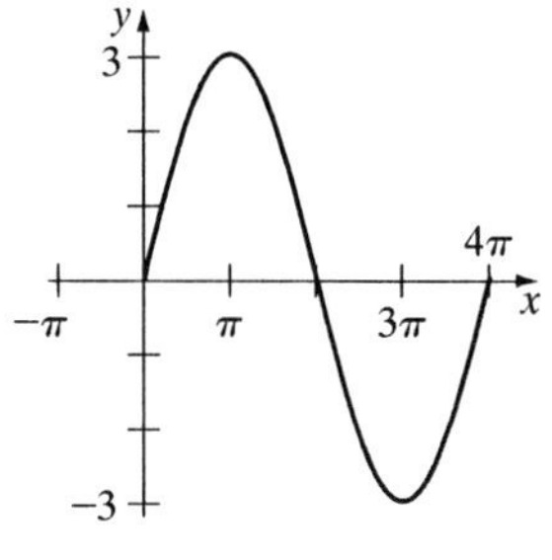

11. 4; 2

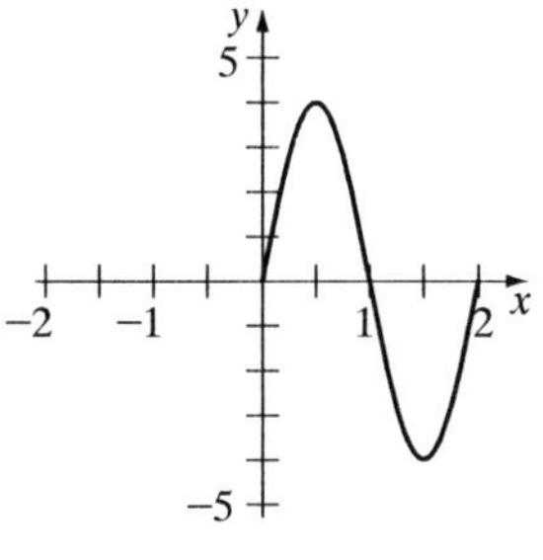

13. 3; 1

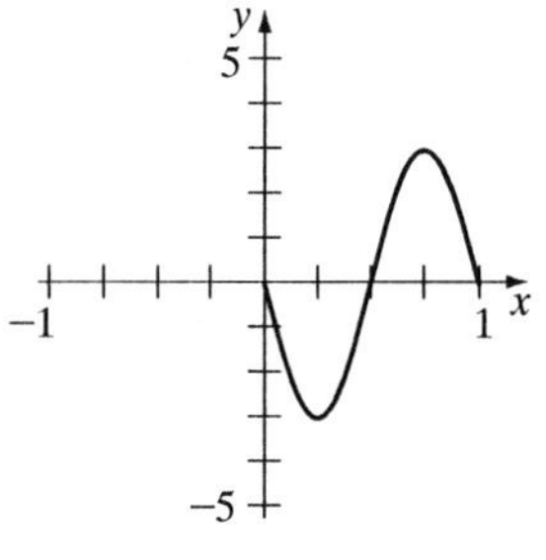

15. 1; 3π

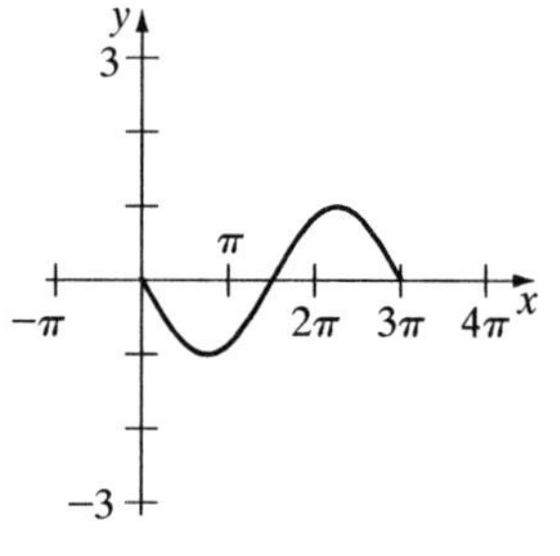

17. $1; 2\pi; \pi$

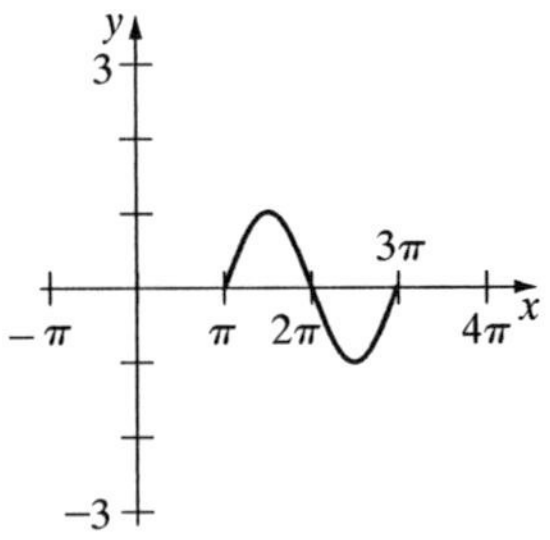

19. $1; \pi; \frac{\pi}{2}$

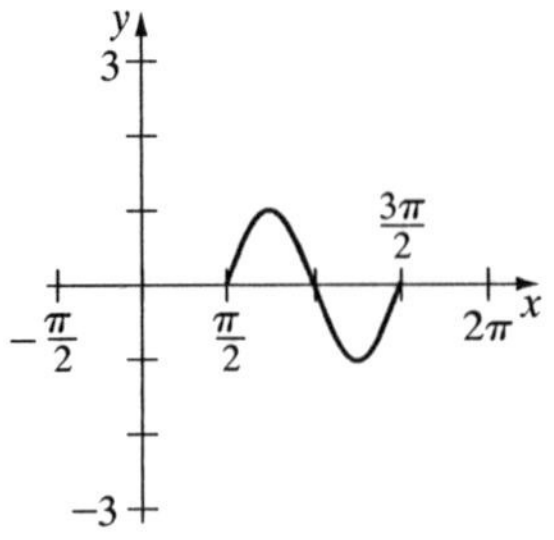

21. $3; \pi; \frac{\pi}{2}$

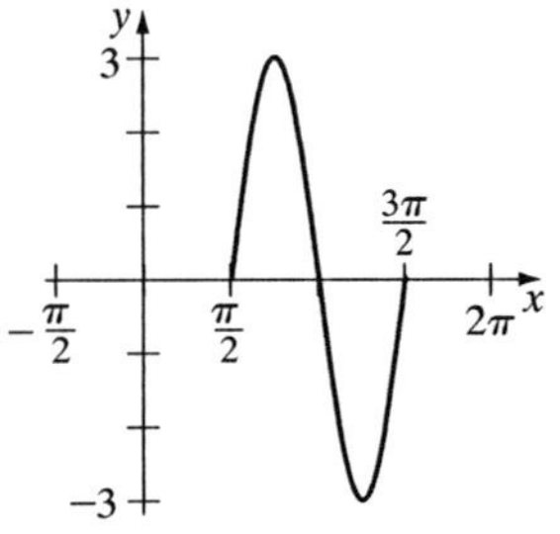

23. $\frac{1}{2}; 2\pi; -\frac{\pi}{2}$

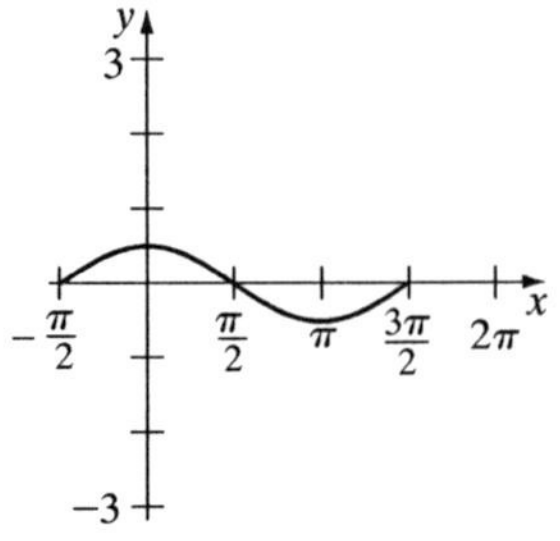

25. $2; \pi; -\frac{\pi}{4}$

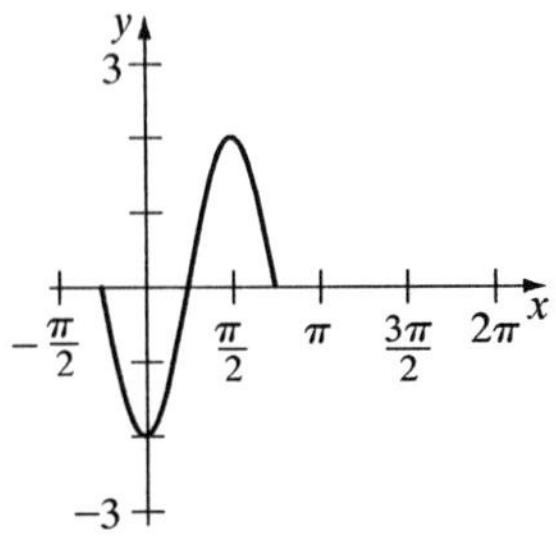

27. $3; 2; -\frac{2}{\pi}$

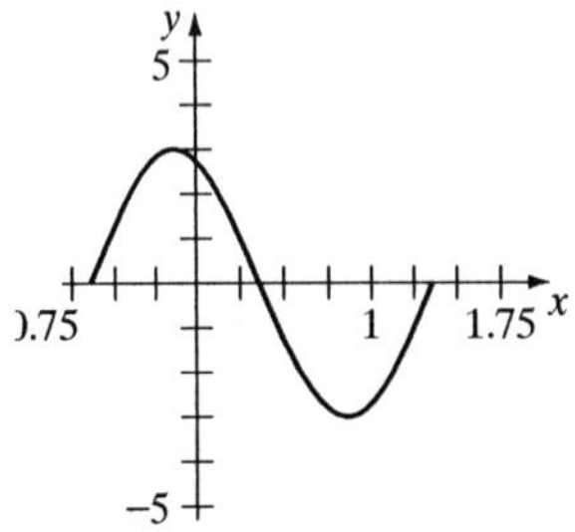

29. $2; 1; -2$

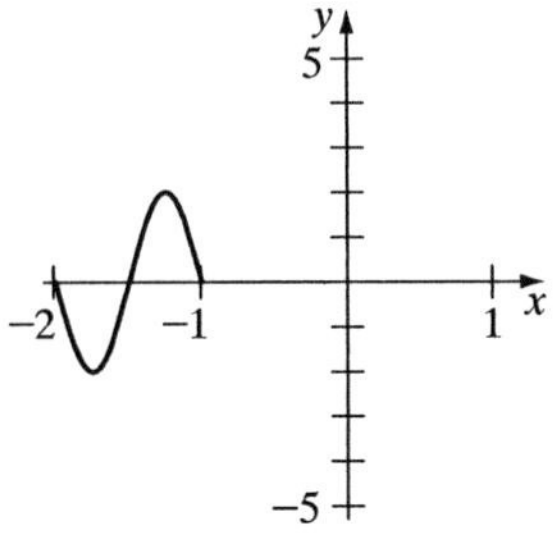

31. 2

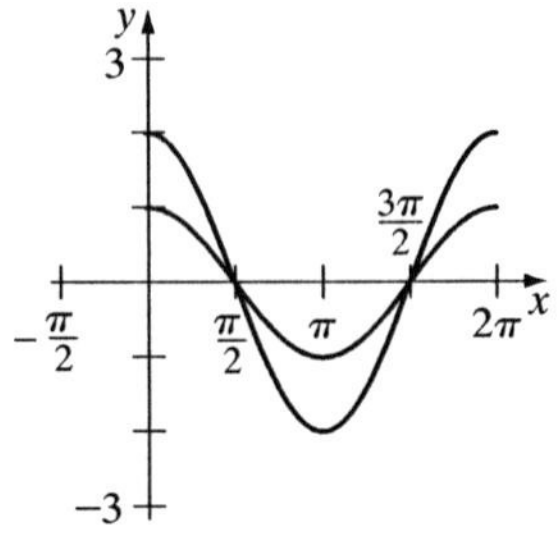

33. 2

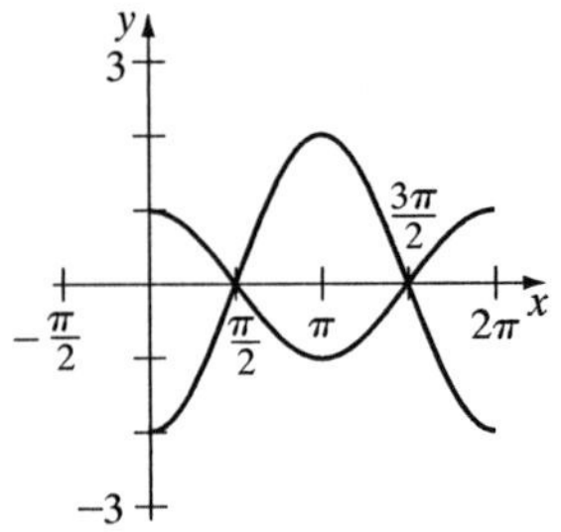

35. $1; \pi$

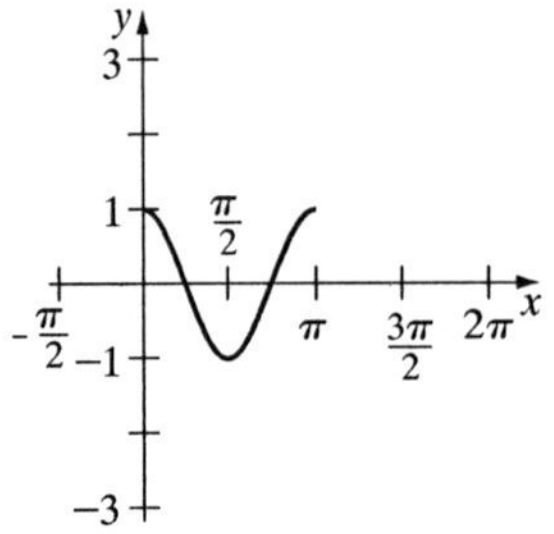

37. $4; 1$

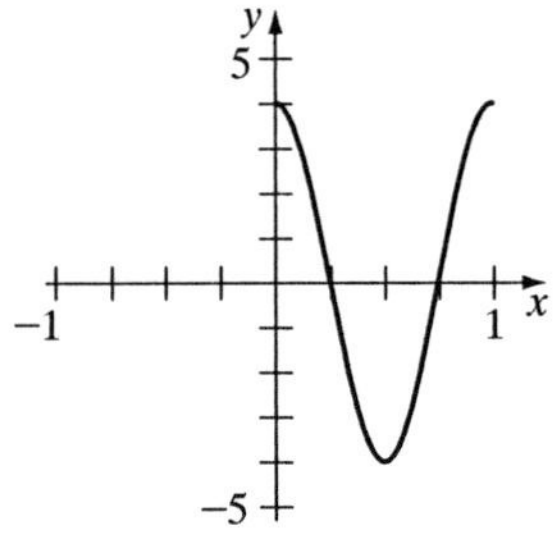

39. $4; 4\pi$

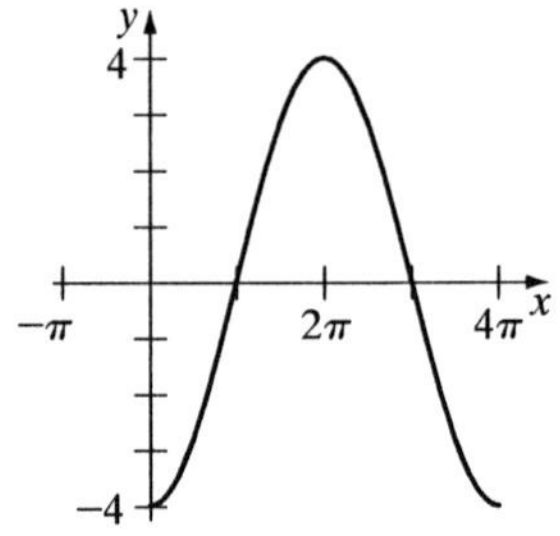

41. $\frac{1}{2}; 6$

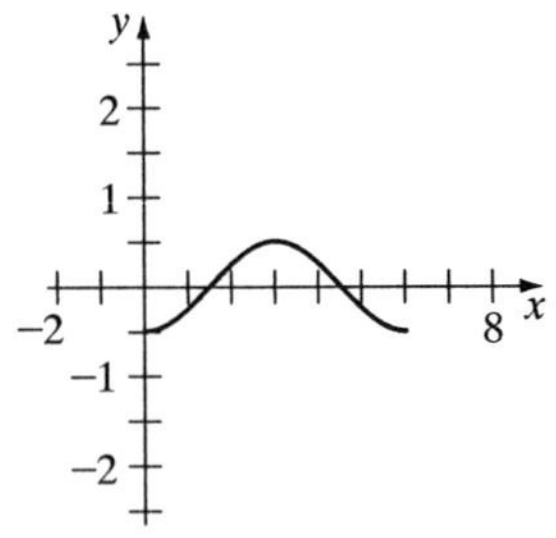

43. $3; \pi; \frac{\pi}{2}$

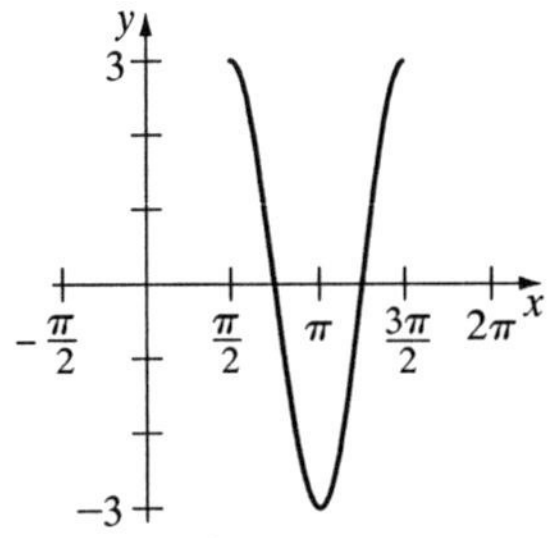

45. $\frac{1}{2}; \frac{2\pi}{3}; -\frac{\pi}{6}$

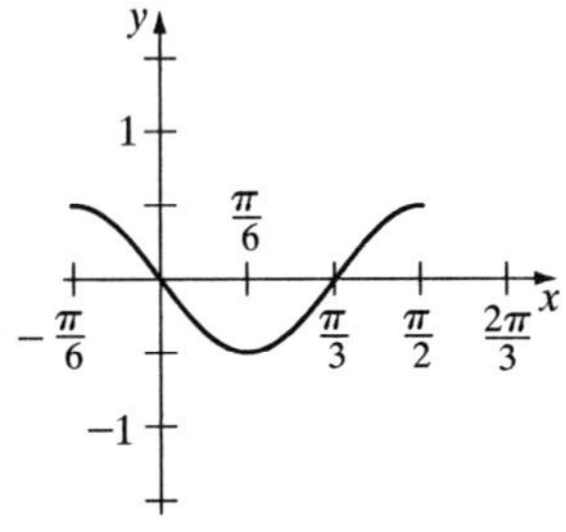

47. $3; \pi; \frac{\pi}{4}$

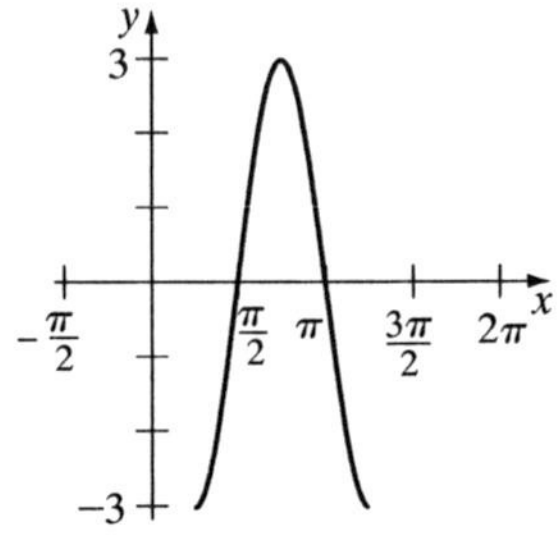

49. 2; 1; −4

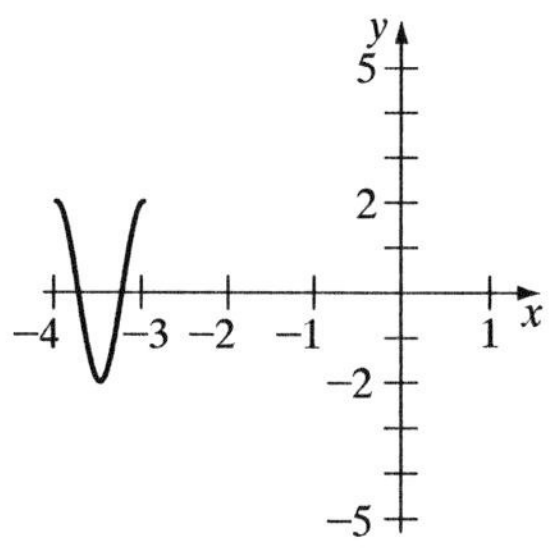

51.

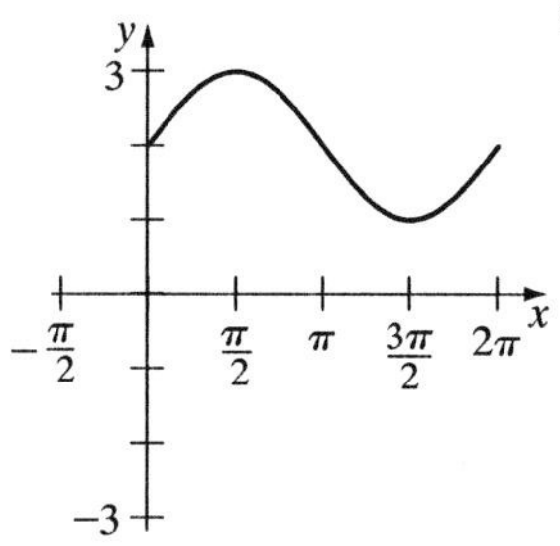

53.

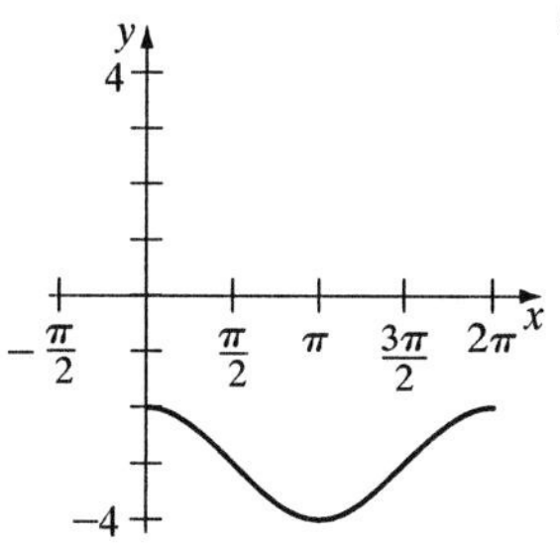

55.

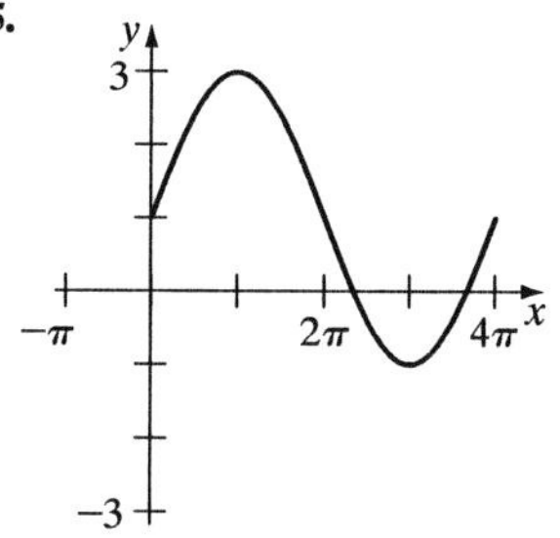

57.

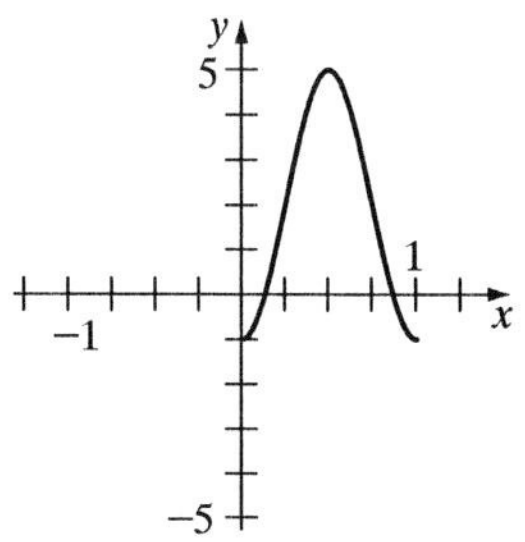

59. 33 days **61.** 23 days **63.** March 21 **65.** No

67.

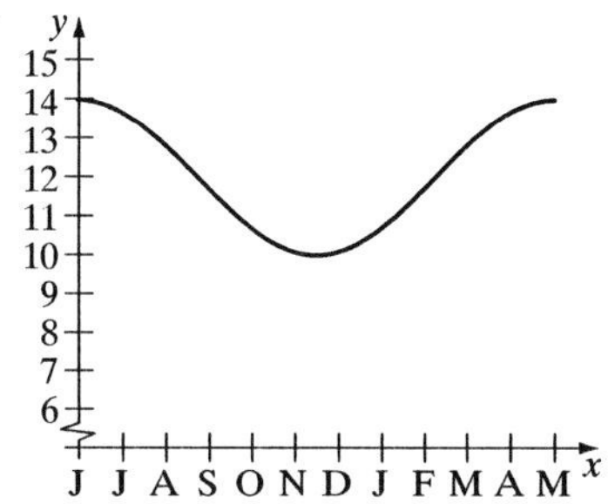

69. a. 3 **b.** 365 days **c.** 15 hours of daylight **d.** 9 hours of daylight **e.**

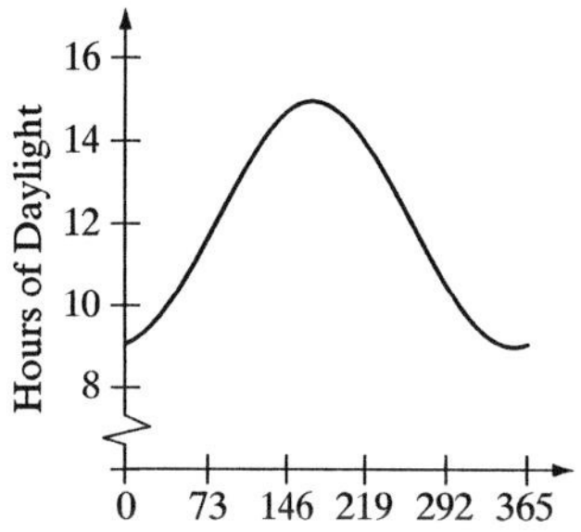

71. $y = 3\cos\dfrac{\pi x}{6} + 9$ **73.** $4; \pi; \dfrac{\pi}{2}$

85.

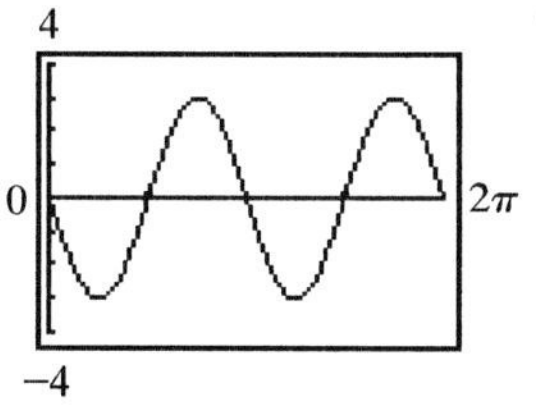

87.

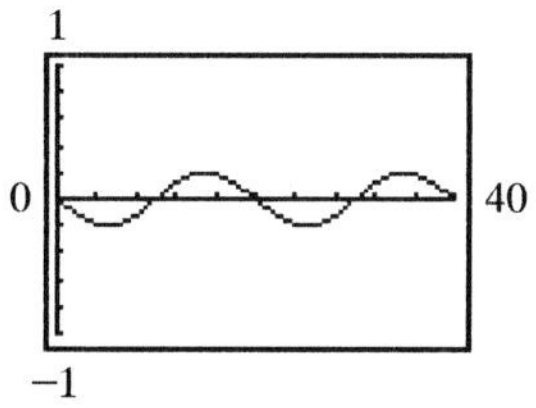

89.

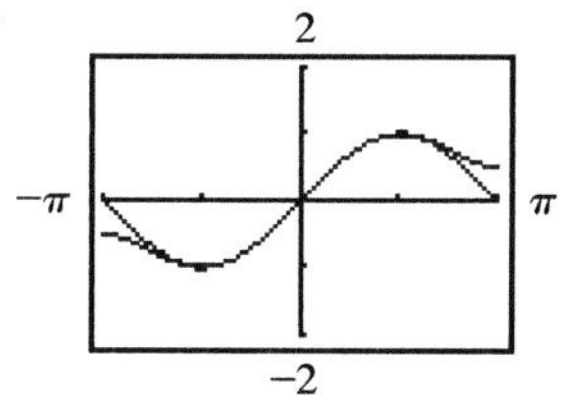

91.

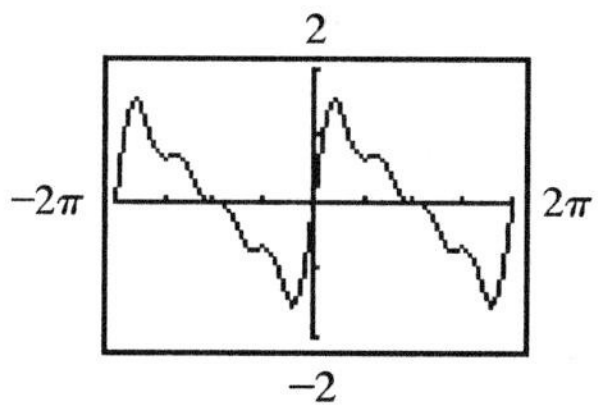

The graph is similar to $y = \sin x$, except the amplitude is greater and the curve is less smooth.

93. a.

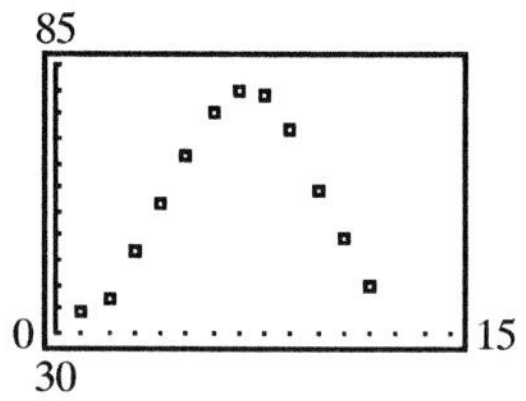

b. $y = 22.61\sin(0.50x - 2.04) + 57.17$

c.

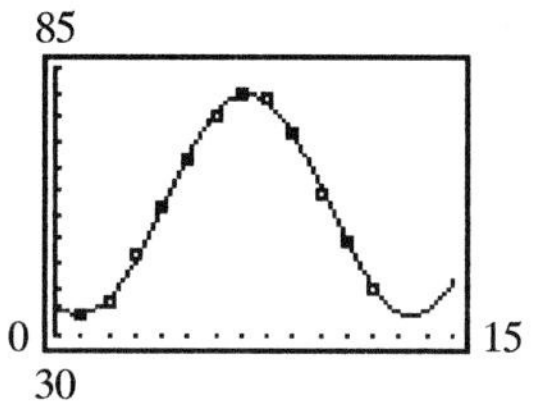

95.

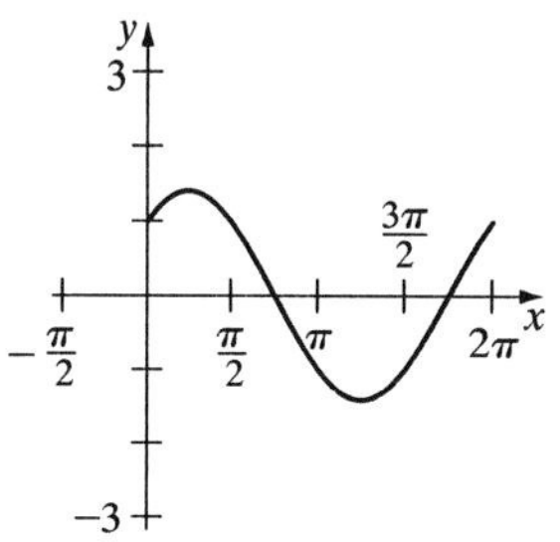

97. $y = 2\cos(4x + \pi)$

Section 5.6

Check Point Exercises

1.

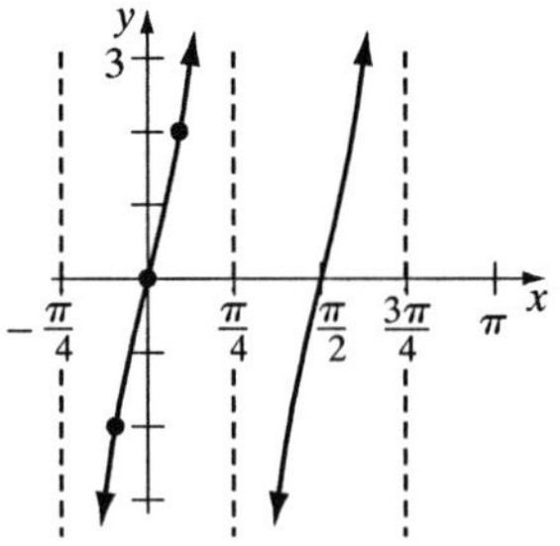

2.

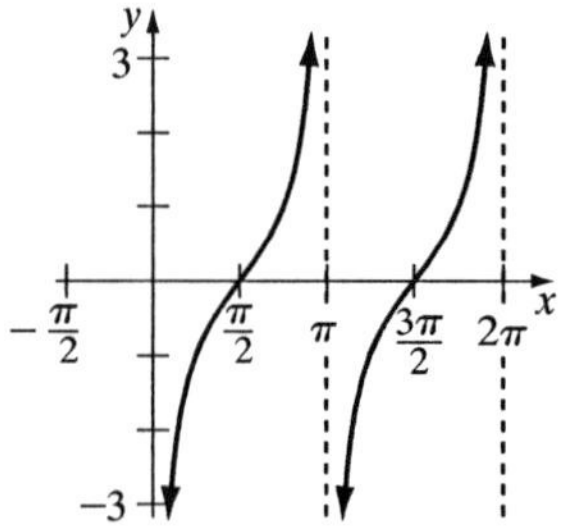

3.

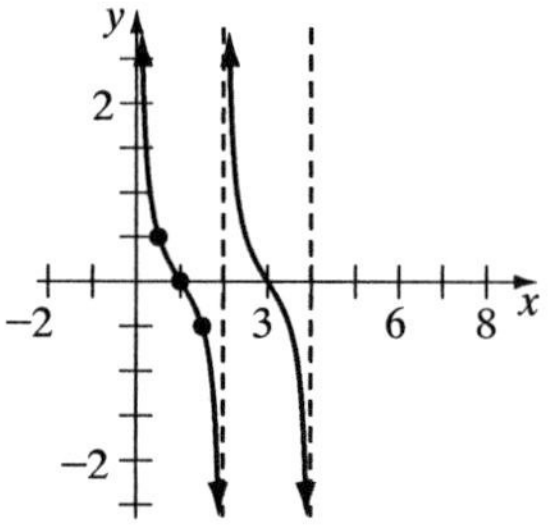

4.

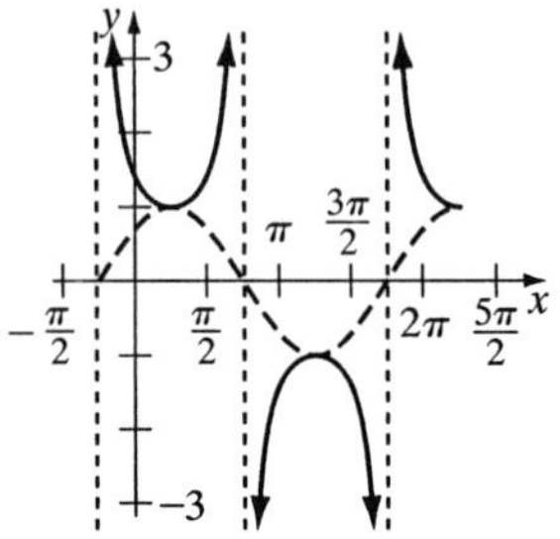

5.

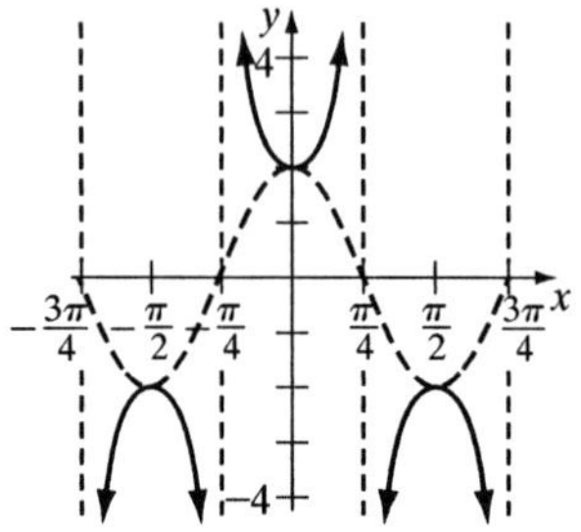

Exercise Set 5.6

1. $y = \tan(x + \pi)$ **3.** $y = -\tan\left(x - \frac{\pi}{2}\right)$

5.

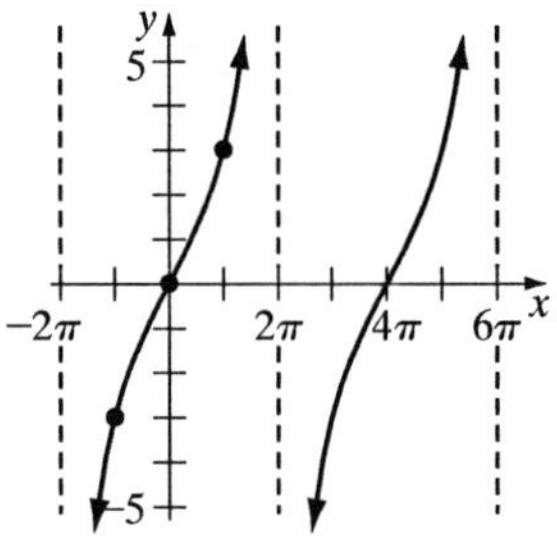

7.

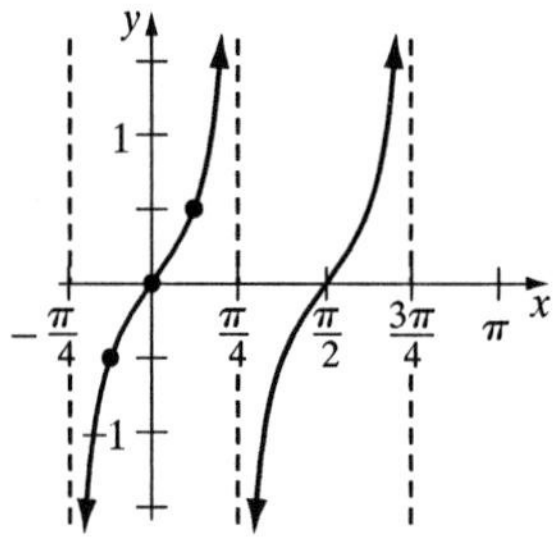

9.

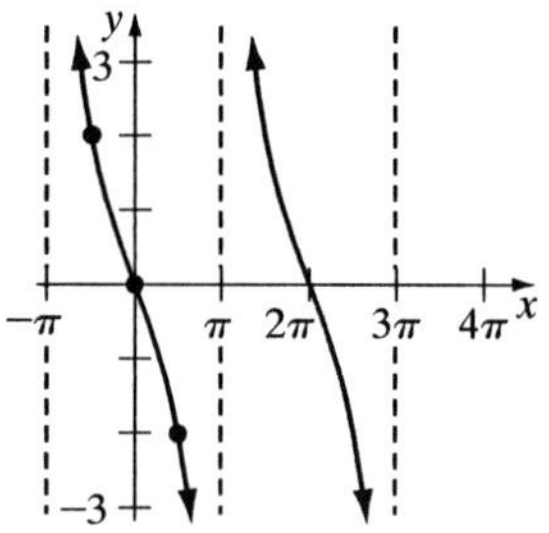

11.

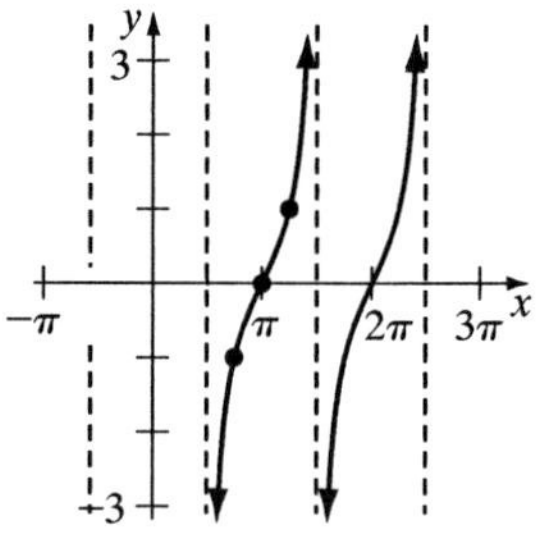

13. $y = -\cot x$

15. $y = \cot\left(x + \frac{\pi}{2}\right)$

17.

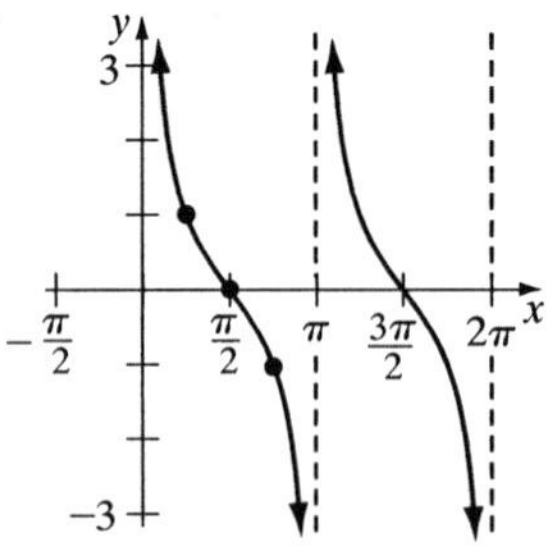

19.

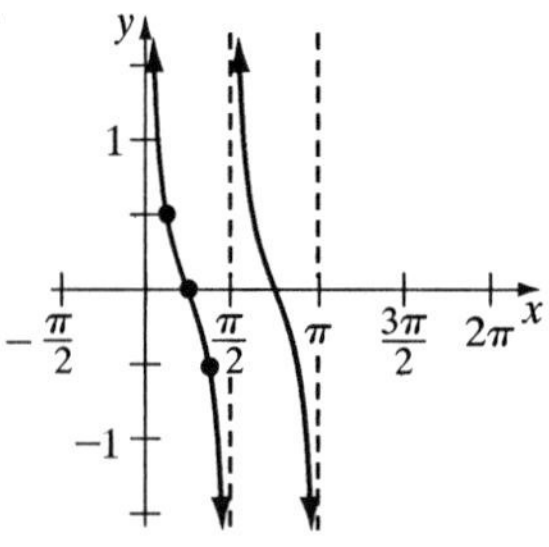

21.

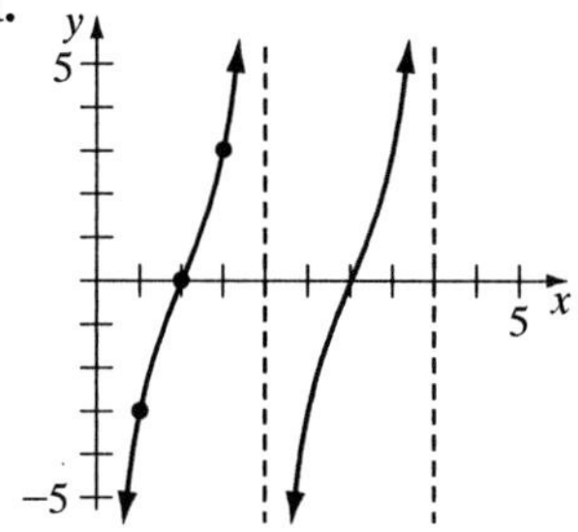

23.

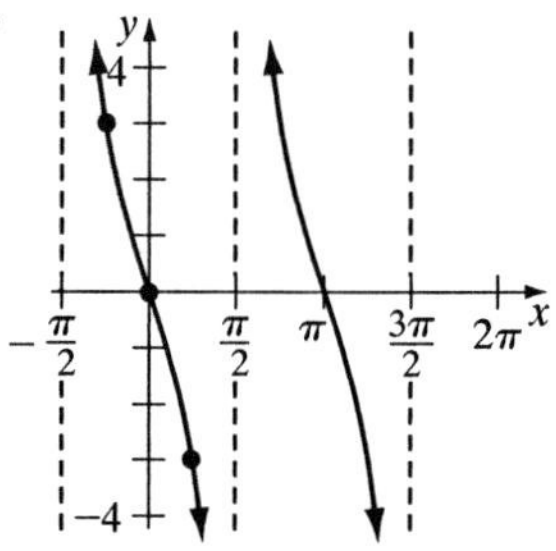

25. $y = -\frac{1}{2}\csc\frac{x}{2}$;

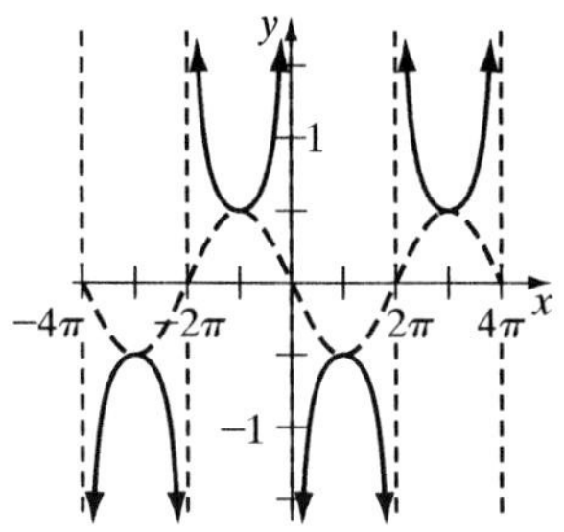

27. $y = \frac{1}{2}\sec 2\pi x$;

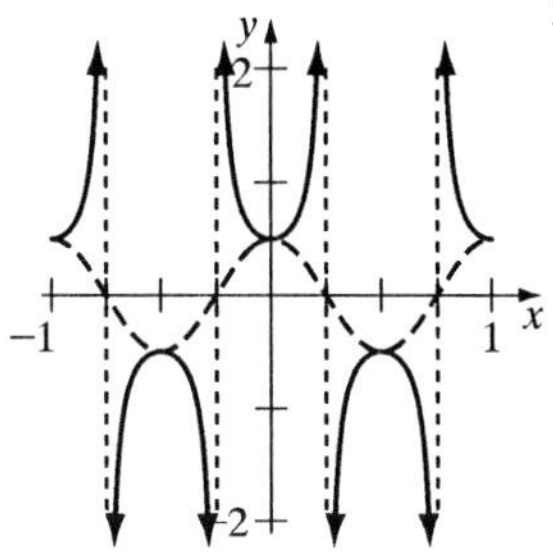

29.

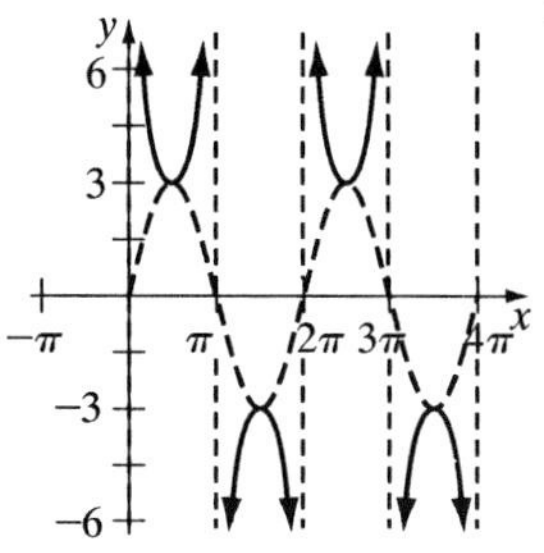

31.

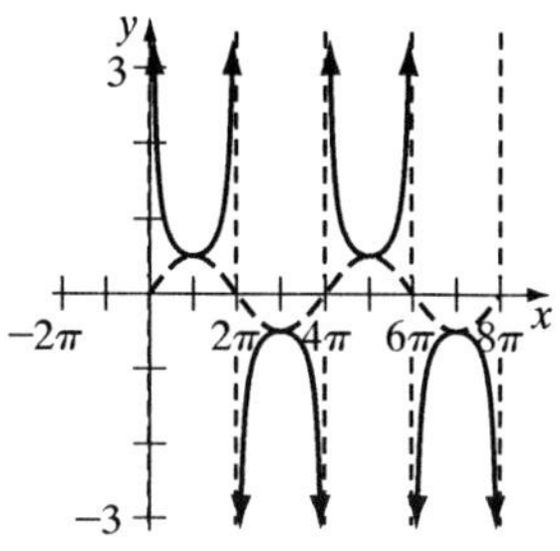

33.

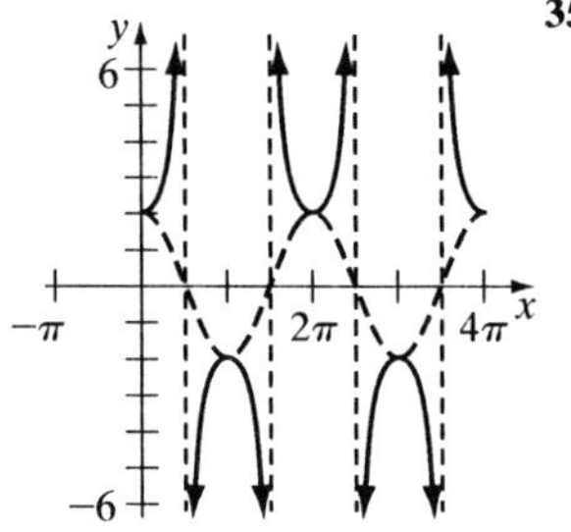

35.

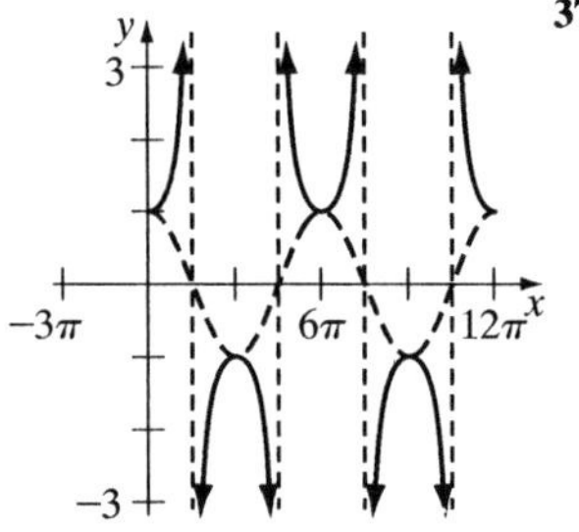

37.

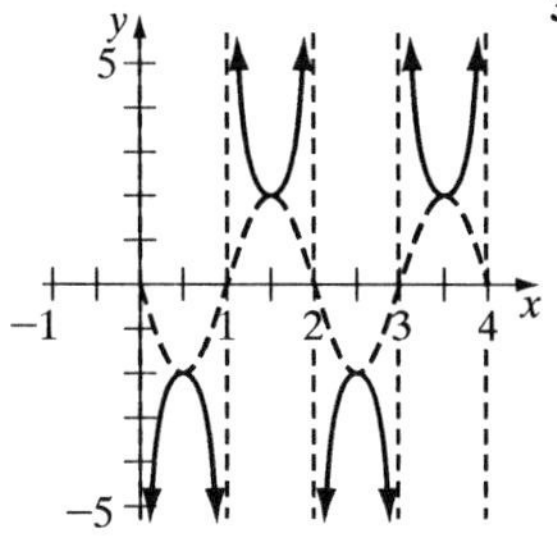

39.

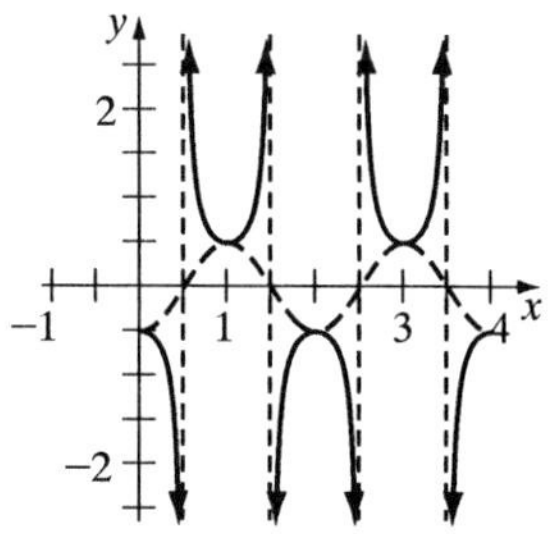

41.

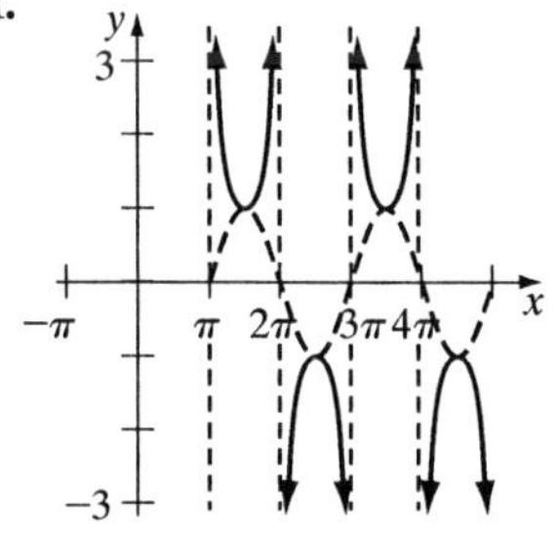

43.

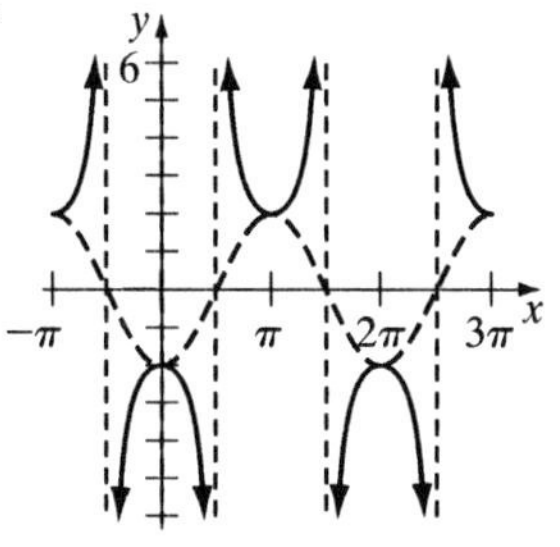

45. a.

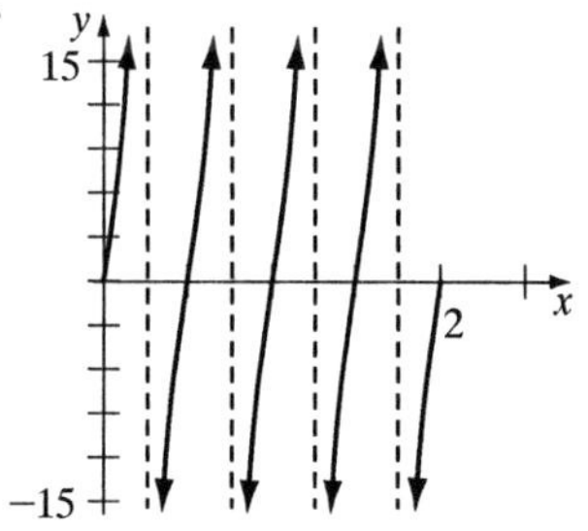

b. 0.25, 0.75, 1.25, 1.75; The beacon is shining parallel to the wall at these times.

47. $d = 10 \sec x$

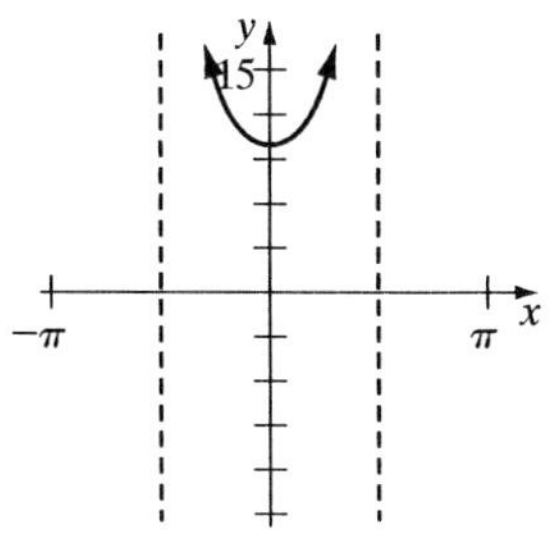

49.

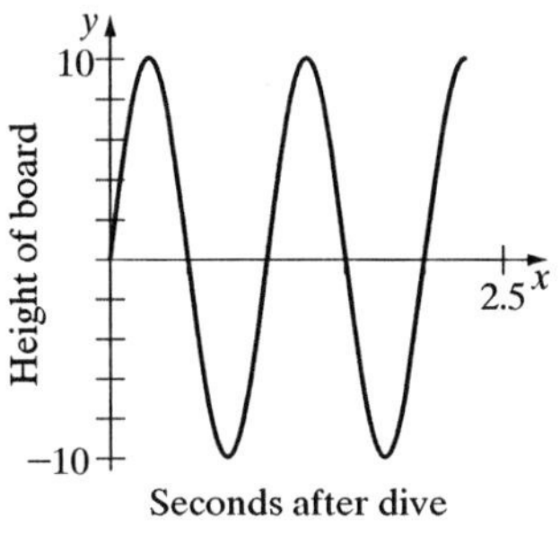

63.

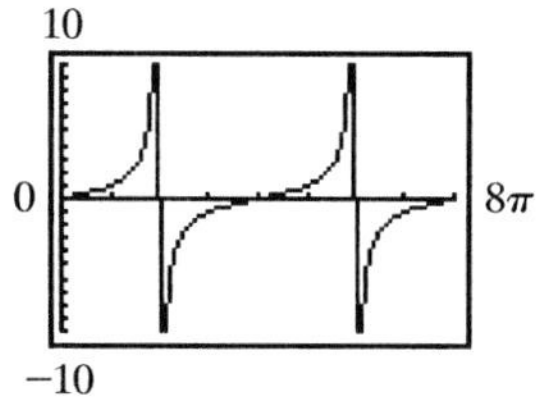

65.

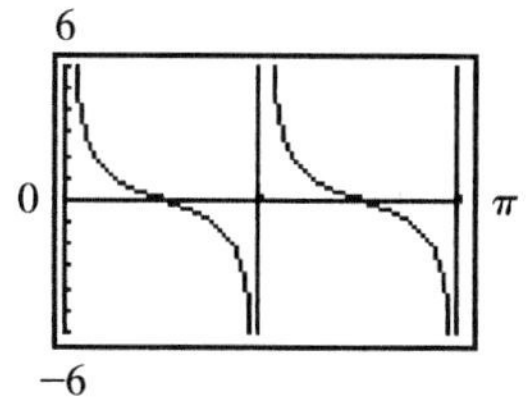

67.

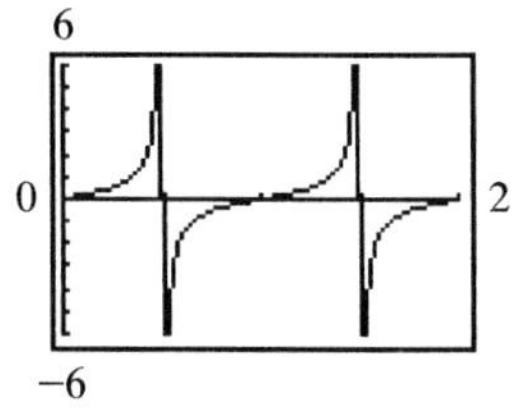

69.

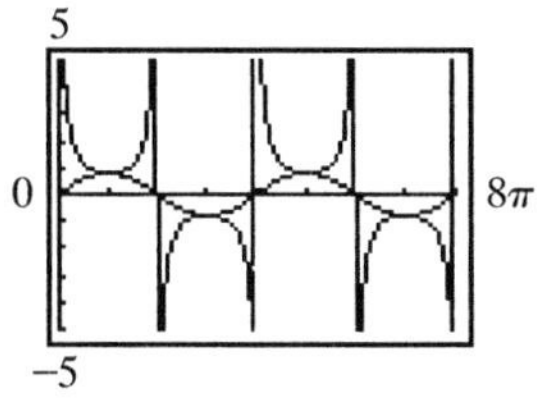

71.

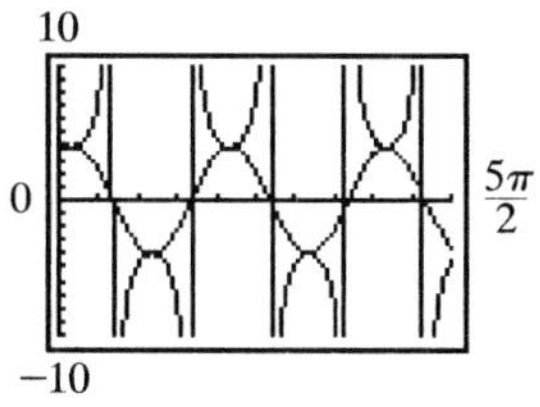

73.

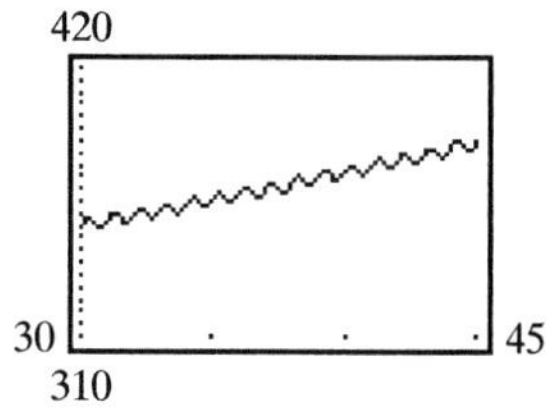

75. $y = \cot\frac{3}{2}x$ **77.** 2^{-x} decreases the amplitude as x gets larger.

Section 5.7

Check Point Exercises

1. $\frac{\pi}{3}$ **2.** $-\frac{\pi}{4}$ **3.** $\frac{2\pi}{3}$ **4.** $-\frac{\pi}{4}$ **5. a.** 1.2310 **b.** −1.5429 **6. a.** 0.7 **b.** 0 **c.** not defined **7.** $\frac{3}{5}$ **8.** $\frac{\sqrt{3}}{2}$ **9.** $\sqrt{x^2+1}$

Exercise Set 5.7

1. $\frac{\pi}{6}$ **3.** $\frac{\pi}{4}$ **5.** $-\frac{\pi}{6}$ **7.** $\frac{\pi}{6}$ **9.** $\frac{3\pi}{4}$ **11.** $\frac{\pi}{2}$ **13.** $\frac{\pi}{6}$ **15.** 0 **17.** $-\frac{\pi}{3}$ **19.** 0.30 **21.** −0.33 **23.** 1.19 **25.** 1.25 **27.** −1.52 **29.** −1.52 **31.** 0.9 **33.** $\frac{\pi}{3}$ **35.** $\frac{\pi}{6}$ **37.** 125 **39.** $-\frac{\pi}{6}$ **41.** $-\frac{\pi}{3}$ **43.** 0 **45.** not defined **47.** $\frac{3}{5}$ **49.** $\frac{12}{5}$ **51.** $-\frac{3}{4}$ **53.** $\frac{\sqrt{2}}{2}$ **55.** $\frac{4\sqrt{15}}{15}$ **57.** $-2\sqrt{2}$ **59.** 2 **61.** $\frac{\sqrt{1-x^2}}{x}$ **63.** $\sqrt{1-4x^2}$ **65.** $\frac{\sqrt{x^2-1}}{x}$ **67.** $\frac{\sqrt{3}}{x}$ **69.** $\frac{\sqrt{x^2+4}}{2}$

71. a.

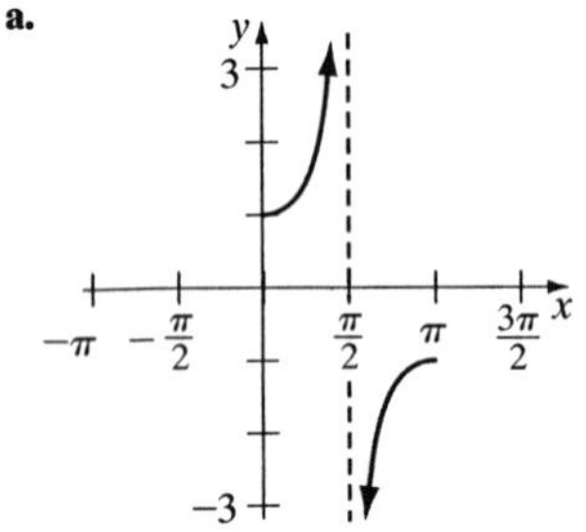

b. No horizontal line intersects the graph of $y = \sec x$ more than once, so the function is one-to-one and has an inverse function.

c.

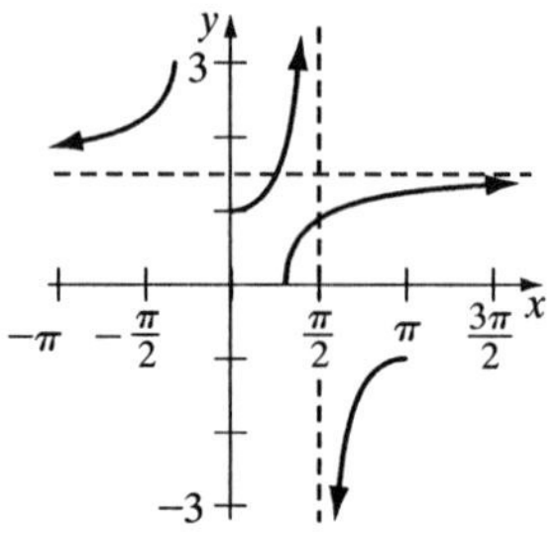

73. 0.408 radians; 0.602 radians; 0.654 radians; 0.645 radians; 0.613 radians **75.** 1.3157 radians or 75.4° **77.** 1.1071 sq units

91.

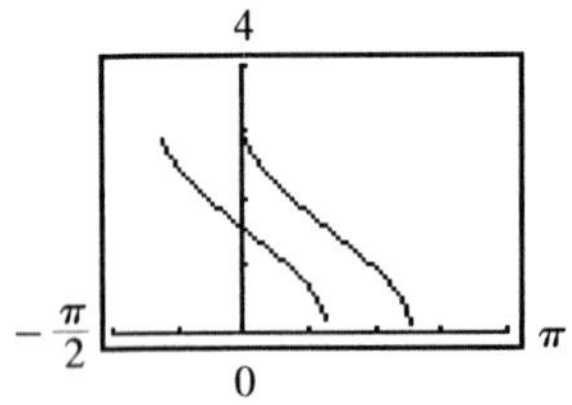

93.

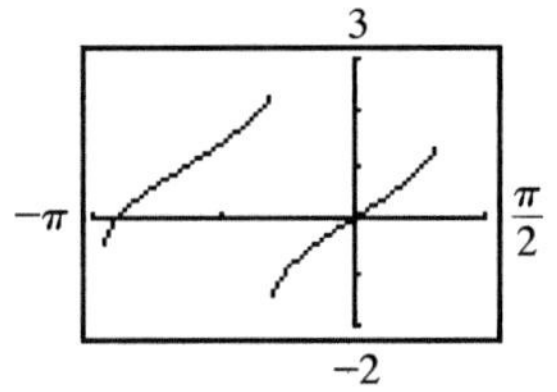

Shifted left 2 units and up 1 unit

95.

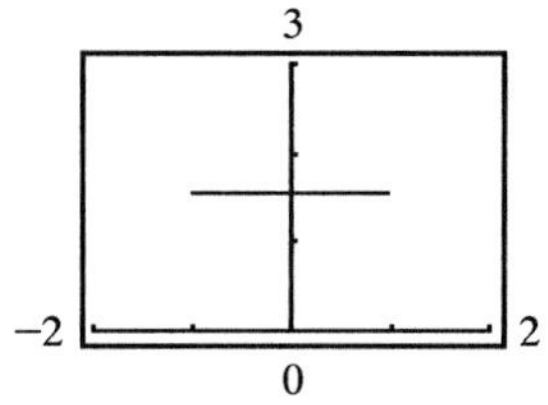

It seems $\sin^{-1} x + \cos^{-1} x = \frac{\pi}{2}$ for $-1 \le x \le 1$

97. $x = \sin\frac{\pi}{8}$ **99.** $\tan\alpha = \frac{8}{x}$, so $\tan^{-1}\frac{8}{x} = \alpha$. $\tan(\alpha+\theta) = \frac{33}{x}$, so $\tan^{-1}\frac{33}{x} = \alpha + \theta$. $\theta = \alpha + \theta - \alpha = \tan^{-1}\frac{33}{x} - \tan^{-1}\frac{8}{x}$.

Section 5.8

Check Point Exercises

1. $B = 27.3°$; $b \approx 4.34$; $c \approx 9.45$ **2.** 994 ft **3.** 29.0° **4.** 60.3 ft **5. a.** S 25°E **b.** S 15°W **6. a.** 4.2 m **b.** S87.7°W **7.** $d = -6\cos\frac{\pi}{2}t$ **8. a.** 12 cm **b.** $\frac{1}{8}$ cm per sec **c.** 8 sec

Exercise Set 5.8

1. $B = 66.5°$; $a \approx 4.35$; $c \approx 10.90$ **3.** $B = 37.4°$; $a \approx 42.90$; $b \approx 32.80$ **5.** $A = 73.2°$; $a \approx 101.02$; $c \approx 105.52$ **7.** $b \approx 39.95$; $A \approx 37.3°$; $B \approx 52.7°$ **9.** $c \approx 26.96$; $A \approx 23.6°$; $B \approx 66.4°$ **11.** $a \approx 6.71$; $B \approx 16.6°$; $A \approx 73.4°$ **13.** N 15° E **15.** S 80° W **17.** $d = -6\cos\frac{\pi}{2}t$ **19.** $d = 3\sin\frac{4\pi}{3}t$ **21. a.** 5 in. **b.** $\frac{1}{4}$ in per sec **c.** 4 sec **23. a.** 6 in. **b.** 1 in. per sec **c.** 1 sec **25. a.** $\frac{1}{2}$ in. **b.** 0.32 in. per sec **c.** 3.14 sec **27. a.** 5 in. **b.** $\frac{1}{3}$ in. per sec **c.** 3 sec **29.** 2059 ft **31.** 695 ft **33.** 1376 ft **35.** 15.1° **37.** 33.7 ft **39.** 90 mi north and 120 mi east **41.** 13.2 mi **43.** N 53° W **45.** N 89.5° E

47. $d = 6 \sin \pi t$ **49.** $d = \sin 528 \pi t$ **59.**

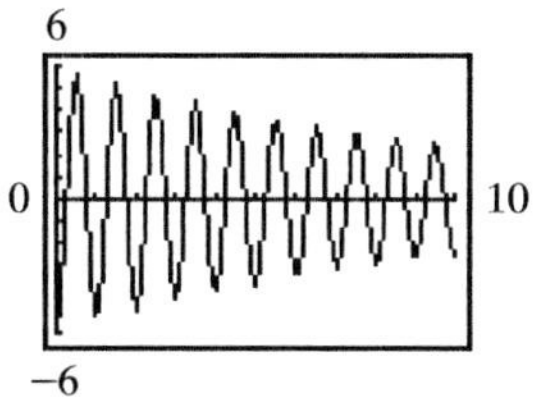

10 complete oscillations

61. 48 ft

Chapter 5 Review Exercises

1.

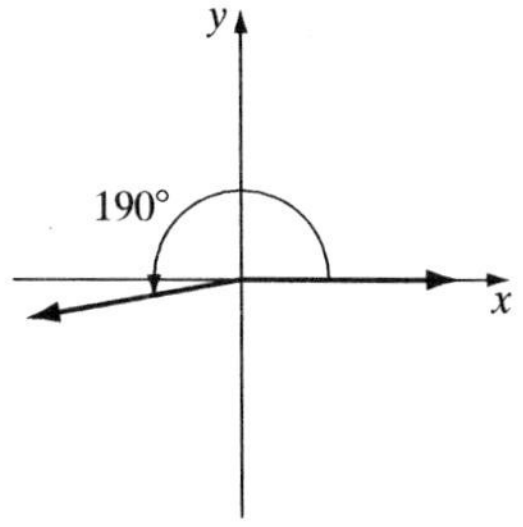

2.

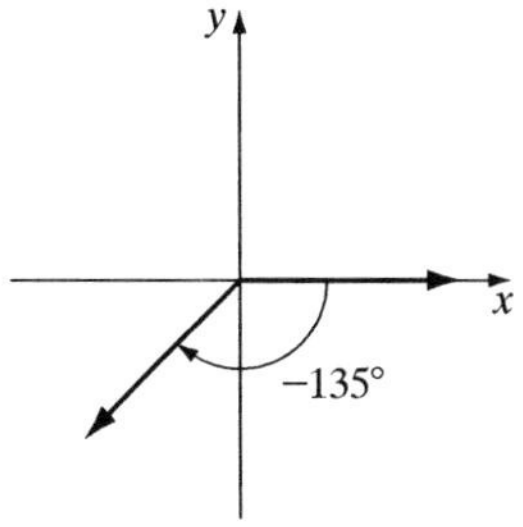

3.

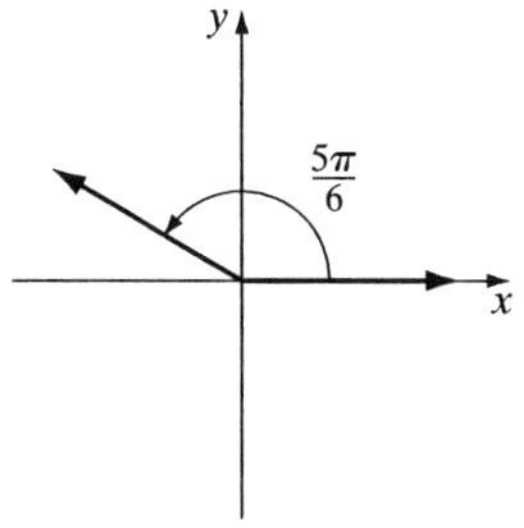

4.

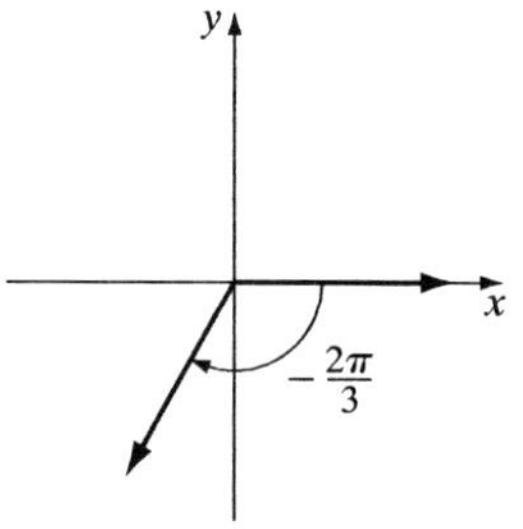

5. 40° **6.** 275° **7.** 17°; 107° **8.** no complement; $\frac{\pi}{3}$ radians **9.** 4.5 radians **10.** $\frac{\pi}{12}$ radians **11.** $\frac{2\pi}{3}$ radians **12.** $\frac{7\pi}{4}$ radians

13. 300° **14.** 252° **15.** −150° **16.** $\frac{15\pi}{2}$ ft ≈ 23.56 ft **17.** 20.6π radians per min **18.** 42,412 ft per min **19.** $\sin\theta = \frac{3}{5}$; $\cos\theta = \frac{4}{5}$; $\tan\theta = \frac{3}{4}$; $\csc\theta = \frac{5}{3}$; $\sec\theta = \frac{5}{4}$; $\cot\theta = \frac{4}{3}$ **20.** $\sqrt{3}$ **21.** $\frac{\sqrt{2}}{2}$ **22.** $\frac{2\sqrt{3}}{3}$ **23.** 1 **24.** $\frac{\sqrt{21}}{7}$ **25.** cos 20°

26. sin 0 **27.** 42 mm **28.** 23 cm **29.** 37 in. **30.** 772 ft **31.** 31 m **32.** 56° **33.** $\sin\theta = -\frac{5\sqrt{26}}{26}$; $\cos\theta = -\frac{\sqrt{26}}{26}$; $\tan\theta = 5$; $\csc\theta = -\frac{\sqrt{26}}{5}$; $\sec\theta = -\sqrt{26}$; $\cot\theta = \frac{1}{5}$ **34.** $\sin\theta = -1$; $\cos\theta = 0$; $\tan\theta$ is undefined; $\csc\theta = -1$; $\sec\theta$ is undefined; $\cot\theta = 0$

35. quadrant I **36.** quadrant III **37.** $\sin\theta = -\frac{\sqrt{21}}{5}$; $\tan\theta = -\frac{\sqrt{21}}{2}$; $\csc\theta = -\frac{5\sqrt{21}}{21}$; $\sec\theta = \frac{5}{2}$; $\cot\theta = -\frac{2\sqrt{21}}{21}$

38. $\sin\theta = \frac{\sqrt{10}}{10}$; $\cos\theta = -\frac{3\sqrt{10}}{10}$; $\csc\theta = \sqrt{10}$; $\sec\theta = -\frac{\sqrt{10}}{3}$; $\cot\theta = -3$ **39.** 85° **40.** $\frac{3\pi}{8}$ **41.** 50° **42.** $-\frac{\sqrt{3}}{2}$ **43.** $-\sqrt{3}$

44. $\sqrt{2}$ **45.** $\frac{\sqrt{3}}{2}$ **46.** $-\sqrt{3}$ **47.** $-\frac{2\sqrt{3}}{3}$ **48.** $-\frac{\sqrt{3}}{2}$ **49.** $\frac{\sqrt{2}}{2}$ **50.** 1

51.

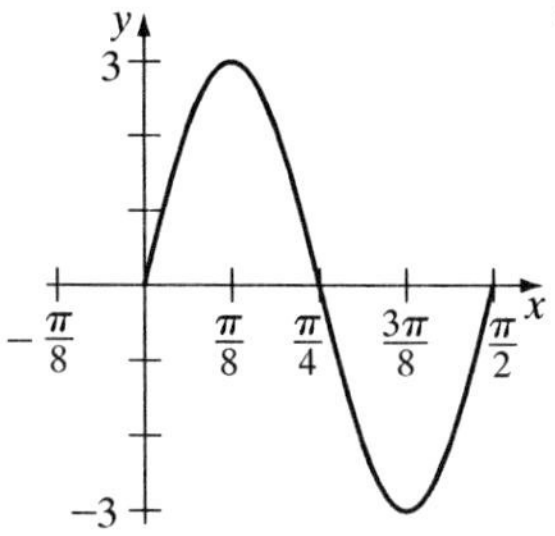

52.

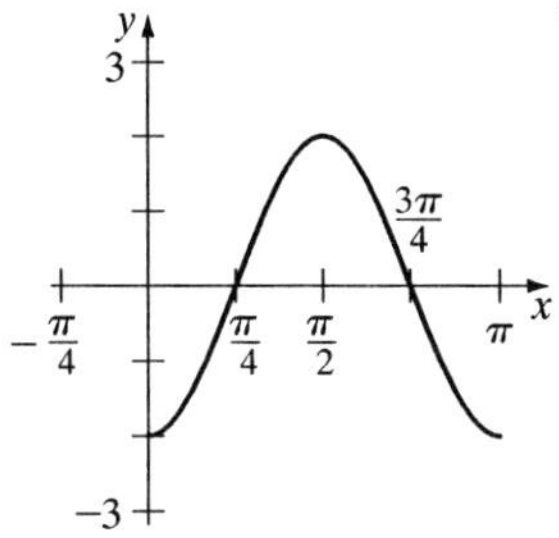

53.

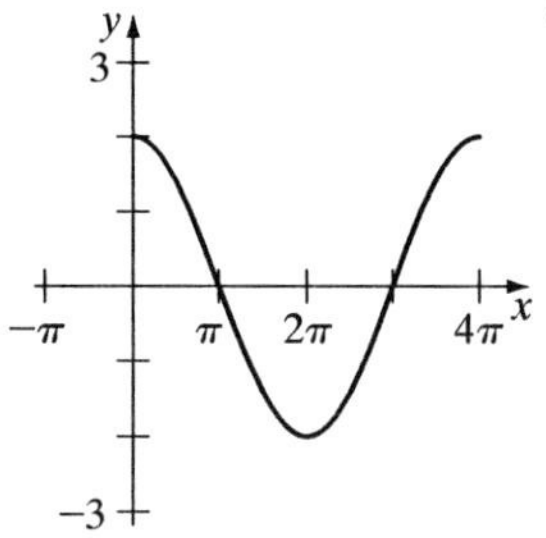

54.

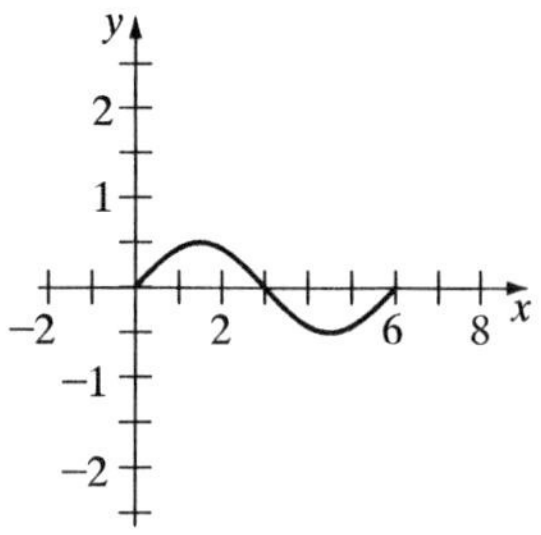

55.

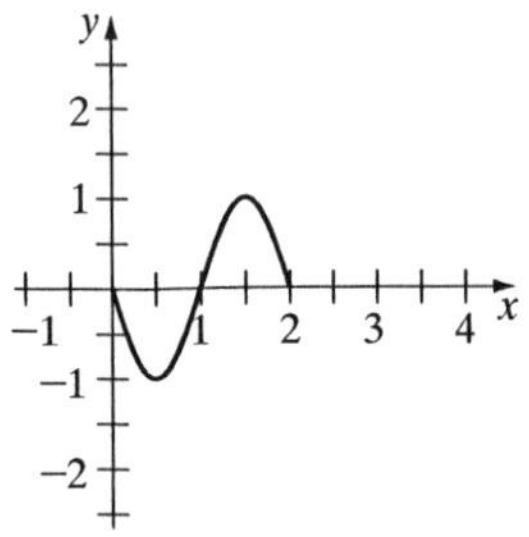

56.

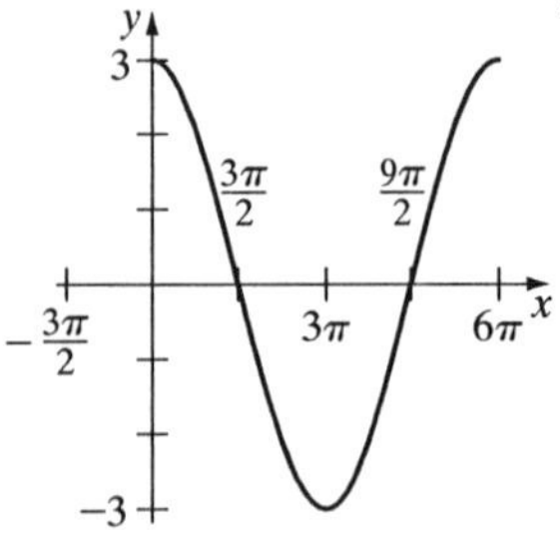

57.

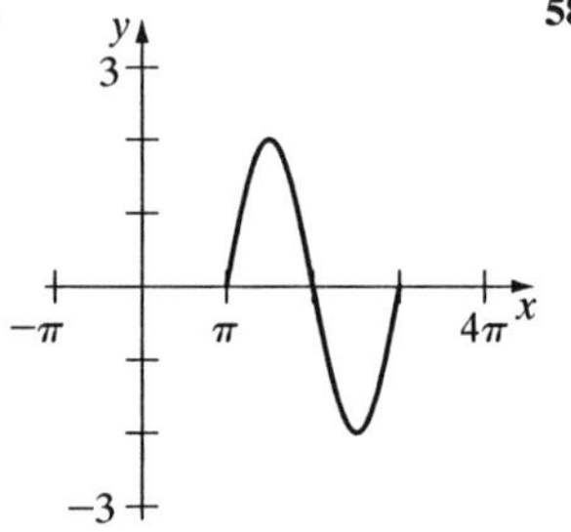

58.

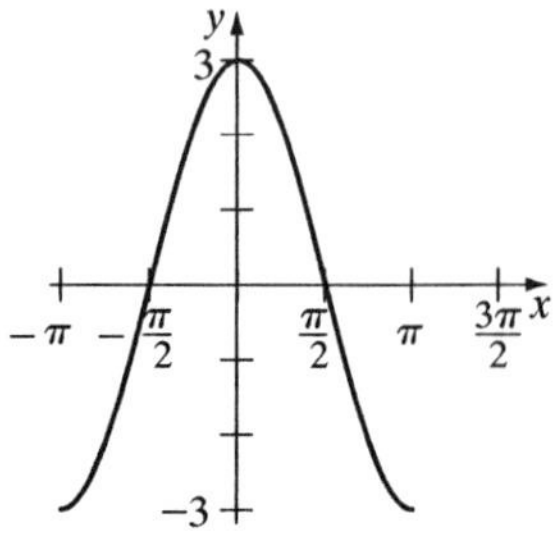

59.

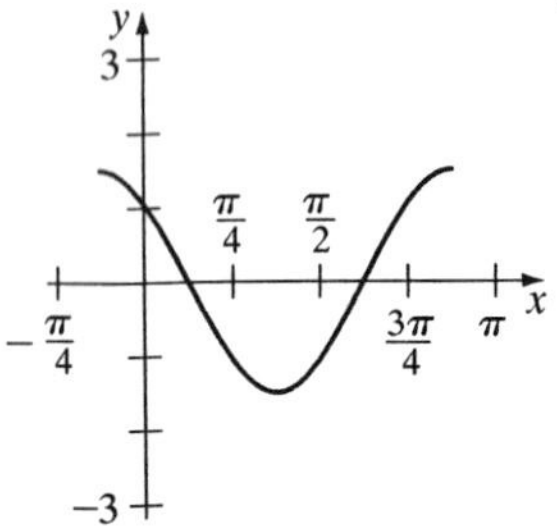

60.

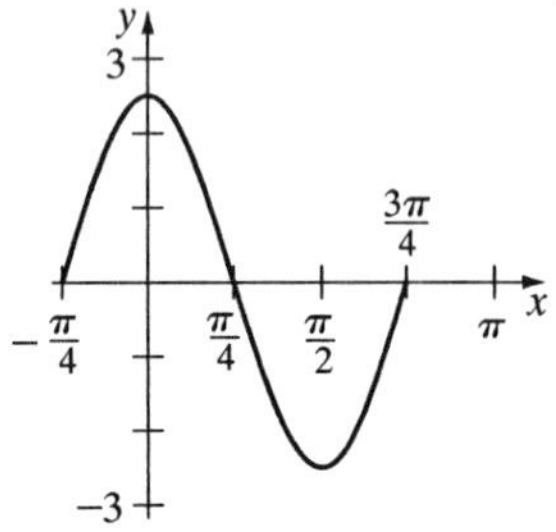

61.

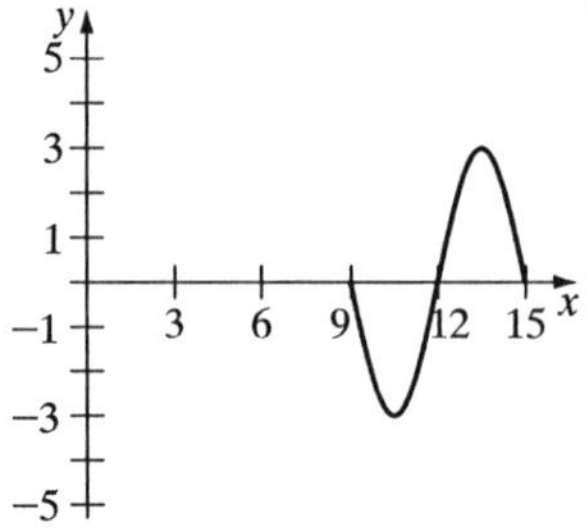

62.

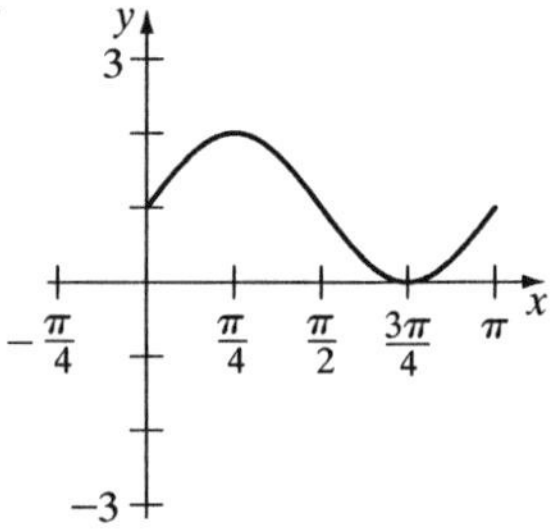

63.

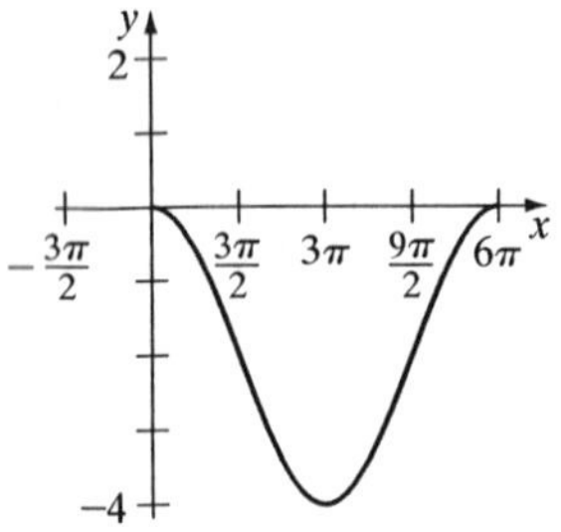

64. a. ≈98.52° **b.** 24 hr **c.** 5:00 P.M.; 98.9° **d.** 5:00 A.M.; 98.3°

e.

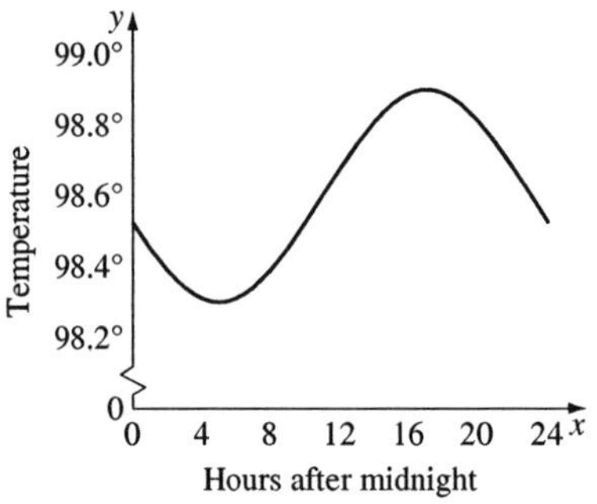

65.

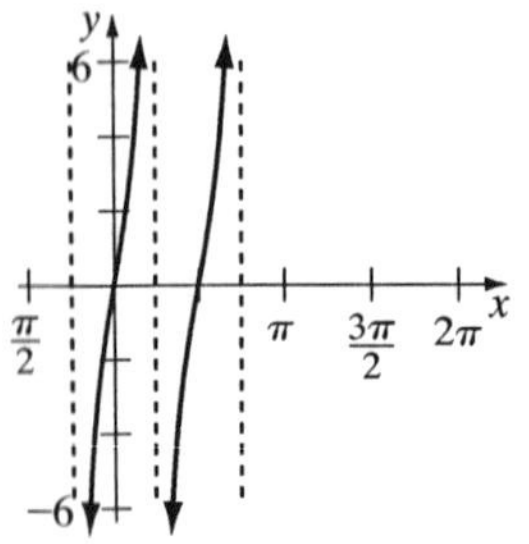

66.

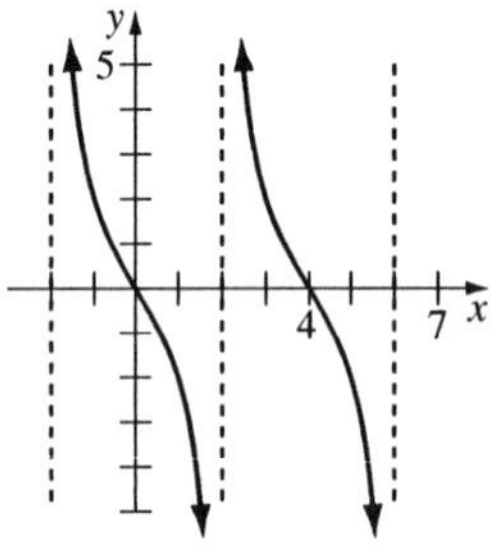

67.

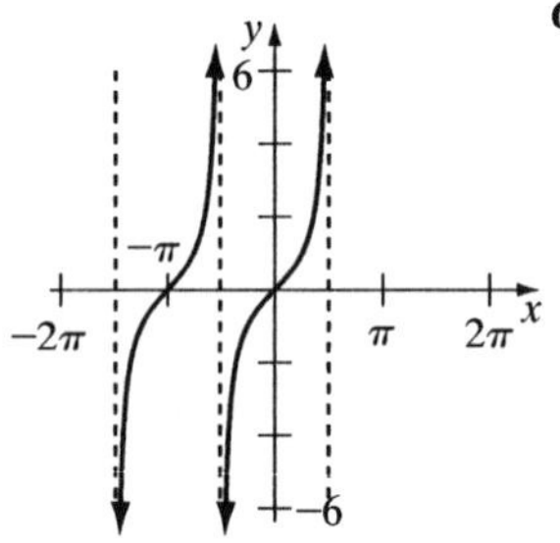

68.

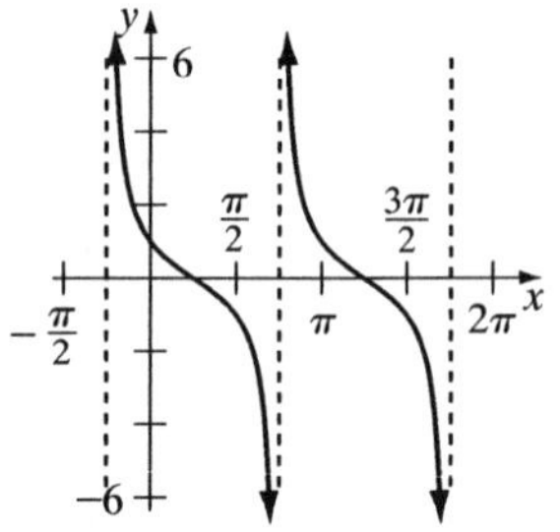

69.

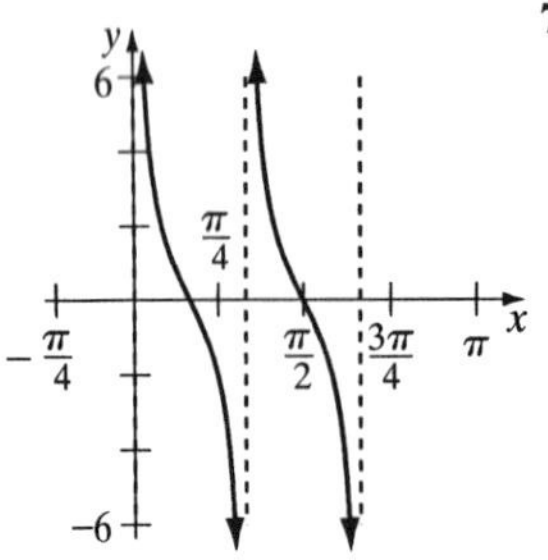

70.

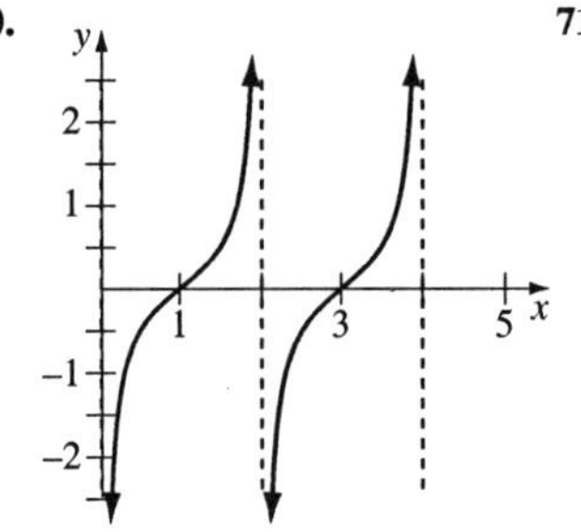

71.

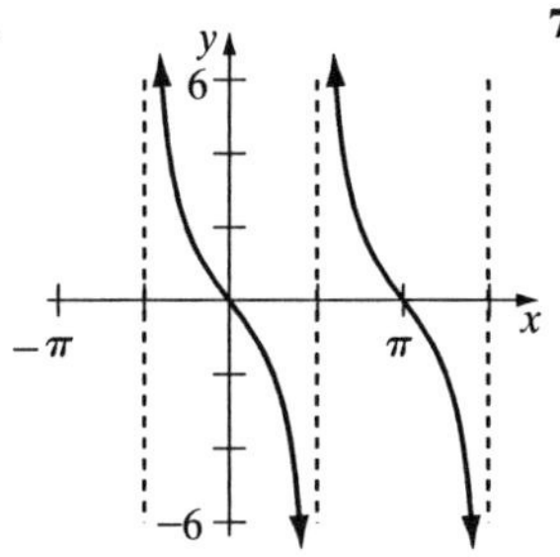

72.

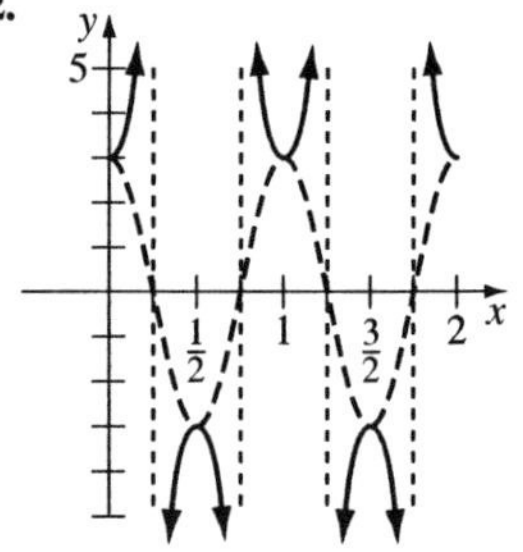

73.

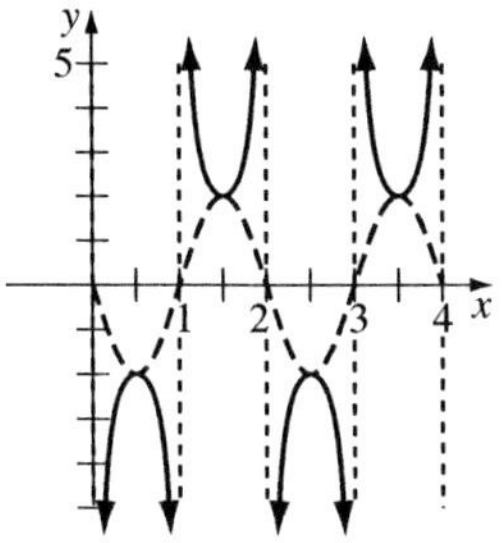

74.

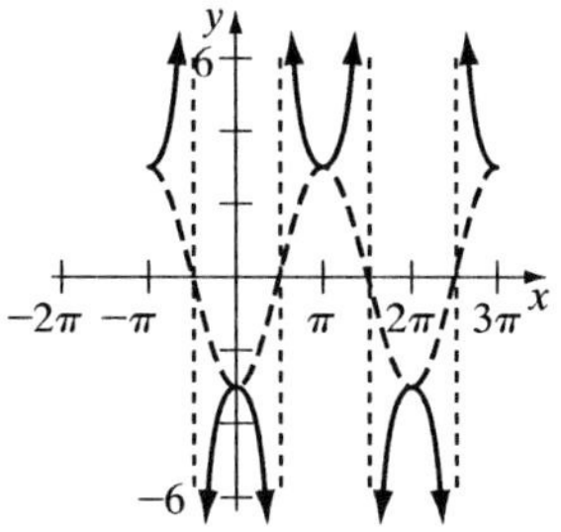

75.

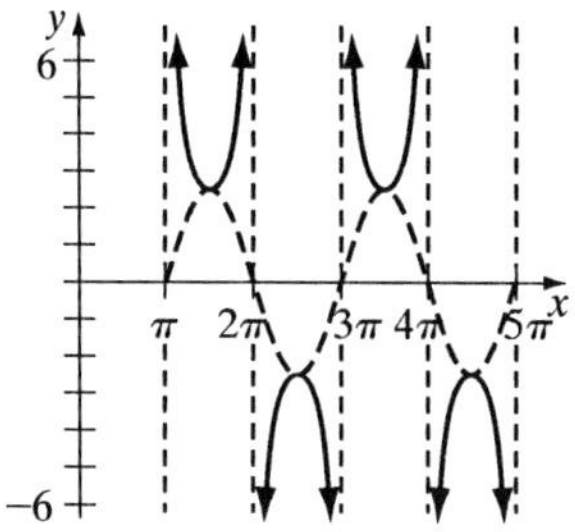

76. $\frac{\pi}{2}$ **77.** 0 **78.** $\frac{\pi}{4}$ **79.** $-\frac{\pi}{3}$ **80.** $\frac{2\pi}{3}$ **81.** $-\frac{\pi}{6}$ **82.** $\frac{\sqrt{2}}{2}$ **83.** 1 **84.** $-\frac{\sqrt{3}}{3}$ **85.** $-\frac{\sqrt{3}}{3}$ **86.** 2

87. $\frac{4}{5}$ **88.** $\frac{4}{5}$ **89.** $-\frac{3}{4}$ **90.** $-\frac{3}{4}$ **91.** $\frac{\pi}{3}$ **92.** $\frac{\pi}{3}$ **93.** $-\frac{\pi}{6}$ **94.** $\frac{2\sqrt{x^2+4}}{x^2+4}$ **95.** $\frac{x\sqrt{x^2-1}}{x^2-1}$ **96.** $B \approx 67.7°; a \approx 37.9; b \approx 9.25$ **97.** $A \approx 52.6°; a \approx 7.85; c \approx 9.88$ **98.** $A \approx 16.6°; B \approx 73.4°; b \approx 6.71$ **99.** $A \approx 21.3°; B \approx 68.7°; c \approx 3.86$ **100.** 38 ft

101. 90 yd **102.** 21.7 ft **103.** N 35° E **104.** S 35° W **105.** 24.6 mi **106. a.** 1282.2 mi **b.** S 74° E **107. a.** 20 cm **b.** $\frac{1}{8}$ cm per sec **c.** 8 sec **108. a.** $\frac{1}{2}$ cm **b.** 0.64 cm per sec **c.** 1.57 sec **109.** $d = -30 \cos \pi t$ **110.** $d = \frac{1}{4} \sin \frac{2\pi}{5} t$

Chapter 5 Test

1. $\frac{3\pi}{4}$ radians **2.** $\frac{4\pi}{13}$ **3.** $\frac{25\pi}{3}$ ft $\approx$ 26.18 ft **4.** $\sin \theta = \frac{5\sqrt{29}}{29}$; $\cos \theta = -\frac{2\sqrt{29}}{29}$; $\tan \theta = -\frac{5}{2}$; $\csc \theta = \frac{\sqrt{29}}{5}$; $\sec \theta = -\frac{\sqrt{29}}{2}$; $\cot \theta = -\frac{2}{5}$

5. quadrant III **6.** $\sin \theta = -\frac{2\sqrt{2}}{3}$; $\tan \theta = -2\sqrt{2}$; $\csc \theta = -\frac{3\sqrt{2}}{4}$; $\sec \theta = 3$; $\cot \theta = -\frac{\sqrt{2}}{4}$ **7.** $\frac{\sqrt{3}}{6}$ **8.** $-\sqrt{3}$ **9.** $-\frac{\sqrt{2}}{2}$

10.

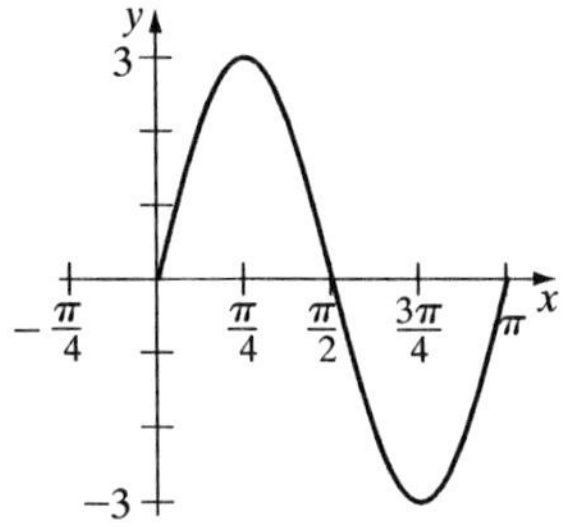

11.

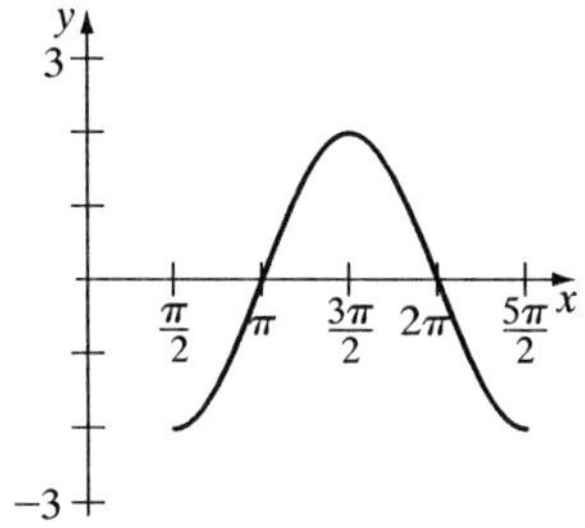

12.

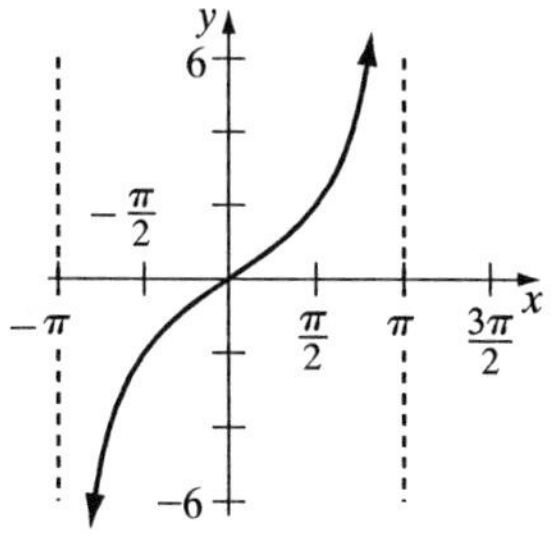

13.

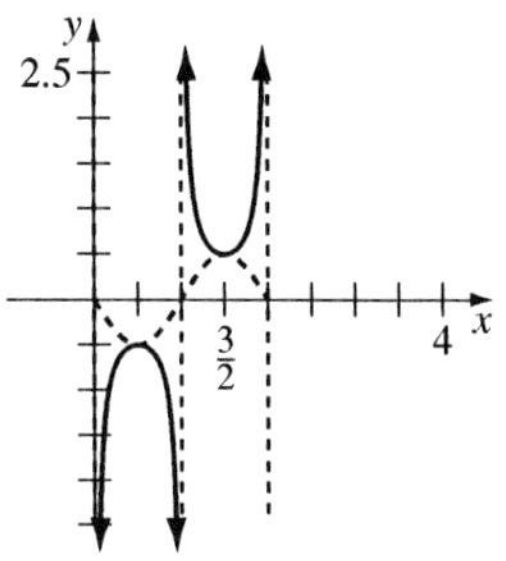

14. $-\sqrt{3}$ **15.** $B = 69°; a = 4.7; b = 12.1$ **16.** 23 yd **17.** 36.1°

18. N 80° W **19. a.** 6 in. **b.** $\frac{1}{2}$ in. per sec **c.** 2 sec

20. Trigonometric functions are periodic.

Cumulative Review Exercises (Chapters 1–5)

1. $\{-3, 6\}$ **2.** $\{-5, -2, 2\}$ **3.** $\{4\}$ **4.** $\{7\}$ **5.** $\{-1, 2, 3\}$ **6.** $-3 \le x \le 8$ **7.** $f^{-1}(x) = x^2 + 6$

8. $4x^2 - \frac{14}{5}x - \frac{17}{25} + \frac{284}{125x + 50}$ **9.** $\log 1000 = 3$ **10.** $280°$ **11.** 3 positive real roots; 1 negative real root

12.

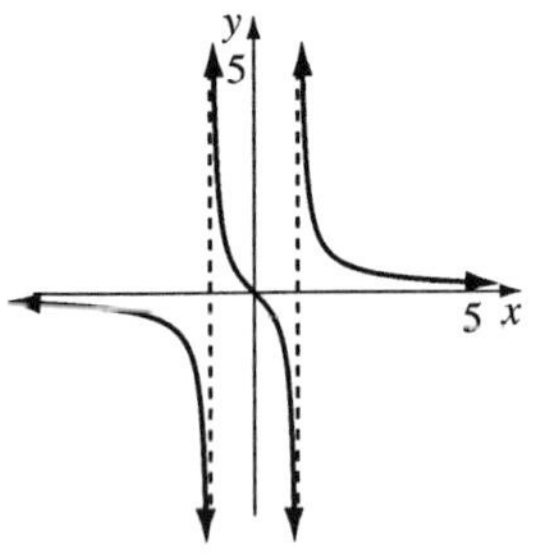

13.

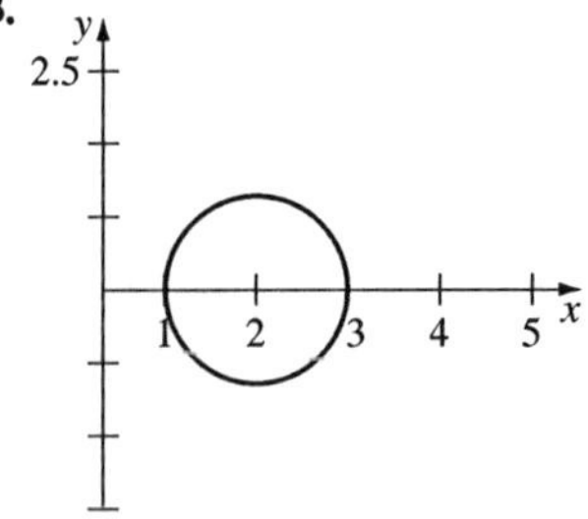

14.

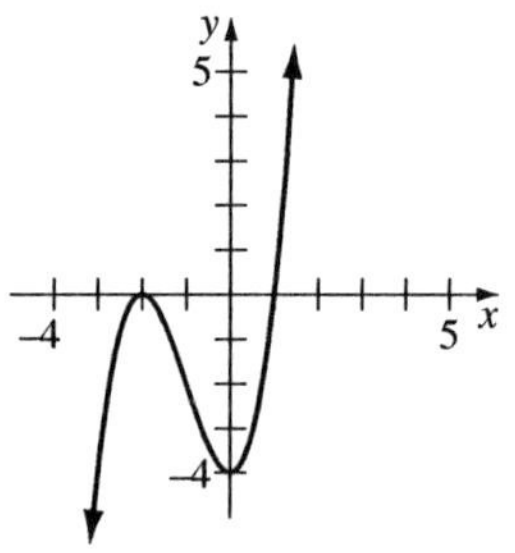

15.

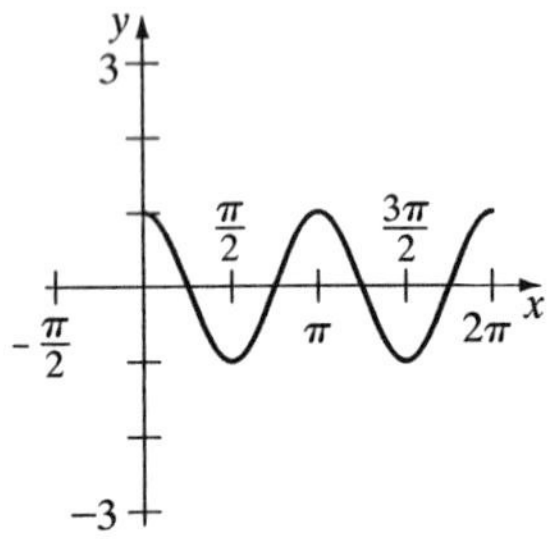

16.

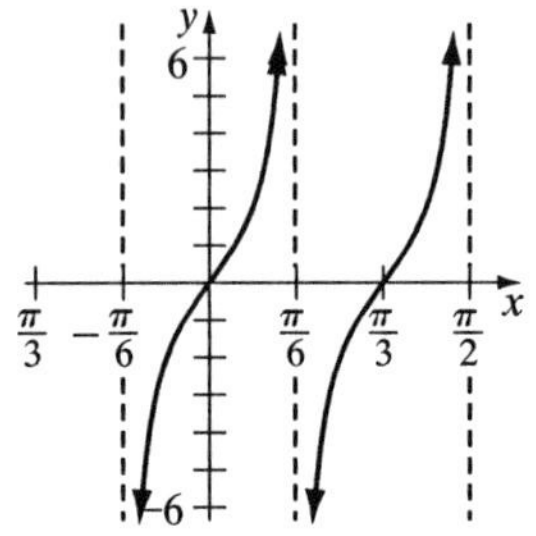

17. 48 performances
18. a. $A = 10.2e^{0.3057t}$
b. 2001
19. 1540 Btu per hr
20. $76°$

CHAPTER 6

Section 6.1

Check Point Exercises

1. $\csc x \tan x = \frac{1}{\sin x} \cdot \frac{\sin x}{\cos x} = \frac{1}{\cos x} = \sec x$

2. $\cos x \cot x + \sin x = \cos x \cdot \frac{\cos x}{\sin x} + \sin x = \frac{\cos^2 x}{\sin x} + \sin x \cdot \frac{\sin x}{\sin x} = \frac{\cos^2 x + \sin^2 x}{\sin x} = \frac{1}{\sin x} = \csc x$

3. $\sin x - \sin x \cos^2 x = \sin x(1 - \cos^2 x) = \sin x \cdot \sin^2 x = \sin^3 x$

4. $\frac{\sin x}{1 + \cos x} + \frac{1 + \cos x}{\sin x} = \frac{\sin x(\sin x)}{(1 + \cos x)(\sin x)} + \frac{(1 + \cos x)(1 + \cos x)}{(\sin x)(1 + \cos x)} = \frac{\sin^2 x + 1 + 2\cos x + \cos^2 x}{(1 + \cos x)(\sin x)}$

$= \frac{\sin^2 x + \cos^2 x + 1 + 2\cos x}{(1 + \cos x)(\sin x)} = \frac{1 + 1 + 2\cos x}{(1 + \cos x)(\sin x)} = \frac{2 + 2\cos x}{(1 + \cos x)(\sin x)} = \frac{2(1 + \cos x)}{(1 + \cos x)(\sin x)} = \frac{2}{\sin x} = 2\csc x$

5. $\frac{\cos x}{1 + \sin x} = \frac{\cos x(1 - \sin x)}{(1 + \sin x)(1 - \sin x)} = \frac{\cos x(1 - \sin x)}{1 - \sin^2 x} = \frac{\cos x(1 - \sin x)}{\cos^2 x} = \frac{1 - \sin x}{\cos x}$

6. $\frac{\sec x + \csc(-x)}{\sec x \csc x} = \frac{\sec x - \csc x}{\sec x \csc x}$

$= \frac{\frac{1}{\cos x} - \frac{1}{\sin x}}{\frac{1}{\cos x} \cdot \frac{1}{\sin x}} = \frac{\frac{\sin x}{\cos x \cdot \sin x} - \frac{\cos x}{\cos x \cdot \sin x}}{\frac{1}{\cos x \cdot \sin x}} = \frac{\frac{\sin x - \cos x}{\cos x \cdot \sin x}}{\frac{1}{\cos x \cdot \sin x}} = \frac{\sin x - \cos x}{\cos x \cdot \sin x} \cdot \frac{\cos x \cdot \sin x}{1} = \sin x - \cos x$

7. Left side: $\frac{1}{1 + \sin\theta} + \frac{1}{1 - \sin\theta} = \frac{1(1 - \sin\theta)}{(1 + \sin\theta)(1 - \sin\theta)} + \frac{1(1 + \sin\theta)}{(1 - \sin\theta)(1 + \sin\theta)} = \frac{1 - \sin\theta + 1 + \sin\theta}{(1 + \sin\theta)(1 - \sin\theta)} = \frac{2}{1 - \sin^2\theta}$;

Right side: $2 + 2\tan^2\theta = 2 + 2\left(\frac{\sin^2\theta}{\cos^2\theta}\right) = \frac{2\cos^2\theta}{\cos^2\theta} + \frac{2\sin^2\theta}{\cos^2\theta} = \frac{2\cos^2\theta + 2\sin^2\theta}{\cos^2\theta} = \frac{2(\cos^2\theta + \sin^2\theta)}{\cos^2\theta} = \frac{2}{\cos^2\theta} = \frac{2}{1 - \sin^2\theta}$

Exercise Set 6.1

For Exercises 1–59, proofs may vary.

65.

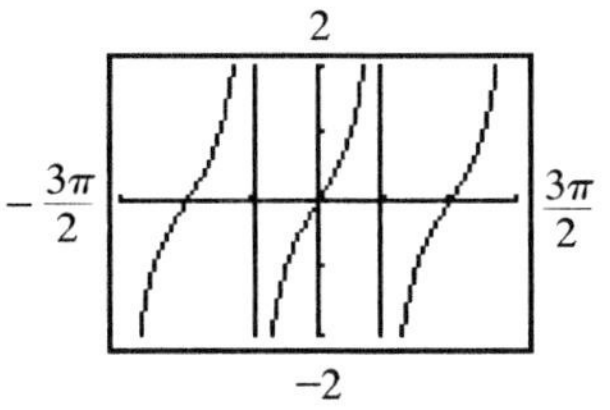

Proofs may vary

67.

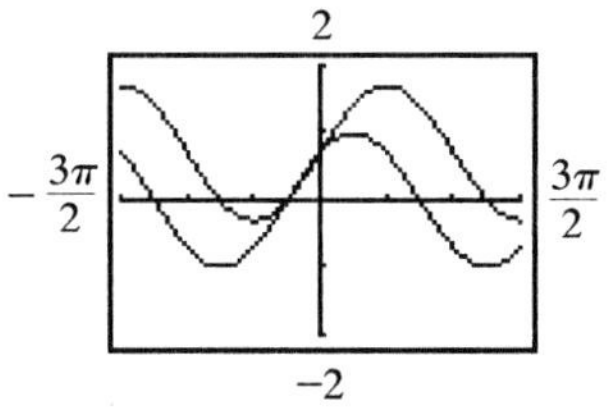

Values for x may vary.

69.

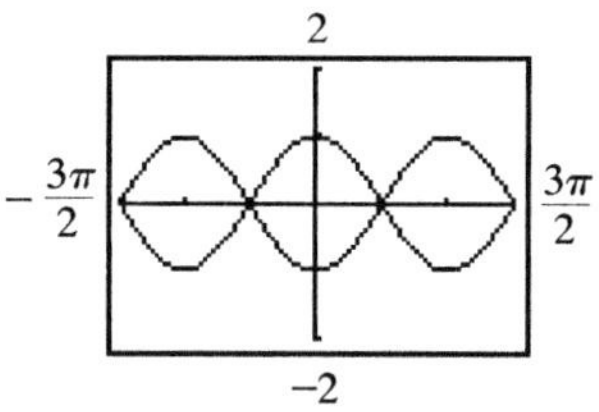

Values for x may vary.

71.

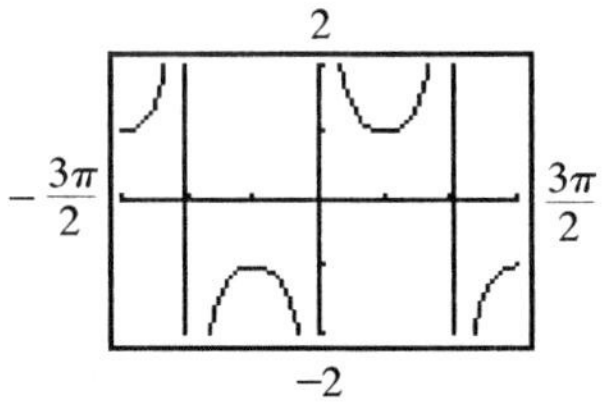

Proofs may vary.

73.

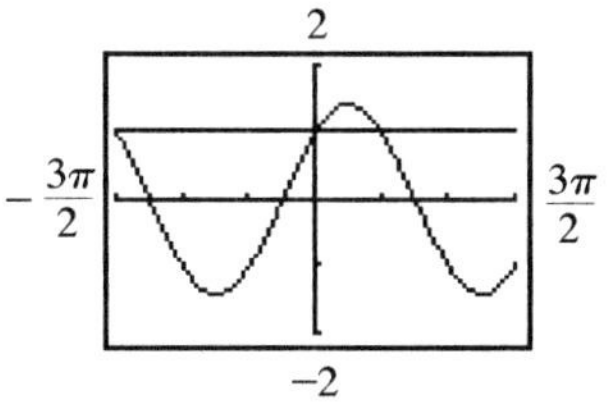

Values for x may vary.

75. Proofs may vary.
77. Answers may vary.

Section 6.2

Check Point Exercises

1. $\frac{\sqrt{3}}{2}$ **2.** $\frac{\sqrt{3}}{2}$ **3.** $\frac{\cos(\alpha-\beta)}{\cos\alpha\cos\beta} = \frac{\cos\alpha\cos\beta + \sin\alpha\sin\beta}{\cos\alpha\cos\beta} = \frac{\cos\alpha}{\cos\alpha}\cdot\frac{\cos\beta}{\cos\beta} + \frac{\sin\alpha}{\cos\alpha}\cdot\frac{\sin\beta}{\cos\beta} = 1 + \tan\alpha\tan\beta$

4. $\frac{\sqrt{2}+\sqrt{6}}{4}$ **5. a.** $\cos\alpha = -\frac{3}{5}$ **b.** $\cos\beta = \frac{\sqrt{3}}{2}$ **c.** $\frac{-3\sqrt{3}-4}{10}$ **d.** $\frac{4\sqrt{3}-3}{10}$

6. a. $y = \sin x$ **b.** $\cos\left(x+\frac{3\pi}{2}\right) = \cos x\cos\frac{3\pi}{2} - \sin x\sin\frac{3\pi}{2} = \cos x\cdot 0 - \sin x\cdot(-1) = \sin x$

7. $\tan(x+\pi) = \frac{\tan x + \tan\pi}{1-\tan x\tan\pi} = \frac{\tan x + 0}{1-\tan x\cdot 0} = \frac{\tan x}{1} = \tan x$

Exercise Set 6.2

1. $\frac{\sqrt{6}+\sqrt{2}}{4}$ **3.** $\frac{\sqrt{2}-\sqrt{6}}{4}$ **5. a.** $\alpha = 50°, \beta = 20°$ **b.** $\cos 30°$ **c.** $\frac{\sqrt{3}}{2}$ **7. a.** $\alpha = \frac{5\pi}{12}, \beta = \frac{\pi}{12}$ **b.** $\cos\left(\frac{\pi}{3}\right)$ **c.** $\frac{1}{2}$

For Exercises 9–11, proofs may vary. **13.** $\frac{\sqrt{6}-\sqrt{2}}{4}$ **15.** $\frac{\sqrt{6}+\sqrt{2}}{4}$ **17.** $\sqrt{3}+2$ **19.** $2-\sqrt{3}$ **21.** $-\frac{\sqrt{6}+\sqrt{2}}{4}$

23. $\frac{\sqrt{6}-\sqrt{2}}{4}$ **25.** $\sin 30°; \frac{1}{2}$ **27.** $\tan 45°; 1$ **29.** $\sin\frac{\pi}{6}; \frac{1}{2}$ **31.** $\tan\frac{\pi}{6}; \frac{\sqrt{3}}{3}$ For Exercises 33–55, proofs may vary.

57. a. $-\frac{63}{65}$ **b.** $-\frac{16}{65}$ **c.** $\frac{16}{63}$ **59. a.** $\frac{4+6\sqrt{2}}{15}$ **b.** $\frac{3-8\sqrt{2}}{15}$ **c.** $\frac{54-25\sqrt{2}}{28}$ **61. a.** $\frac{8\sqrt{3}+15}{34}$ **b.** $\frac{15\sqrt{3}-8}{34}$

c. $\frac{480-289\sqrt{3}}{33}$ **63. a.** $y = \sin x$ **b.** $\sin(\pi - x) = \sin\pi\cos x - \cos\pi\sin x = 0\cdot\cos x - (-1)\sin x = \sin x$

65. a. $y = 2\cos x$ **b.** $\sin\left(x+\frac{\pi}{2}\right) + \sin\left(\frac{\pi}{2}-x\right) = \sin x\cos\frac{\pi}{2} + \cos x\sin\frac{\pi}{2} + \sin\frac{\pi}{2}\cos x - \cos\frac{\pi}{2}\sin x$

$= \sin x\cdot 0 + \cos x\cdot 1 + 1\cdot\cos x - 0\cdot\sin x = \cos x + \cos x = 2\cos x$ **67.** Proofs may vary.; amplitude is $\sqrt{13}$; period is 2π

77.

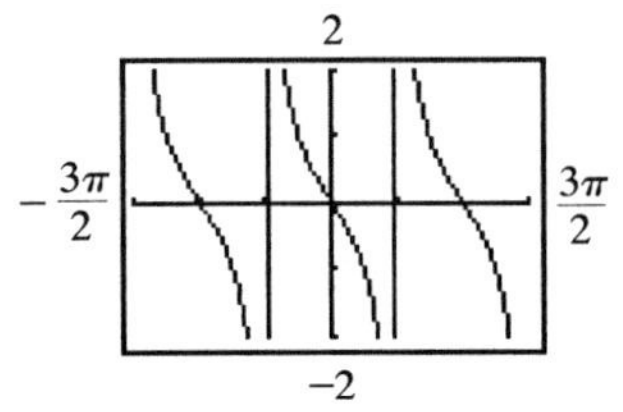

Proofs may vary.

79.

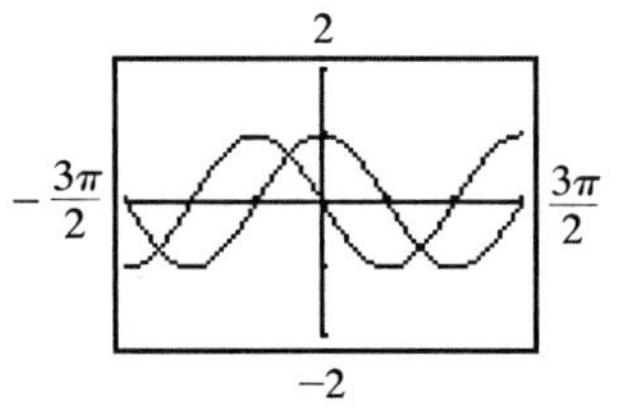

Values for x may vary.

81.

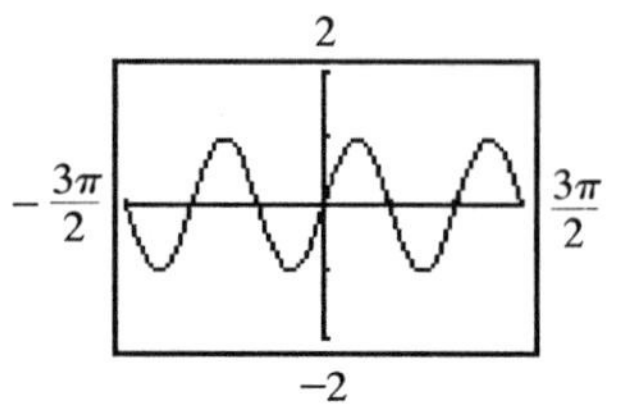

Proofs may vary.

83. Proofs may vary. **85.** $-\frac{24}{25}$ **87.** −1 **89.** $\frac{x\sqrt{1-y^2}-y}{\sqrt{1+x^2}}$

Section 6.3

Check Point Exercises

1. a. $\frac{24}{25}$ **b.** $-\frac{7}{25}$ **c.** $\frac{24}{7}$ **2.** $\frac{\sqrt{3}}{2}$ **3.** $\sin 3\theta = \sin(2\theta + \theta) = \sin 2\theta \cos\theta + \cos 2\theta \sin\theta = 2\sin\theta\cos\theta\cos\theta$
$= 2\sin\theta\cos\theta\cos\theta + (2\cos^2\theta - 1)\sin\theta = 2\sin\theta\cos^2\theta + 2\sin\theta\cos^2\theta - \sin\theta = 4\sin\theta\cos^2\theta - \sin\theta = 4\sin\theta(1-\sin^2\theta) - \sin\theta$
$= 4\sin\theta - 4\sin^3\theta - \sin\theta = 3\sin\theta - 4\sin^3\theta$

4. $\sin^4 x = (\sin^2 x)^2 = \left(\frac{1-\cos 2x}{2}\right)^2 = \frac{1 - 2\cos 2x + \cos^2 2x}{4} = \frac{1}{4} - \frac{1}{2}\cos 2x + \frac{1}{4}\cos^2 2x = \frac{1}{4} - \frac{1}{2}\cos 2x + \frac{1}{4}\left(\frac{1+\cos 2(2x)}{2}\right)$
$= \frac{1}{4} - \frac{1}{2}\cos 2x + \frac{1}{8} + \frac{1}{8}\cos 4x = \frac{3}{8} - \frac{1}{2}\cos 2x + \frac{1}{8}\cos 4x$

5. $-\frac{\sqrt{2+\sqrt{3}}}{2}$ **6.** $\frac{\sin 2\theta}{1+\cos 2\theta} = \frac{2\sin\theta\cos\theta}{1+(1-2\sin^2\theta)} = \frac{2\sin\theta\cos\theta}{2-2\sin^2\theta} = \frac{2\sin\theta\cos\theta}{2(1-\sin^2\theta)} = \frac{2\sin\theta\cos\theta}{2\cos^2\theta} = \frac{\sin\theta}{\cos\theta} = \tan\theta$

7. $\frac{\sec\alpha}{\sec\alpha\csc\alpha + \csc\alpha} = \frac{\frac{1}{\cos\alpha}}{\frac{1}{\cos\alpha}\cdot\frac{1}{\sin\alpha} + \frac{1}{\sin\alpha}} = \frac{\frac{1}{\cos\alpha}}{\frac{1}{\cos\alpha\sin\alpha} + \frac{\cos\alpha}{\cos\alpha\sin\alpha}} = \frac{\frac{1}{\cos\alpha}}{\frac{1+\cos\alpha}{\cos\alpha\sin\alpha}} = \frac{1}{\cos\alpha}\cdot\frac{\cos\alpha\sin\alpha}{1+\cos\alpha} = \frac{\sin\alpha}{1+\cos\alpha} = \tan\frac{\alpha}{2}$

Exercise Set 6.3

1. $\frac{24}{25}$ **3.** $\frac{24}{7}$ **5.** $\frac{527}{625}$ **7. a.** $-\frac{240}{289}$ **b.** $-\frac{161}{289}$ **c.** $\frac{240}{161}$ **9. a.** $-\frac{336}{625}$ **b.** $\frac{527}{625}$ **c.** $-\frac{336}{527}$ **11. a.** $\frac{4}{5}$ **b.** $\frac{3}{5}$ **c.** $\frac{4}{3}$
13. a. $\frac{720}{1681}$ **b.** $\frac{1519}{1681}$ **c.** $\frac{720}{1519}$ **15.** $\frac{1}{2}$ **17.** $-\frac{\sqrt{3}}{2}$ **19.** $\frac{\sqrt{2}}{2}$ **21.** $\frac{\sqrt{3}}{3}$ For Exercises 23–33, proofs may vary.
35. $\frac{9}{4} - 3\cos 2x + \frac{3}{4}\cos 4x$ **37.** $\frac{1}{8} - \frac{1}{8}\cos 4x$ **39.** $\frac{\sqrt{2-\sqrt{3}}}{2}$ **41.** $\frac{\sqrt{2+\sqrt{2}}}{2}$ **43.** $2+\sqrt{3}$ **45.** $-\sqrt{2}+1$
47. $\frac{\sqrt{10}}{10}$ **49.** $\frac{1}{3}$ **51.** $\frac{7\sqrt{2}}{10}$ **53.** $\frac{3}{5}$ **55. a.** $\frac{2\sqrt{5}}{5}$ **b.** $-\frac{\sqrt{5}}{5}$ **c.** −2 **57. a.** $\frac{3\sqrt{13}}{13}$ **b.** $\frac{2\sqrt{13}}{13}$ **c.** $\frac{3}{2}$
For Exercises 59–67, proofs may vary. **69. a.** $\frac{v_0^2}{32}\cdot\sin 2\theta$ **b.** $\theta = \frac{\pi}{4}$ **71.** $\sqrt{2-\sqrt{2}}\cdot(2+\sqrt{2}) \approx 2.6$

85.

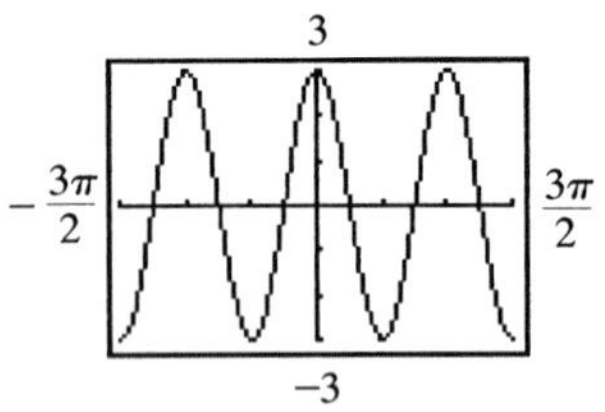

Proofs may vary.

87.

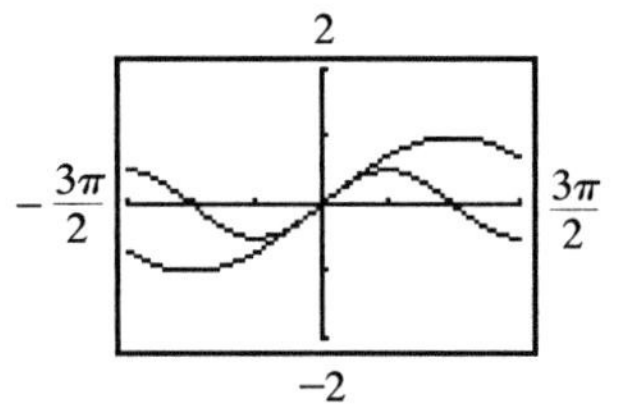

Values for x may vary.

89.

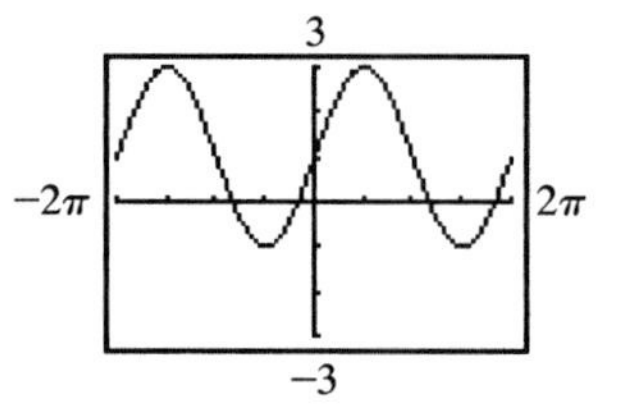

a. $y = 1 + 2\sin x$
b. Proofs may vary.

91.

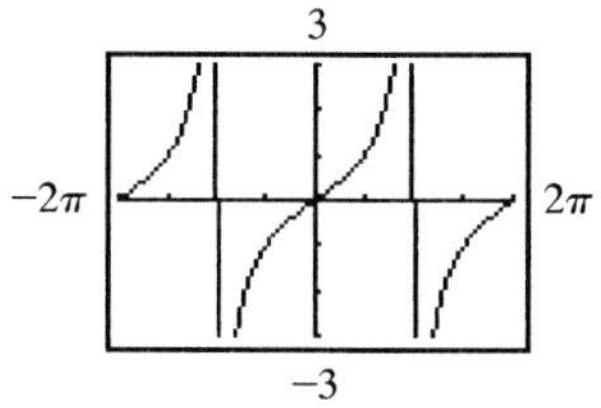

a. $y = \tan\frac{x}{2}$ **b.** Proofs may vary.

93. $\frac{5}{16} - \frac{7}{16}\cos 2x + \frac{3}{16}\cos 4x - \frac{1}{16}\cos 2x \cos 4x$

Section 6.4

Check Point Exercises

1. a. $\frac{1}{2}[\cos 3x - \cos 7x]$ **b.** $\frac{1}{2}[\cos 6x + \cos 8x]$ **2. a.** $2 \sin 5x \cos 2x$ **b.** $2\cos\left(\frac{5x}{2}\right)\cos\left(\frac{x}{2}\right)$

3. $\frac{\cos 3x - \cos x}{\sin 3x + \sin x} = \frac{-2\sin\left(\frac{3x+x}{2}\right)\sin\left(\frac{3x-x}{2}\right)}{2\sin\left(\frac{3x+x}{2}\right)\cos\left(\frac{3x-x}{2}\right)} = \frac{-2\sin\left(\frac{4x}{2}\right)\sin\left(\frac{2x}{2}\right)}{2\sin\left(\frac{4x}{2}\right)\cos\left(\frac{2x}{2}\right)} = \frac{-2\sin 2x \sin x}{2\sin 2x \cos x} = \frac{\sin x}{\cos x} = -\tan x$

Exercise Set 6.4

1. $\frac{1}{2}[\cos 4x - \cos 8x]$ **3.** $\frac{1}{2}[\cos 4x + \cos 10x]$ **5.** $\frac{1}{2}[\sin 3x - \sin x]$ **7.** $\frac{1}{2}[\sin 2x - \sin x]$ **9.** $2 \sin 4x \cos 2x$

11. $2 \sin 2x \cos 5x$ **13.** $2\cos 3x \cos x$ **15.** $2\sin\frac{3x}{2}\cos\frac{x}{2}$ **17.** $2\cos x \cos\frac{x}{2}$ **19.** $\frac{\sqrt{6}}{2}$ **21.** $-\frac{\sqrt{2}}{2}$

For Exercises 23–29, proofs may vary. **31. a.** $y = \cos x$ **b.** Proofs may vary. **33. a.** $y = \tan 2x$ **b.** Proofs may vary.

35. a. $y = \sin 1704\pi t + \sin 2418\pi t$ **b.** $2 \sin 2061\pi t \cdot \cos 357\pi t$

43.

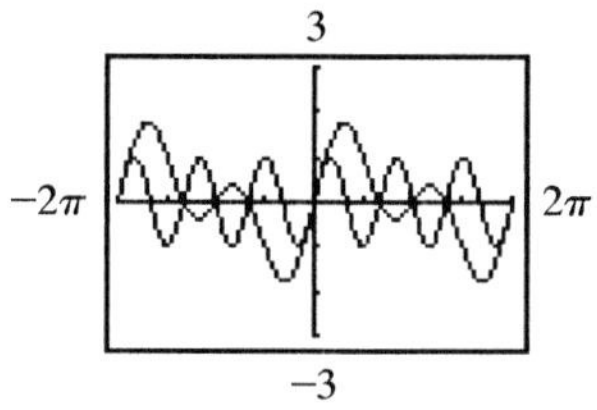

Values for x may vary.

45.

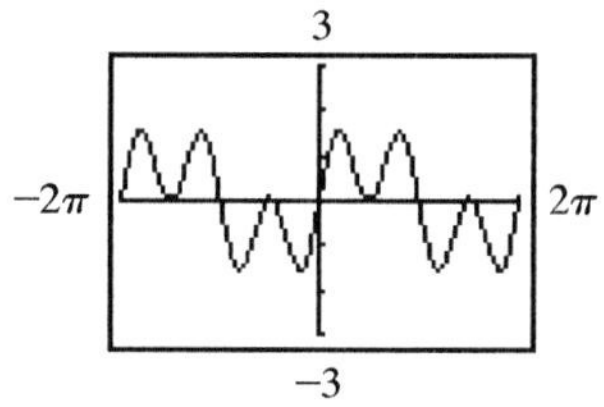

Proofs may vary.

47.

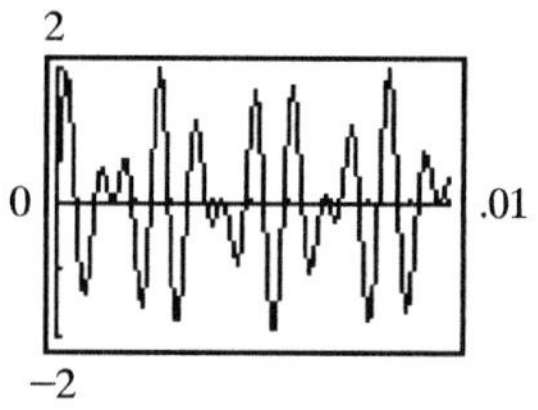

49. a.

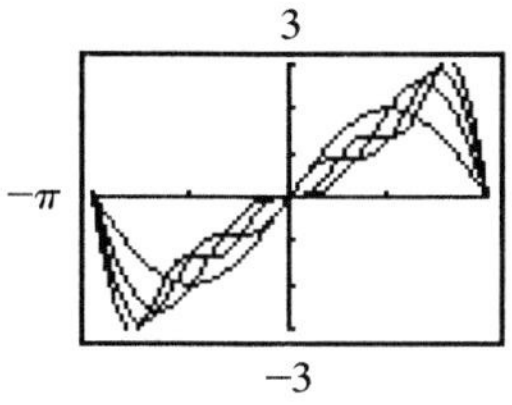

b.

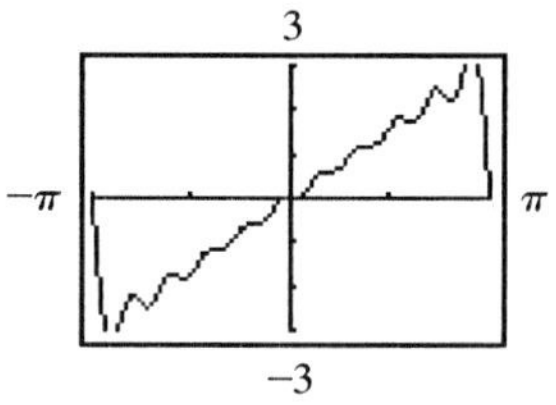

c. $\pi = 4 - \frac{4}{3} + \frac{4}{5} - \frac{4}{7} + \cdots$

For Exercises 51–55, proofs may vary.

Section 6.5

Check Point Exercises

1. $x = \frac{\pi}{3} + 2n\pi$ or $x = \frac{2\pi}{3} + 2n\pi$, where n is any integer. **2.** $\frac{\pi}{6}, \frac{2\pi}{3}, \frac{7\pi}{6}, \frac{5\pi}{3}$ **3.** $\frac{\pi}{2}$ **4.** $\frac{\pi}{6}, \frac{\pi}{2}, \frac{5\pi}{6}$ **5.** $0, \frac{\pi}{4}, \pi, \frac{5\pi}{4}$ **6.** $\frac{\pi}{2}, \frac{7\pi}{6}, \frac{11\pi}{6}$ **7.** $\frac{3\pi}{4}, \frac{7\pi}{4}$ **8.** $\frac{\pi}{2}, \pi$

Exercise Set 6.5

1. Solution **3.** Not a solution **5.** Solution **7.** Solution **9.** Not a solution **11.** $x = \frac{\pi}{3} + 2n\pi$ or $x = \frac{2\pi}{3} + 2n\pi$, where n is any integer. **13.** $x = \frac{\pi}{4} + n\pi$, where n is any integer. **15.** $x = \frac{2\pi}{3} + 2n\pi$ or $x = \frac{4\pi}{3} + 2n\pi$, where n is any integer. **17.** $x = n\pi$, where n is any integer. **19.** $x = \frac{5\pi}{6} + 2n\pi$ or $x = \frac{7\pi}{6} + 2n\pi$, where n is any integer. **21.** $\theta = \frac{\pi}{6} + 2n\pi$ or $\theta = \frac{5\pi}{6} + 2n\pi$, where n is any integer. **23.** $\theta = \frac{3\pi}{2} + 2n\pi$, where n is any integer. **25.** $\frac{\pi}{6}, \frac{\pi}{3}, \frac{7\pi}{6}, \frac{4\pi}{3}$ **27.** $\frac{5\pi}{24}, \frac{7\pi}{24}, \frac{17\pi}{24}, \frac{19\pi}{24}, \frac{29\pi}{24}, \frac{31\pi}{24}, \frac{41\pi}{24}, \frac{43\pi}{24}$ **29.** $\frac{\pi}{18}, \frac{7\pi}{18}, \frac{13\pi}{18}, \frac{19\pi}{18}, \frac{25\pi}{18}, \frac{31\pi}{18}$ **31.** 0 **33.** no solution **35.** $\frac{4\pi}{9}, \frac{8\pi}{9}, \frac{16\pi}{9}$ **37.** $0, \frac{\pi}{3}, \pi, \frac{4\pi}{3}$ **39.** $\frac{\pi}{2}, \frac{7\pi}{6}, \frac{11\pi}{6}$ **41.** $\frac{2\pi}{3}, \pi, \frac{4\pi}{3}$ **43.** $\frac{3\pi}{2}$ **45.** $\frac{\pi}{2}, \frac{3\pi}{2}$ **47.** $\frac{\pi}{4}, \pi, \frac{5\pi}{4}$ **49.** $\frac{5\pi}{6}, \frac{7\pi}{6}, \frac{11\pi}{6}$ **51.** $\frac{\pi}{4}, \frac{5\pi}{4}$ **53.** $0, \frac{2\pi}{3}, \pi, \frac{4\pi}{3}$ **55.** $0, \pi$ **57.** $\frac{\pi}{2}, \frac{7\pi}{6}, \frac{11\pi}{6}$ **59.** π **61.** $\frac{\pi}{6}, \frac{5\pi}{6}$ **63.** $\frac{\pi}{6}, \frac{\pi}{2}, \frac{5\pi}{6}, \frac{3\pi}{2}$ **65.** $0, \frac{2\pi}{3}, \frac{4\pi}{3}$ **67.** $\frac{2\pi}{3}, \frac{4\pi}{3}$ **69.** $\frac{\pi}{8}, \frac{3\pi}{8}, \frac{9\pi}{8}, \frac{11\pi}{8}$ **71.** $0, \frac{\pi}{2}$ **73.** $\frac{\pi}{4}, \frac{3\pi}{4}$ **75.** $\frac{\pi}{12}, \frac{\pi}{4}, \frac{3\pi}{4}, \frac{11\pi}{12}, \frac{17\pi}{12}, \frac{19\pi}{12}$ **77.** 0 **79.** 0.4 sec and 2.1 sec **81.** 49 days and 292 days **83.** $t = 2 + 6n$ or $t = 4 + 6n$ where n is any nonnegative integer. **85.** 21° or 69°.

97.

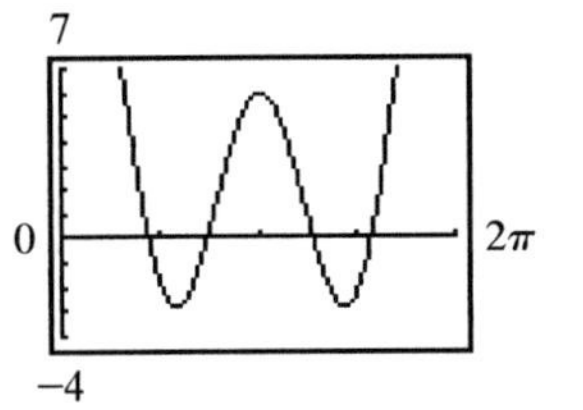

$x = 1.37, x = 2.30, x = 3.98,$ or $x = 4.91$

99.

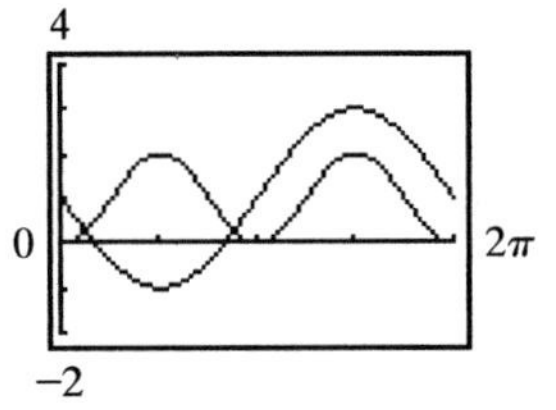

$x = 0.37$ or $x = 2.77$

101.

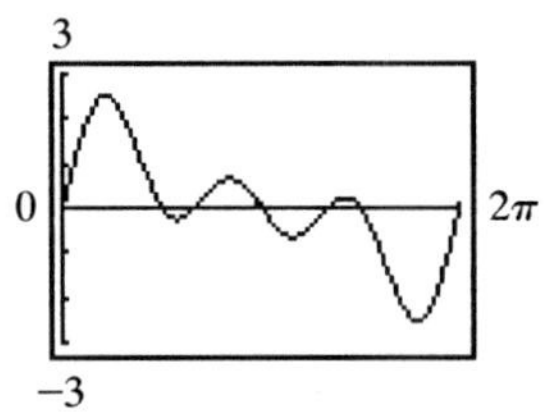

$x = 0, x = 1.57, x = 2.09, x = 3.14,$ $x = 4.19,$ or $x = 4.71$

103. no solution **105.** $\frac{\pi}{3}, \frac{5\pi}{3}$

Chapter 6 Review Exercises

For Exercises 1–12, proofs may vary. For Exercises 19–30, proofs may vary. **13.** $\frac{\sqrt{6} - \sqrt{2}}{4}$ **14.** $\frac{\sqrt{2} - \sqrt{6}}{4}$ **15.** $2 - \sqrt{3}$ **16.** $\sqrt{3} + 2$ **17.** $\frac{1}{2}$ **18.** $\frac{1}{2}$

31. a. $y = \cos x$ **b.** $\sin\left(x - \frac{3\pi}{2}\right) = \sin x \cos\frac{3\pi}{2} - \cos x \sin\frac{3\pi}{2} = \sin x \cdot 0 - \cos x \cdot -1 = \cos x$

32. a. $y = -\sin x$ **b.** $\cos\left(x + \frac{\pi}{2}\right) = \cos x \cos\frac{\pi}{2} - \sin x \sin\frac{\pi}{2} = \cos x \cdot 0 - \sin x \cdot 1 = -\sin x$

33. a. $y = \tan x$ **b.** $y = \frac{\tan x - 1}{1 - \cot x} = \frac{\frac{\sin x}{\cos x} - 1}{1 - \frac{\cos x}{\sin x}} = \frac{\frac{\sin x - \cos x}{\cos x}}{\frac{\sin x - \cos x}{\sin x}} = \frac{\sin x - \cos x}{\cos x} \cdot \frac{\sin x}{\sin x - \cos x} = \frac{\sin x}{\cos x} = \tan x$

34. a. $\frac{33}{65}$ **b.** $\frac{16}{65}$ **c.** $-\frac{33}{56}$ **d.** $\frac{24}{25}$ **e.** $\frac{2\sqrt{13}}{13}$ **35. a.** $-\frac{63}{65}$ **b.** $-\frac{56}{65}$ **c.** $\frac{63}{16}$ **d.** $\frac{24}{25}$ **e.** $\frac{5\sqrt{26}}{26}$

36. **a.** 1 **b.** $-\frac{3}{5}$ **c.** undefined **d.** $-\frac{3}{5}$ **e.** $\frac{\sqrt{10+3\sqrt{10}}}{2\sqrt{5}}$ **37.** **a.** 1 **b.** $\frac{4\sqrt{2}}{9}$ **c.** undefined **d.** $\frac{4\sqrt{2}}{9}$ **e.** $-\frac{\sqrt{3}}{3}$

38. $\frac{\sqrt{3}}{2}$ **39.** $-\frac{\sqrt{3}}{3}$ **40.** $\frac{\sqrt{2-\sqrt{2}}}{2}$ **41.** $2-\sqrt{3}$ **42.** $\frac{1}{2}[\cos 2x - \cos 10x]$ **43.** $\frac{1}{2}[\sin 10x + \sin 4x]$ **44.** $-2 \sin x \cos 3x$

45. $\frac{\sqrt{6}}{2}$ **46.** Proofs may vary. **47.** Proofs may vary. **48.** **a.** $y = \cot x$ **b.** Proofs may vary.

49. $x = \frac{2\pi}{3} + 2n\pi$ or $x = \frac{4\pi}{3} + 2n\pi$, where n is any integer. **50.** $x = \frac{\pi}{4} + 2n\pi$ or $x = \frac{3\pi}{4} + 2n\pi$, where n is any integer.

51. $x = \frac{7\pi}{6} + 2n\pi$ or $x = \frac{11\pi}{6} + 2n\pi$, where n is any integer. **52.** $x = \frac{\pi}{6} + n\pi$, where n is any integer **53.** $\frac{\pi}{2}, \frac{3\pi}{2}$ **54.** $\frac{\pi}{6}, \frac{5\pi}{6}, \frac{9\pi}{6}$

55. $\frac{3\pi}{2}$ **56.** $0, \frac{\pi}{3}, \pi, \frac{5\pi}{3}$ **57.** π **58.** $\frac{\pi}{6}, \frac{5\pi}{6}, \frac{3\pi}{2}$ **59.** $\frac{\pi}{6}, \frac{5\pi}{6}, \frac{7\pi}{6}, \frac{11\pi}{6}$ **60.** $0, \pi, \frac{7\pi}{6}, \frac{11\pi}{6}$ **61.** $0, \frac{\pi}{6}, \pi, \frac{11\pi}{6}$ **62.** $0, \pi$

63. $t = \frac{2}{3} + 4n$ or $t = \frac{10}{3} + 4n$, where n is any integer. **64.** 12° or 78°

Chapter 6 Test

1. $-\frac{63}{65}$ **2.** $\frac{56}{33}$ **3.** $-\frac{24}{25}$ **4.** $\frac{3\sqrt{13}}{13}$ **5.** $\frac{\sqrt{6}+\sqrt{2}}{4}$ **6.** $\cos x \csc x = \cos x \cdot \frac{1}{\sin x} = \frac{\cos x}{\sin x} = \cot x$

7. $\frac{\sec x}{\cot x + \tan x} = \frac{\frac{1}{\cos x}}{\frac{\cos x}{\sin x} + \frac{\sin x}{\cos x}} = \frac{\frac{1}{\cos x}}{\frac{\cos^2 x + \sin^2 x}{\sin x \cos x}} = \frac{1}{\cos x} \cdot \frac{\sin x \cos x}{1} = \sin x$

8. $1 - \frac{\cos^2 x}{1 + \sin x} = 1 - \frac{(1 - \sin^2 x)}{1 + \sin x} = 1 - \frac{(1 + \sin x)(1 - \sin x)}{1 + \sin x} = 1 - (1 - \sin x) = \sin x$

9. $\cos\left(\theta + \frac{\pi}{2}\right) = \cos\theta \cos\frac{\pi}{2} - \sin\theta \sin\frac{\pi}{2} = \cos\theta \cdot 0 - \sin\theta \cdot 1 = -\sin\theta$

10. $\frac{\sin(\alpha - \beta)}{\sin\alpha\cos\beta} = \frac{\sin\alpha\cos\beta - \cos\alpha\sin\beta}{\sin\alpha\cos\beta} = \frac{\sin\alpha\cos\beta}{\sin\alpha\cos\beta} - \frac{\cos\alpha\sin\beta}{\sin\alpha\cos\beta} = 1 - \cot\alpha\tan\beta$

11. $\sin t \cos t(\tan t + \cot t) = \sin t \cos t\left(\frac{\sin t}{\cos t} + \frac{\cos t}{\sin t}\right) = \sin^2 t + \cos^2 t = 1$ **12.** $\frac{7\pi}{18}, \frac{11\pi}{18}, \frac{19\pi}{18}, \frac{23\pi}{18}, \frac{31\pi}{18}$, and $\frac{35\pi}{18}$

13. $\frac{\pi}{2}, \frac{7\pi}{6}, \frac{3\pi}{2}, \frac{11\pi}{6}$ **14.** $0, \frac{\pi}{3}, \frac{5\pi}{3}$ **15.** $0, \frac{2\pi}{3}, \frac{4\pi}{3}$

Cumulative Review Exercises (Chapters 1–6)

1. $-3, 1 + 2i$, and $1 - 2i$ **2.** $x = \frac{\log 125}{\log 11} + 1$ or $x \approx 3.01$ **3.** $(-\infty, -4] \cup [2, \infty)$ **4.** $\frac{\pi}{3}, \frac{5\pi}{3}$

5.

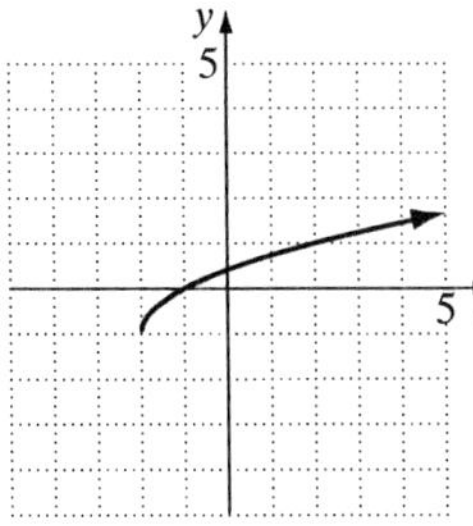

6.

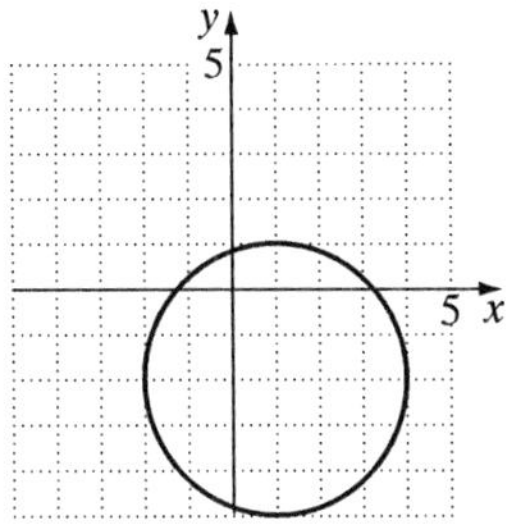

7.

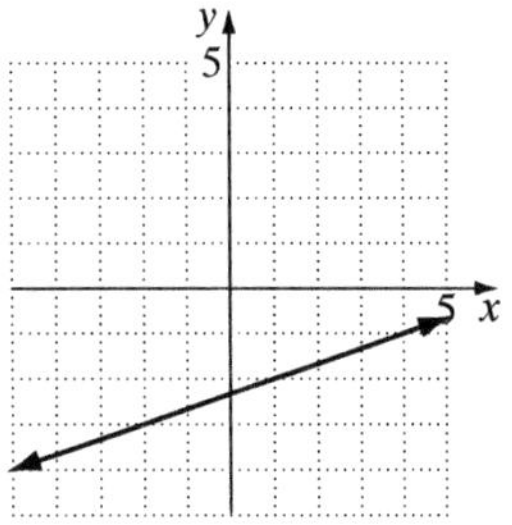

8.

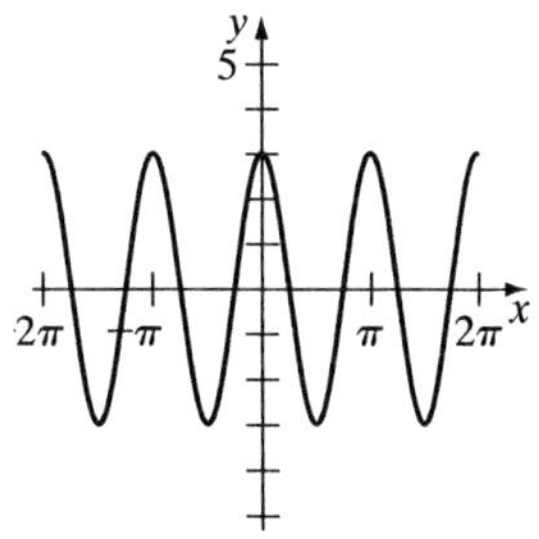

9.

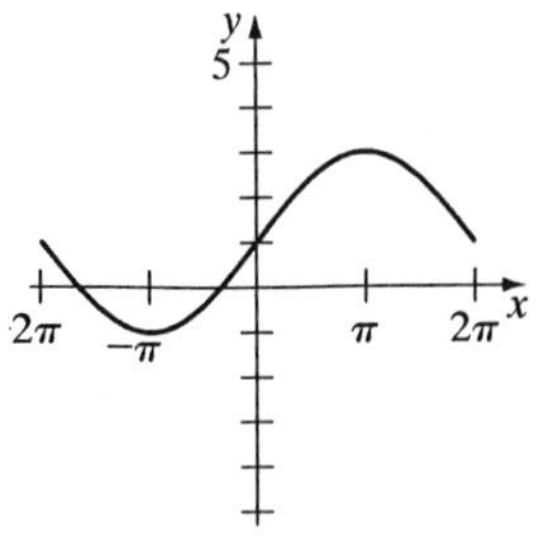

10.

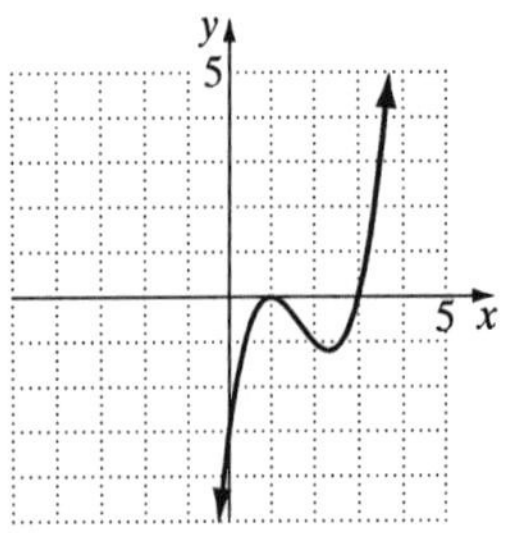

11. $2a + h + 3$ **12.** $-\frac{\sqrt{2}}{2}$ **13.** Proofs may vary. **14.** $\frac{16}{9}\pi$ radians **15.** $t \approx 19.1$ yr **16.** $f^{-1}(x) = \frac{3x + 1}{x - 2}$ **17.** $B = 67°, b = 28.27, c = 30.71$ **18.** 106 mg **19.** $h \approx 15.9$ ft **20.** Answers may vary.